Lecture Notes in Artificial Intelligence 565

Subseries of Lecture Notes in Computer Science
Edited by J. Siekmann

Lecture Notes in Computer Science

Edited by G. Goos and J. Hartmanis

J. D. Becker   I. Eisele   F. W. Mündemann (Eds.)

Parallelism, Learning, Evolution

Workshop on Evolutionary Models and Strategies
Neubiberg, Germany, March 10-11, 1989
Workshop on Parallel Processing: Logic,
Organization, and Technology – WOPPLOT 89
Wildbad Kreuth, Germany, July 24-28, 1989
Proceedings

Springer-Verlag
Berlin Heidelberg New York
London Paris Tokyo
Hong Kong Barcelona
Budapest

Series Editor

Jörg Siekmann
Institut für Informatik, Universität Kaiserslautern
Postfach 3049, W-6750 Kaiserslautern, FRG

Volume Editors

J. D. Becker*
I. Eisele
Institut für Physik, Fakultät für Elektrotechnik, Universität der Bundeswehr
Werner-Heisenberg-Weg 39, W-8014 Neubiberg, FRG

F. W. Mündemann*
Fakultät für Informatik, Universität der Bundeswehr
Werner-Heisenberg-Weg 39, W-8014 Neubiberg, FRG

* Also at:
ICAS, Institute for Cybernetic Anthropology Starnberg
Maisingerschluchtstraße 4a, W-8130 Starnberg, FRG

CR Subject Classification (1991): C.1.2-3, F.1.3, F.4.1, I.2.3, I.2.6

ISBN 3-540-55027-5 Springer-Verlag Berlin Heidelberg New York
ISBN 0-387-55027-5 Springer-Verlag New York Berlin Heidelberg

Typesetting: Camera ready by author
Printing and binding: Druckhaus Beltz, Hemsbach/Bergstr.
45/3140-543210 - Printed on acid-free paper

Preface

Parallel and distributed processing raises many new questions which did not play a role in the classical von Neumann pradigm. Thus, as in two preceding volumes[+], also here we try to explore other areas which may be important for parallel processing, in addition to presenting material directly concerned with the field.

In search of such relevant concepts, we have included papers on aspects of
- space and time,
- representations of systems,
- non-Boolean logics,
- metrics,
- dynamics and structure, and
- superposition and uncertainties.

In particular we should like to point out that distributed representations of information may share many features with quantum physics, such as the superposition principle and the uncertainty relations. Another important issue is logics; if Boolean logic has not developed as a natural ability of the human race in the course of evolution, this might mean that there should be an advantage in using protologics for problem solving. The reason for this could be that Boolean logic is too rigid for situations in which information is not conserved.

The central part of this volume contains material on general parallel processing machines, neural networks, and system-theoretic aspects.

We have added to the papers presented at WOPPLOT 89 the material of another workshop on Evolutionary Strategies, because we feel that these will yield very powerful and important applications for parallel processing machines, and because this section represents a kind of completion of this volume. Furthermore, evolutionary and genetic algorithms not only speed up considerably the search for suboptimal solutions to hard problems; it seems that they also open the pathway to new problem classes to be treated by computers.

[+] WOPPLOT 83, Springer Lecture Notes in Physics, Vol. 196

WOPPLOT 86, Springer Lecture Notes in Computer Science, Vol. 253

For technical and personal reasons it has not been easy to edit and complete this volume, and we apologize very much to the authors as well as to the publishing company for the delay.

It is a pleasure to thank the publisher, Springer-Verlag, for giving us again the opportunity to present this collection of papers to the scientific community.

Neubiberg, Summer 1991   J. Becker, I. Eisele, F. Mündemann

Contents

+) Reprinted from: Peter J. Plath (ed.): Optimal Structures in Heterogeneous Reaction Systems
  Springer Series in Synergetics (1989)

On Parallel Consciousness

Peter Molzberger

Fakultät für Informatik

Institut für Systemorientierte Informatik

Universität der Bundeswehr München

Werner-Heisenberg-Weg 39

D - 8014 Neubiberg

Abstract

I assume that our current difficulties in modelling self-organizing systems (in A.I., genetics etc.) are deeply rooted in our way of thinking. Our current paradigms are like hypnotic spell making it impossible for us to see the open door out of the box we have locked ourselves in. Instead we are running against the solid walls.

The first steps to get out of that box are the recognition that we are in a box and the moment we are realizing that the door is open and that there is no trap.

This paper is intended to contribute a little bit to recognize the box. Our understanding of parallelity - especially our inability to hold contradicting mental structures in our minds - seems to be a key to that.

Motivation

I'm a computer scientist, coming from electrical engineering. When I first 'met' the computer 25 years ago, I was fascinated. I switched my job. Since that time my dream has been to create artificial intelligence.

Today everybody speaks about A.I. and there is a tremendous amount of research and commercial interest in the field. Nevertheless I feel frustrated by the way we approach the challenge. I see that we are stuck; we are fighting windmills. We do it with great commitment and with great bravery, but there is no chance to come to a real break-through, as long as we do it the way we do it.

I'm looking for a totally different solution, a new paradigm. Now, as we know from history, especially Kuhn /1/, it is dangerous to break paradigms. It evokes a lot of emotions in the people's minds and I may very well end up being burned on a stake. (In earlier days that was a real physical stake; today our stakes are more subtil, but - nevertheless - they serve the same purpose.) So, if you as the reader, will become upset about this article - this is a very natural reaction. Preparing a break-through in A.I. is my official goal. Behind that there is a deeper motivation I'm not paid for: giving a real contribution to life on this planet in a very critical situation. To me understanding our stuckness in A.I. means the same stuckness we meet everywhere in science and in the thinking of people outside of science. In my understanding it is a matter of survival for mankind to find solutions we seem to be unable to see at the moment.

Feeling to Be in the Trap

We are surrounded by self-organizing systems:

- cells;
- animals, plants;
- brains;
- social systems, etc. .

Whenever we look at such systems, we find operationally closed loops. A very simple loop would be:

A modifies B;
B modifies C;
C modifies A.

As an example, take a living cell. Every element in the cell not only has to perform a function (as operator), but must also be produced (as operand) by other elements. Imagine a 'hierarchy' of machines (enzymes) finishing at its top the machines needed at the bottom. Actually it is not a hierarchy: the top of the hierarchy is bent back to the bottom. Scientists have found a nice name for this kind of structures: autopoietic systems.

Now, if we try to formalize this model or program it on a computer, as a self-organizing system, we run into difficulties. We create antinomies. Our model breaks down, it trivializes itself. In computer science we have a very advanced theory of computability. It says, in a nutshell:

"It is not possible to model operationally closed systems."

To me this looks like a gigantic joke. We know very well that we are surrounded by this kind of systems and they work pretty well. At the same time we can prove scientifically that it is principally impossible to model them with our most powerful tools.

We do not deny that we can model the behavior of a single atom or molecule to any degree of accuracy desired. Furthermore we do not deny that we can model - at least in principle - any number of interacting molecules. (This is a matter of computing power, of course.)

As we will hardly find anybody in science today ready to discuss mysterious vital forces from outside the molecular system, there is no reason, why we should not be able to model a living cell or any self-organizing system. This paradoxical situation does not only apply to computer simulation. We are in a similar trap everywhere we deal with self-organizing systems: Biologists do not really understand what life means, economists have no appropriate models for (living!) economic systems, etc. etc. . Even a (living!) body of law cannot be represented without antinomies, as was found recently by trying to create large expert systems. So, even on a very practical level, in applying today's technology, we are confronted with apparently unsur-mountable difficulties. Where do the difficulties come from? They are neither to be found in false analysis of the system nor anywhere in the way of computing. They cannot be met by more complex models nor by more powerful computers or better pro-gramming languages. They are far deeper; they are at the roots of our thinking, at our perception of the world. The whole situation is very much like that at the final state of other paradigms. Contradictions within the accepted framework of assumptions about reality can no longer be neglected or suppressed. The whole building has become shaky.

At the beginning of this century we had a similar situation, for example, in physics: The models of ether - a substance assumed

between the celestial bodies, responsilbe for transmitting gra-
vity and electro-magnetism - became more and more complicated and
contradictory, until Einstein replaced them by a set of very
simple, but abstract assumptions - called special relativity.

The revolution we are now waiting for will be different in so
far, as not only some physicists are concerned, but it goes to
the 'bones' of everyone.

The Aristotelian Paradigm

On WOPPLOT 86 the philosopher von Müller gave a talk on this
topic. He said that our current way of experiencing reality, the
Aristotelian Paradigm (A.P.), is founded on four pillars - our
assumptions about

- space;
- time (causality);
- logic;
- identity.

Each of these sub-paradigms is very powerful in itself, but
additionally they support one another. We cannot break down one
of them without making the whole building tumble down. We cannot
break it down piecewise without creating tremendous trouble in
our minds. Only when we replace all four pillars simultaneously,
we come on firm ground again /2/.

This is what makes us scared. We have to face too much 'magic',
too much 'incomprehensible' things at the same time. We would be
totally confused and somehow we know that sub-consciously.

In my view, one of the restrictions of the A.P. is that we cannot
really think parallelism. Of course, we think that we can think

it, but we don't. In our minds we replace anything that works in
parallel by piecewise sequential processes.

Computer scientists are masters in that. We know how to sequen-
tialize any powerful parallel activity to a string of miniscule
micro-instructions. We have made the principal sequentialibility
of any parallelism to one of our basic laws in theoretical
computer science. We believe that everything that can be computed, can be computed on a universal Turing machine, which works
sequentially. Everything that does not fit into that scheme is
simply denied, just not recognized. We are so used in doing so
that we cannot even find a plausible example for irreducible
parallelism.

We break the wholeness and go on working with its parts, taking
them for the whole. In case of irreducible parallelism we run
into difficulties showing up in

- antinomies,
- strange loops, and
- infinite regressions.

The sequential parts do not form a consistent system. During the
last years it has become a fashion to play around with this
stuff, as Hofstadter's success shows /3/.

Once again, we are not the only ones behaving that way. Von
Weizsäcker sees that clearly in physics, when he says /4/:

1.   Breaking reality into parts is always faulty.

2.   Breaking reality into parts is a precondition for
     conceptual thinking. That means that conceptional
     thinking is always an approximation to reality.

3.   This is also true for these statements, because
     conceptional thinking is being used.

My assumption is: We cannot see a solution, as long as we are in
the A.P. We are just blinded. The moment we get out of that, we
will be able to think irreducible parallelism. We will be able to
understand, why the whole is more than the sum of its parts: We
will understand that two parallel processes - communicating to
each other - generate a synergistic effect resulting in a
behavior that cannot be explained out of the behavior of the two
processes. So we will be able to understand the very nature of
self-organizing processes, of living systems of all kind, of
evolution. We may even be able to understand creativity which -
to me - is a synergistic effect between thoughts as mental
structures, and on this way we might be able to take a big step
into real artificial intelligence. We will also make a gigantic
step in understanding evolution. As WOPPLOT is a famous forum for
the representation of advanced genetic algorithms, I'm going to
point that out once again in detail:

In some talks it became very clear that evolution is a function
of organization. It takes years for a most intelligent human
being to create the appropriate organization, in order to show
similar effects as we know from nature. What is still beyond any
modelling is the fact that this organization is created by
evolution itself. We can easily see that trying to model this
leads to an infinite regress: Imagine, ten years from now,
somebody will find a super-algorithm explaining how evolution was
able to create and optimize the genetic algorithms we are playing
with now. Then we would ask: How did evolution find this super-
algorithm? Etc. etc. ! The way to avoid the outside programmer or
the infinite regress seems to bend the sequence of super-
programs into a circular structure. So we come to an operationally closed system, withdrawing formal representation on a
computer.

Whatever we do, we find ourselves caught in paradoxes, trapped in
a box.

How to Get out of the Box

The old paradigm, the A.P., defends itself as all beliefs and paradigms do. So we are caught in a box we have constructed ourselves collectively. There is no logical way out of the box, because our logical system is valid only inside the box. A good metaphor for this situation is: The instructions how to leave the box are written on the outside.

Clearly, when we have left the box, we will find it all very simple. "Why haven't we seen that before?", we will say. Why haven't the sailors seen that the earth cannot be flat by simply looking at a sailing-ship vanishing on the horizon, with its top as the last part?

When there is no way to get out of the box, the only way is to **be** outside of the box. Some day we will find ourselves outside, tunnelled out. To our today's minds this sounds like magic, until we become used to it and find a formal description like in quantum mechanics. We experience a very analogous situation every night when going to sleep. There is no logical way how to go to sleep. On the contrary! The more we try, the less we'll succeed. But - nevertheless - somehow it happens and we don't even realize. We are usually more successful in observing the opposite transition - waking up. Sometimes we experience, how the old reality - our last dream - is becoming inconsistent and fading away. Falling asleep and waking up are not just metaphors for a shift of paradigm. They are analogous transitions from one state of consciousness to another one. Leaving the A.P. will certainly throw us into a grossly altered state. In both states we have a seemingly consistent perception of reality, but they are inconsistent to one another, as most people remember solutions for their problems which they found in the dream state.

Before going on, I would like to give one more analogon wich is very convincing to me: hypnosis. People in hypnosis - say in a TV demonstration on stage - behave totally consistent in their reality. When someone hypnotized is told to be a bird, he is totally convinced to be a bird. He sees feathers, where the spectators around him see clothes, and it is impossible to tell him that these are not feathers.

To me paradigms are not like hypnotic spells, they are hypnotic spells. It is exactly the same mechanism with the only difference that we have created the suggestion collectively and have perpetuated it since thousands of years (Aristotle hasn't done it to us!).

So getting out of the box is getting out of a hypnosis. You have little chance, when you do not know that you are hypnotized or don't believe so. The A.P. is currently becoming weaker and will eventually break down. Otherwise we would partially not be aware of the fact that we are in that box. I believe, we are not only leaving that box, but we are on the way to understand how to deal with such boxes generally. Otherwise we wouldn't have become aware of the mechanisms of paradigms some 20 years ago and wouldn't have invented the word 'paradigm' in its current meaning.

What can we do to accelerate the process of leaving the box?

1. by becoming more and more aware of the existence of that box;
2. by inviting antinomies into our lives.

In the normal course of events people try to suppress contradictions as long as possible, because these are very uncomfortable. The moment we know that they can help us, we may be able to invite even uncomfortable guests.

The recipe is not a new one. Zen people use koans - paradoxical sentences. The student is advised to meditate about the contradiction, until his mental system, containing his paradigmatic barriers, breaks down. Zen people go in search of confusion deliberately.

How Do I Know?

All this is a speculation, but one shared by more and more people around the world. I've been engaged in several fields very close to this topic

- Constructivism;
- NLP - Neuro Linguistic Programming;
- Polycontextural Logic.

To me constructivism is the philosophical base. Famous authors in this field are von Förster /5/, von Glasersfeld /6/ and Maturana /7/. The main point is constructivism claims that we construct our subjective realities. Radical constructivism even denies the existence of an outside absolute reality.

NLP was founded at Santa Cruz University about 15 years ago, as a new approach to psychology, by a mathematician (Bandler), a linguist (Grinder), and a computer scientist (Dilts) /8/. What makes NLP   and its derivations so fascinating to me is the practical demonstration of the ideas of constructivism about the nature of reality. During the last years I've done myself many things with people and groups, which I would have classified as 'magic' or 'miracles' a few years ago.

As this is not the place to give introductions into that field, I'm presenting a fairy tale an NLP practitioner showed me some day with the remark: "This is all about NLP." The reader may choose to enjoy it or feel scared of it /9/.

The Prince and the Magician

Once upon a time there was a young prince who believed in all things but three. He did not believe in princesses, he did not believe in islands, he did not believe in God. His father, the king, told him that such things did not exist.

As there were no princesses or islands in his father's domains, and no sign of God, the prince believed his father.

But then, one day, the prince ran away from his palace and came to the next land. There, to his astonishment, from every coast he saw islands, and on these islands, strange and troubling creatures whom he dared not name. As he was searching for a boat, a man in full evening dress approached him along the shore.

"Are those real islands?" asked the yound prince.
 "Of course they are real islands," said the man in evening dress.
"And those strange and troubling creatures?"
"They are all genuine and authentic princesses."
"Then God must also exist!" cried the prince.
"I am God," replied the man in evening dress, with a bow.

The young prince returned home as quickly as he could.

"So, you are back," said his father, the king.
"I have seen islands, I have seen princesses, I have seen God," said the prince reproachfully.

The king was unmoved.

"Neither real islands, nor real princesses, nor a real God exist."
"I saw them!"
"Tell me how God was dressed."
"God was in full evening dress."
"Were the sleeves of his coat rolled back?"
The prince remebered that they had been. The king smiled.
"That is the uniform of a magician. You have been deceived."

At this, the prince returned to the next land and went to the same shore. where once again he came upon the man in full evening dress.

"My father, the king, has told me who you are," said the prince indignantly. "You deceived me last time, but not again. Now I know that those are not real islands and real princesses, because you are a magician."

Tho man on the shore smiled.
"It is you who are deceived, my boy. In your father's kingdom, there are many islands and many princesses. But you are under your father's spell, so you cannot see them."

The prince pensively returned home. When he saw his father, he looked him in the eye.
"Father, is it true that you are not a real king, but only a magician?"

The king smiled and rolled back his sleeves.
"Yes, my son, I'm only a magician."

"Then the man on the other shore was God."
"The man on the other shore was another magician."
"I must know the truth, the truth beyond magic."
"There is no truth beyond magic," said the king.

The prince was full of sadness. He said, "I will kill myself." The king by magic caused Death to appear. Death stood in the door and beckoned to the prince. The prince shuddered. He remembered the beautiful but unreal islands and the unreal but beautiful princesses.

"Very well," he said, "I can bear it"
"You see, my son," said the king, "you, too, now begin to be a magician."

In Search of Confusion

Coming back to my motivation, I would like to give some practical advice for the reader to become aware that there is more in this universe than the A.P. lets us know.

I start with the provocative statement saying that

"parallelism is synergy".

What I mean by that is: When antonomous systems start to communicate freely (without being bound to a common given system of reference), unexpected high-order structures will pop up spontaneously. This can easily be observed in groups of human beings. From group dynamics we have known for a long time that social structures organize themselves spontaneously, within seconds. With so-called synergistic teams /13,14/ we can come to a performance of work and learning that is very difficult to explain on the base of individual faculties. A high trust level in the team is necessary for this effect or - in other words - absence of resistance of the people against one another.

My claim is now that the same mechanisms apply to the thoughts in a mind: as soon as a person is able to hold conflicting thoughts in her/his mind and to be 'in peace' with them, a synergistic effect will occur and new thoughts, reconciling the conflicting ones, will pop up. This is what we call 'creativity'. It is the process which Hegel described as

thesis + antithesis = synthesis.

To my understanding it is essential that there is no resistance against the thoughts. When the person is fully relaxed, without emotions against the ideas, this synthesis will take place very fast. It seems to me that it will take place in no time or out of time, in one sudden flash. (The last section will mainly deal with the question how to cope with resistance.)

This is the way I propose for getting out of the box. My assumption is that there is no principal difference between a small thought (belief) and a big one (paradigm). Both are self-fulfilling statements about reality or statements creating their own local reality. So, instead of speaking of parallel thoughts, we can as well talk about the co-existence of parallel conflicting realities in our minds.

Experimenting with Contradicting Mental Structures

The moment we are creating a statement about reality (belief, paradigm) in our minds, we are also creating the opposite statement (very much like an electron/positron pair in microphysics). We choose one of the two positions to be 'TRUE', the other one 'FALSE'.

The 'TRUE' position will become part of our conscious reality, the other one will show up as doubt in our own minds or as opposition in other people's minds. This may be conscious or subconscious. We seem to have no other choice to create our (subjective) reality than working with dualismus: "tertium non datur!"

Contradicting beliefs fight each other. They are exclusive to our mind system. Each of them has survival character. This survival character may be stronger than that of its carrier: some people even prefer to die before giving up their convictions. Such mental structures defend themselves violently, as we know from psychology and history. Max Planck said that a new physical theory would be accepted to the same degree that the physicists opposing it died out. So we can say that statements about reality in our individual or collective mind(s) behave very much like living systems. The question is how to 'come to peace' between them? When we manage to hold the two contradicting ideas in our minds simultaneously, see them as complementary, something

strange takes place: A new point of view will be born, trans-
cending the opposites (without destroying them). Something takes
place in our consciousness beyond our normal scope of understanding. It is a synergistic leap, a creative process: something
is created spontaneously on a new dimension. We will be surprised
by the sudden illumination and may say "aha!"

Everyone knows this effect that happens frequently when you are
really willing to listen to other people. What I'm claiming for
here is to systemize that to come to a 'technology of creativity'.

What I'm proposing here is an emotional shift in our reaction to
opposing ideas. Our normal behavior is mainly the fight-or-flight
reaction or - very rarely - the surrender to the other idea. Very
often we show more or less violent aggressive reactions (fight)
or we pretend to ignore it: We suppress our emotional reactions
(flight). I am talking of a third way "tertium datur", i.e. the
parallel coexistence of seemingly irreconcilable statements.
Attempting this will lead to all forms of emotional resistance,
the fight for survival of the old tertium-non-datur paradigm and
our position "I am right and you are wrong". You may feel very
uncomfortable; you may produce anger, hatred, fear, pain, feel
silly, embarrassed and display many kinds of bodily reactions,
even cramps and nausea. The moment you are able to confront all
these reactions as normal resistance of a belief system, they
will vanish or become pleasant. From my - not yet always
reproducible - personal experience I know that looking at all
this turmoil with the attitude of a scientist, doing a most
fascinating experiment with myself, enables me just to observe
the symptoms with interest and finally to enjoy them.

The theory is extremely simple, but the difficulty are our
emotional reactions, which seem to be much stronger than our
mental willingness. (Jesus, who said in the Bible (Matthew 26,41)
"The spirit indeed is willing, but the flesh is weak", must have
known about that.)

On workshops I use to perform experiments with people, making it much easier for them to experience the mechanics of their mind structures in defending the A.P. Anyway, you as the reader, will get a glimpse of it by the following Gedankenexperiment: You are likely to be the carrier of some great idea, something in science that really moves you, something you see as your personal contribution, your message for the scientific world. Can you identify that, please!

Now there will certainly be opposition to that idea, people who do not agree and who cause some emotional trouble to you. I mean a competing idea, not one you have given up in favor of your new one. Perhaps you find a person you consider as incompetent or a little bit crazy, someone you do not like to discuss with, because is doesn't make any sense. Take that person!

Now, just as an exercise, imagine that this person sees him-/herself in his/her world as just as honorable as you do in yours and has the same pure intention to bring his/her message over. (He/she doesn't know that he/she is incompetent or crazy!) As a next step, can you imagine that person's feelings towards you? He/she may see you as incompetent or crazy! Perhaps you can manage now to step fully into the other person's moccasins. Would you be able to write his/her papers, to give his/her talks from a point of view fully convinced of being right? Try it a little bit in your mind's eye! Perhaps you can experience in your imagination a symmetry of your points of view. When you can do that, you will see that there does not exist an absolute reference system for 'TRUE' and 'FALSE'. There is no reason not to say: "Oh, both points of view are equally possible and as a matter of personal preference I have chosen this one and he that one."

Now we are playing a game and for having more fun out of it, let's both assume "I am right" and forget that it was an assumption.

In other words: We have agreed on a social game with a subconscious rule.

Well, as a final step, you may feel tired of playing the game and remember the rule you agreed to forget. Notice that I dropped the words "in your imagination". Do it really! The moment you are able to hold both opinions in your mind simultaneously and take the position (outside!) that both are equally right or wrong, you will most likely create a fruitful sythesis containing both points of view.

I suppose you have not followed the exercise. You have given up sooner or later, because it was too much trouble or you experienced it just as nonsense. Notice that you have found a very good reason why to give up. The A.P. is in charge of our mental system and as scientists we are especially well trained in finding rational explanations. Notice now, how you reacted to this last sentence!

An other possibility is that you said "so what!" No objections to the theory, but unfortunately - just in my case - it does not apply. The person opposing me is incompetent, because ... whatever ...

Don't be discouraged! In my view, the very fact that you have been reading this paper up to this point is a clear indication that you are really on the way to parallel consciousness and of leaving the box.

Now we are clearly not in the field of magic but of science; what may be experienced as magic has a structure to it which leads to reproducible results and is teachable.

This is not enough for a computer scientist committed to contribute to a break-through in A.I. What we need here is a formal system able to be processed on a computer. We believe that we have found that instrument in Günther's Polycontextural Logic, /10,11/, a logical system that cannot be reduced to our common Aristotelian Logic. 'Polycontextural' can be translated as 'multi-reality', which means that the world is seen in terms of multitude of subjective realities. So a variable TRUE in one reality may well be FALSE in another, corresponding, for example, to different opinions of people.

In my view, these three fields come together very closely, but they do say very much about the actual break-down of the A.P. When there is a general shift of consciousness going on these years, we should expect for statistical reasons that we find certain people already in the new state, others swinging back and forth. During the last years I was highly interested in meeting or reading about people definitely living in naturally altered states (i.e. without use of drugs). Beside the mystical litera- ture I found some more or less accurate descriptions of so-called higher states which - to my understanding - correspond to going out of the A.P. (or other collective spells in other cultures). The most profound protocol of such a permanent transition was written some 40 years ago by Merrel-Wolff, a scientist working in mathematics and physics as well as in philosophy /12/.

Most of the literature is hardly suitable to be quoted, but putting many puzzle stones together, I came to a very profound and consistent image of what it means to be out of the A.P.

19

References

1. Kuhn, T.S.: Die Struktur wissenschaftlicher Revolutionen, Suhrkamp tw 25, 5. Aufl., Frankfurt 1981 (The Structure of Scientific Revolutions, Univ. of Chicago 1962)

2. Müller, Albrecht v.: Zeit und Logik, W. Baur Verlag 1983

3. Hofstadter, D.R.: Gödel, Escher, Bach: An Eternal Golden Braid, Penguin, Harmondsworth 1980

4. Weizsäcker, C.F. v.: Vortrag auf Symposion "Raum und Zeit", Grainau 9.5.1986

5. Foerster, H.v.: Über das Konstruieren von Wirklichkeiten. In: ib.: Sicht und Einsicht: Versuche zu einer operativen Erkenntnistheorie, Braunschweig 1985

6. Glasersfeld, E.v.: Einführung in den radikalen Konsstruktivismus. In: Watzlawick, P.(Hrsg.): Die erfundene Wirklichkeit, München 1981

7. Maturana, H., Varela, F.: Der Baum der Erkenntnis - Die biologischen Wurzeln des menschlichen Erkennens, Bern 1987

8. Dilts, R. et al.: Neuro-Linguistic Programming, Meta Publications, Cupertino 1980

9. Fowles, J.: The Magus, Dell Publishing Co., Inc.

10. Günther, G.: Beiträge zur Grundlegung einer operationsfähigen Dialektik, Hamburg 1976

11. Günther, G.: Idee und Grundriß einer nicht-Aristotelischen Logik, Hamburg 1978

12. Merrell-Wolff, F.: Pathways Through the Space,
    The Julian Press, N.Y. 1983

13. Molzberger, P.: Auf den Spuren unserer kreativen Kräfte, in:
    Leidlmair, Neumair (Hrsg.): Wozu künstliche Intelligenz,
    VMGÖ Wien 1988

14. Molzberger, P.: Synergistic Teams - Talking to the Right
    Brain, Vortrag auf 1. Intern. Congress on Cerebral
    Dominances, München, Sept. 1988

COMPLEMENTARITY AND NON-LOCALITY IN COMPLEX SYSTEMS

Walter v. Lucadou

Hildastr. 64, D - 7800 Freiburg

Abstract

Many of the early founders of quantum theory expected to find
"quantum-like" features also within the domain of living
systems. But these ideas have not found much support in biology
or psychology until now. However, modern system theory dealing
with complex self-referential systems might cast a new light
upon this question. Since quantum mechanical observables cannot
be applied directly to complex systems, a categorical analysis
of those systems is necessary. After discussing some general
features of quantum theory a phenomenological non-classical
model to describe highly complex systems in developed.

It is argued that structure and function can be seen as
complementary categories of complex systems which lead to the
formulation of uncertainty relations. Furthermore non-locality
is regarded as a crucial criterion for the underlying
non-classical structure of complex systems.

The outcome of a psychological experiment is reported which is
suggestive of non-local correlations within man-machine
interaction. The results of this experiment support the model
that a self-referential complex system may be able to select
certain noise fluctuations from a causally separated system via
non-local correlations if a feedback-loop generates an
"organizational closure" of the "observer" and the "observed".

The results further suggest that these constraints can be adequately described by the concept of "pragmatic information" which was proposed by von Weizsäcker to "measure" the "meaning" of a given information.

1. How general is quantum theory?

The idea that quantum theory may have also some relevance for biology and psychology was already formulated by its founders such as N. Bohr, E. Schrödinger, W. Pauli, and P. Jordan. Especially the latter argued that quantum mechanics would be the only possibility to reconcile the personal experience of "free will" with the physical world view. He (Jordan 1947) assumed that quantum mechanical indeterministic processes inside the human brain may be amplified in such a way that they result in unpredictable behaviour of the person which is then interpreted as manifestation of free will.

Practically, however, these ideas did never play an important role neither in biology nor in psychology. Some biologists even argued that for principal reasons quantum effects cannot show up in biology (Bünning 1943).

However, the systemtheoretic paradigm has changed the situation insofar as it is assumed that theoretical constructs need not be bound to a specific substratum and thus do not require a reductionistic point of view (see v. Lucadou & Kornwachs 1983) in the sense that physics is more fundamental than psychology. The advantage of system theory is that it can be applied to psychological problems as well as to physical problems simply by the fact that it only deals with the structure of descriptions and the way how information drawn from experiments is represented and mapped in the context of this description. The process of operationalization and interpretation describes the interface between the "real world" (whatever this may be) and the description of it.

It is often said that quantum mechanical effects only occur in microscopic domain where single elementary particles are involved. However, this is not true in general. There exist several macroscopic phenomena such as superconductivity or superfluidity, which can only be understood on the conceptual basis of quantum theory and which must be considered as macroscopic quantum effects (see Ristik 1981). These systems are macroscopic since they include a macroscopic number of particles. Other systems such as the EPR-experiments (see Paty 1977) involve single particle conditions, but the relevant distances (between the two spin measurement devices) are in a macroscopic range.

However, all these macroscopic quantum effects can only be detected under highly artificial conditions, which do not exist in everyday live situations. These systems are completely isolated from their environment (see Clough 1988) and/or involve highly coherent conditions and/or very low temperatures (see e.g. Kawamoto 1987) and/or extreme measurement resolutions. Usually it is argued that the correspondence principle describes the limitations of the classical description: If the action E*dt is in the order of magnitude of h, or the distances in order of magnitude of the wavelength and if the quantum numbers n are small enough, the correct description of the system is given by quantum theory. If these conditions are not fulfilled the classical theory seems to be a good approximation. However, under highly artificial conditions, as mentioned above, these criteria cannot be applied blindly. As an example may serve the postcollision interaction, where interference patterns can be detected even with very high quantum numbers n if the resolution of the measurement is high enough (Morgenstern et al. 1977).

From this point of view we cannot rule out the possibility that quantum effects may underlie "genuine" macroscopic systems involving long distances, many particles, not isolated from their environment, at high temperatures and transmitting large amounts of action. However, in these "natural" systems we have

no experimental possibility in physics to measure the difference
between the quantum mechanical and the classical description.
From a positivistic point of view the assumption of an
underlying "real" quantum mechanical structure of "natural"
classical systems is therefore an additional assumption which
cannot be justified by experiments (see Walker, May,
Spottiswoode, Piantanida 1987). The so-called "Copenhagen
interpretation" takes this into account by its distinction of
microphysical objects and macrophysical measurement devices.

From an epistemological point of view this situation might be
unsatisfactory but - as we will see later - the assumption that
quantum effects play a role in biological and psychological
systems involves much more difficult questions than the correct
interpretation of quantum physics. Nevertheless from a
phenomenological point of view there exist several good
arguments why at least highly complex systems may have some
similarities to quantum physical systems (see Finkelstein &
Finkelstein 1984, Kornwachs 1987, Kornwachs & Lucadou 1985,1989,
Lucadou 1989b)

1. The axioms of QT seem to be of a very general nature since
they do not contain information about physics itself but merely
about the way how measured data are linked with the theory, and
how the states of a system develop in time (Schrödinger
equation). The properties of physical observables are contained
in the mathematical formulation of the corresponding operators,
which, however are not specified in the axioms. Thus they can
equally be applied to elementary particle physics, quantum
chemistry, solid state physics or even cosmology.

2. According to Jammer (1974) the basic axiom of quantum theory
can be formulated in the following way: Each physical system can
be represented by a Hilbertspace H. The basis vectors of this
Hilbertspace completely describe the state of the system. (This
implies the so-called "superposition principle" which says that
each possible state of the system can be described as a

superposition of a set of basis vectors.) It could be argued
that this axiom is too specific for a general description of
other natural systems. Furthermore it seems unclear why the
Hilbertspace structure should play such a predominant role in
the description of natural systems. However, this point becomes
more plausible if it is taken into account that the Hilbertspace
is a generalization of the Fourier transform which describes the
transformation of a sequential measurement protocol (serial
data) to a "parallel" data representation (structural data). At
least in the context of the black box model sequential data are
the only source of information which can be obtained from a
system. Later we will see that structure and function
(behaviour) of a system can be conceived as complementary
concepts which could be properly described in a Hilbertspace. In
regard to the Schrödinger equation it could again be argued that
it is too specific to serve as a general model to describe any
natural system. However, as far as the Hamiltonian H is not
further specified the Schrödinger equation simply states that
the temporal change of the state vector depends on the present
state of the system which indeed is a fairly general statement.
Furthermore there exists indeed a formal similarity between the
Schrödinger equation describing stationary quantum mechanical
systems and the eigenequations describing self-referential
complex systems (see Varela 1981).

3. It is known that any measurement of a sufficient complex
system causes perturbations. At least for practical purpose it
seems to be wishful thinking to expect that such perturbations
can be suppressed deliberately. E.g.: Any questionnaire in
psychology exerts an influence on the assessment of a subject
which cannot be prevented.

4. The results of different measurements A and B depend on their
temporal order: $[AB - BA] \neq 0$. This is a more precise
formulation of the previous statement (3)

5. Some features of macroscopic complex systems seem to be

governed by uncertainty relations.

6. Some features of macroscopic complex systems exhibit stepwise rather than continuous changes, which resemble quantum leaps.

7. Macroscopic complex systems can show chaotic behaviour that cannot operationally be distinguished from the stochastical behaviour of a quantum mechanical system.

8. In some macroscopic (complex) systems "anomalous" correlations have been observed between psychological conditions (or variables) of a human "observer" and variables of an independent physical process. These correlations exist in spite of careful shielding against any know physical signal transfer which could have produced them (Jahn 1981, Jahn & Dunne 1987). It is assumed that these correlations are non-local in nature like the EPR-correlation (see paragraph 4).

The arguments 1. to 7. cannot be regarded as sufficient conditions for the assumption that complex systems can be considered as "non-classical" since in any case a suitable classical model can be given to describe the effect. However the situation resembles very much to the situation of atomic and molecular physics before the invention of quantum theory (see Heisenberg 1976): The spectra of specific atoms or molecules could be calculated with an enormous degree of precision on the basis of classical models. However, the parameters of the corresponding equations had to be determined individually and post hoc and no "explanation" for these specific values could be given. In psychology, in a similar way very distinct classical models can be found, for instance, for learning curves of individuals, which, however, do not allow a generalization of the adapted special system parameters (Geert 1989, see also Hill 1986).

In contrast, if argument 8 could be experimentally demonstrated, this would be a necessary and sufficient condition for a "non-classical" structure. The existence of "genuine" non-local correlations is not compatible with any classical model. This is the essence of Bell's theorem (Bell 1966).

Until now we have not specified what is meant with the notion of complex systems. In the moment we use the concept of complexity in a purely phenomenological way. Attempts to quantify complexity are still rather unsatisfactory (see Kornwachs & Lucadou 1975, 1976, 1978, 1982) because - and this is the point we want to stress here - they still adhere to classical models.

2. Categorical analysis of complex systems

In order to find an appropriate "non-classical" description of complex systems it is not useful to expect that physical concepts could be applied directly to such systems or that the description language which is normally used to describe these systems could be derived from physics. This could only be done if the reductionism problem would have been solved, but for most natural complex systems this is not the case. For instance, until now it is not possible to describe psychological constructs on the basis of physical concepts (see Büchel 1975). The idea that complex systems show a structure which could better be described by a non-classical model with a similar structure as quantum theory does therefore not imply that quantum theory could directly be applied to those systems, it rather implies that a "categorical analysis" is needed to find out which constructs on the description level of the complex system exhibit the underlying quantum-theoretical structure and may act as corresponding observables. In some cases it may be possible to identify these observables with certain physical quantities in a "metaphorical" way (see Jahn, Dunne 1986).

From such a phenomenological point of view the hierarchical structure and the self-reference of natural complex systems seem to be prominent features of complexity. Open systems, dissipative structures, thermodynamic systems far from the equilibrium, chaotic systems, strange attractors are not regarded as complex systems as such but as prerequisites for the latter. To describe natural complex systems Varela (1981) has introduced the notion of "organizational closure": "An organizationally closed unity is defined as a composite unity by a network of interactions of components that (1) through their interactions recursively generate the network of interactions that produce them, and (2) realize the network as a unity in the space in which the components exist by constituting and specifying the unity's boundaries as a cleavage from the background."

In the case of quantum physics the "interactions" are described by the specific form of the Hamiltonian H or any other operator and the "network", namely the set of eigenfunctions is recursively generated by the Schrödinger equation. The constraints that produce "unity's boundaries as a cleavage from the background" turn out to be conservation laws or more generally speaking symmetries of the system.

From this point of view an atom with its spectrum of eigenfunctions can be considered as an organizationally closed system as long as it is not disturbed too much by other processes such as collisions or chemical reactions, which means that certain conservation rules or symmetries remain intact. Metaphorically, one could also say that an atom is the outcome of a "self-observation" or in the context of psychology that for instance self-awareness of a person can be regarded as a kind of "psychological atom" (see also Jahn & Dunne 1986).

The concept of organizational closure introduces a natural way to define a boundary between the "inside" and the "outside" of the organizationally closed system and herewith a possibility to define hierarchical nested systems which are organizationally closed and contain other organizationally closed subsystems.

Another result of such a categorical analysis is the concept of "pragmatic information" as it was proposed by E.U. v. Weizsäcker (1974, see also Gernert 1985, Kornwachs & v. Lucadou 1982, 1984, 1985): The basic idea is that the "meaning" of any given information depends on the context of the whole system which processes the pragmatic information. It can be quantified by the (potential) action it exerts on a given subsystem. It is assumed that it has a dimension analogous to the physical dimension of an action, namely (m * l**2 * t**-1) and that similar to the observables in quantum theory it cannot be formalized as a scalar function, but only as an operator in a Hilbertspace.

He further proposed that pragmatic information should be written as a product of two observables which he called "Erstmaligkeit" E (novelty) and "Bestätigung" B (confirmation):

$$I = E * B \tag{1}$$

This approach takes into account that each portion of meaningful information must presuppose a certain prestructure (confirmation) - for instance, a native language - in order to be understood by the (receiving) system but also something new in order to produce a change. For instance, a joke in a foreign language which is not known would not cause anybody to laugh (no confirmation), and a joke which is already known would not do so either (no novelty).

The concept of action plays a predominant role in quantum physics because it can only exist in portions of h (Planck's constant), which is the minimum action which is transfered in any kind of measurement. Consequently it is assumed that there exists also a minimum action of pragmatic information i on a subsystem if pragmatic information is processed. This is a consequence of the assumption that the perturbation of a system in a measurement cannot be made arbitrarily small (see 3. paragraph 1.). Until now it is not clear whether h = i holds. Furthermore it should be possible to formulate uncertainty relations for systemtheoretic observables which take into account the minimum action of pragmatic information (see Kornwachs 1987) in an analogous way to quantum theory.

3. Complementarity in complex systems

Some authors (see Patee 1987) have argued that structure and function may be complementary categories, however they failed to give a precise formulation of the concept of complementarity. In analogy to quantum theory we require that complementary categories should be observables A, B which fulfil the following conditions:

1) [AB - BA] ‡ 0 which means that the information on the system after the measurement of A and B will be different if A is measured previous to B or vice versa.

2) Dim(A*B)=G(m * l**2 * t**-1) which means that the dimension of the product of the observables A and B should be a mathematical function G of the equivalent to the physical dimension of an action (m=mass, l=length, t=time). In the most simple case the function G is an equality.

For the categories of structure S and function (or behaviour) F it can indeed be made plausible that they meet the above stated two conditions. Any input-, output- or state transition function which describes F must be given as a function either in real time f(t) or in terms of system steps fi. In both cases they contain information on changes per time interval such as information rates, counting rates, switching rates and so on. The dimension of a rate is (t**-1). The category of a structure is more complicated. By definition we assume that it does not contain the concept of time. This does not mean that structures may not depend on time. Certainly, structures may be spread out in space but it is not clear how to give a more precise description in general. A good starting point may be to look at a connectivity matrix where the structure of a system is described by its elements and their connections among each

other. The connectivity matrix MiJ is fully determined by the
three values M,i,j where M may describe the nature of the
connections and i,j the two elements which are connected among
each other. It is obvious that i and j have the same categorical
status which means that it is useful to assume that they
contribute to the same dimension. M defines not only the nature
of the connection but also the nature or a property of the
elements. Thus the categorical equivalent of M is a property
like the mass of a solid and therefore we could describe its
dimension as a generalized mass. The categorical meaning of i
and j contains the notion of a distance otherwise it would not
be necessary to "connect" two different elements. Therefore it
is useful to attribute to i and j independently the dimension of
generalized length. As a result we get for the dimension of
structure Dim (S) = (m * l**2). If we take the product of S and
F we can indeed fulfil condition 2.

The question whether a measurement on the structure S of a
system perturbes a measurement on the function F of the same
system seems to be trivial in any practical case of biological
or psychological systems. However the important question will be
whether such a perturbation can be avoided within a deliberately
chosen accuracy. This is indeed still an unsolved problem and
the proponents of classical ontology still hope to demonstrate
that this is possible. At least they cannot go beyond Planck's
constant h but for practical reasons h is far too small to
detect direct macroscopic consequences of the underlying
uncertainty relations. It seems much more plausible, however,
that the action which is transfered to the complex system during
a measurement depends on the context of the specific system in
question. Thus we can argue that any measurement implies a
minimum amount i of pragmatic information which is transfered to
the system. Formally we can express this idea by the following
uncertainty relation:

$$\Delta F * \Delta S \geq i \qquad\qquad\qquad (2)$$

Let us now assume that we transmit a portion of pragmatic
information to a system using two different representations. If
the system is able to "understand" both representations we can
assume that the system has got more confirmation B than if only
one representation had been "perceived". From the point of view
of the receiving system, this means that it has a higher
structural variance ΔS because it can potentially react to (or
understand) different prestructures (for instance, different
languages). If the system is also able to incorporate much
novelty E this would mean that it is able to process much
pragmatic information. Of course much novelty can only be
incorporated if the system is able potentially to react to it
or, in other words, has a sufficient (potential) functional
variance ΔF. As a result we can categorically connect structural
variance ΔS with confirmation B and functional variance ΔF with
novelty E of an incoming pragmatic information I. Thus we can
rewrite inequality (2) in the following way:

$$\Delta F * \Delta S = E * B \geqq i \tag{3}$$

If we consider the (potential) reaction of the system to the
given pragmatic information, it may lead either to a change in
structure S or behaviour (function) F. Potential changes in
behaviour can be described as autonomy A of the system. If, on
the other hand, the potential action of the system results in
pure changes of the structure this can be described as
reliability R since the system has compensated the pragmatic
information without changing its function or output. In most
cases, however, both S and F may change. However, in any case
the "action" of the pragmatic information I has to be preserved.
This leads to the equation:

$$I = R * A = n * i, \quad n \in \mathbb{N} \tag{4}$$

Equation (1) can be considered as the description from "outside"
and equation (4) from "inside" of the system. If we put
(1),(2),(3) and (4) together we get:

$$B * E = R * A = B' * E' = n * 1 \qquad (5)$$

in which B' and E' belong to a different receiving system.

This equation describes the action of a given pragmatic information on a system. Since R and A or B and E are corresponding pairs of complementary observables nothing can be said upon a partitioning of an outside pragmatic information I to reliability R and autonomy A "inside" the system without further specifying the "measurement" and the surface of the system which is the boundary between the "inside" and the "outside" of the system.

From equation (5) it can be concluded that any portion of pragmatic information I = B*E which interacts with a system (R*A) also produces pragmatic information, but with a new partitioning of novelty E' and confirmation B' which can be interpreted as the "reaction" of the (sub-)system. Therefore, it is an important question to specify the subsystem and its surface.

Without going too far into detail we now assume that pragmatic information can be considered as a constraint that may produce organizational closure on a subsystem. Assume that we do not interact (by measurements) with a subsystem which processes a certain amount of pragmatic information. If the system is rich or complex enough to be described as a self-referential or self-steering system the "self-measurement" of the system produces perturbations as well in relation to the function (behaviour) as in relation to the structure of the system which again causes changes in the "self-measurement". It can be argued that the permanent and unpredictable reorganizations of structure and function within the system would lead to chaotic states but the real behaviour of such systems shows that also ordered structures and behaviour may occur as it is observed in even more simple chaotic systems (see Schuster 1984). It can be

called the "eigenbehaviour" respectively the "eigenstructure" of
the system. Changes between different "eigenconfigurations" of
the system often exhibit a quantum-like character, insofar, as
such eigenconfigurations represent stable "attractors" whilst
the intermediate states belong to chaotic regions of the state
space.

Thus self-reference becomes a criterion for the occurrence of
quantumlike effects in complex systems. In paragraph 2 we argued
that quantum systems like atoms can be regarded as
"self-observing" systems. However, macroscopic systems
necessarily need a certain degree of complexity to become
self-referential (see v. Neumann 1966). There seems to be a gap
in the degree of complexity between microscopic and macroscopic
self-referential systems. Hence, we expect that the appearance
of quantum effects in macroscopic systems depends on their
complexity and only show up if a certain degree of complexity is
given.

In psychology many qualitative examples for uncertainty
relations as described above could be given (see for instance
Devereux 1972, Fahrenberg 1979), however, they cannot be
regarded as sufficient criteria for an underlying quantum
theoretical structure. Uncertainty relations also exist in
classical physics (for instance Küpfmüller's uncertainty
relation: $\Delta x * \Delta k \geq 1$ for a classical wave packet). In this
case, however the uncertainty depends on practical
considerations (for instance on the finite length of a
measurement which restricts the accuracy of the measurement of
the corresponding "complementary" observable). In this case
however the accuracy can be increased deliberately.

The only feature of quantum physics that has no classical
equivalent is "non-locality". Non-locality is a direct
consequence of the superposition-principle and the
quantum-theoretical axiom that the "collapse" of the
superposition occurs in the moment of the measurement.

4. Non-locality in complex systems

The well-known EPR-effect (see Einstein, Podolsky & Rosen 1935)
in quantum physics which describes the spontaneous decay of a
two particle system into two single particles is a specific
example of the socalled "measurement problem". It demonstrates
that the "collapse of the state vector" is not only a problem of
separability but also of non-locality. Whenever the statevector
is collapsed the whole spacial distribution of the "wave
function" changes immediately. In the case of the EPR-effect the
measurement causes the change of a singlet state

$$|si\rangle = (1/sqr(2)) (|u+\rangle |v-\rangle - |u-\rangle |v+\rangle) \tag{6}$$

which cannot be factorized into two independent states u and v
to one of two finite factorized states

$$|sf1\rangle = |u+\rangle |v-\rangle \quad or \quad |sf2\rangle = |u-\rangle |v+\rangle \tag{7}$$

which occurs with equal probability. It can be shown (see Aspect
et al. 1982, Clausner & Schimony 1978) that equation (6) holds
even if the two parts u and v of the system are space-like
separated which explicitly means that the transition from si to
sf does not occur at the moment of separation! If we adopt the
realist's interpretation of the wavefunction we have to admit
that the change of the state function occurs instantaneously in
the spacelike separated system and thus violates the principle
of locality. At first glance this seems also to contradict the
requirement of Lorentz invariance.

But the seemingly contradiction to the theory of special
relativity is meaningless because an actual information transfer
is not possible by means of the non-local change of the
statevector because it cannot be measured directly. The only
measurable quantities are non-local correlations between two
spacelike separated measurement devices. At each separate

device, however, it can only be measured an equal distributed
ensemble of the two possible final states (see eq. 7). To
establish the existence of the non-local correlation a
coincidence measurement has to be made after which the results
of the two measurement devices are compared to each other. This
procedure, however, requires "normal" signal transfer. Thus the
discrepancy with special relativity vanishes if it is taken into
account that the non-local correlation can only be measured post
hoc. On the other hand it is important to emphasize that such
non-local correlations cannot occur in classical systems and
indeed Bell's inequality shows that any classical theory would
deliver predictions that are at variance with those of quantum
theory. In Bell's inequality the only assumption which is made
in relation to the classical system is that the final state of
the system is already reached at the moment of separation but
not at the moment of the measurement. Experimentally the
predictions of quantum theory could be confirmed (see Aspect et
al. 1982, Patty 1977).

It is not easy to find an analogous situation to the EPR-effect
in the realm of macroscopic complex systems because it is
difficult to define symmetries or corresponding conservation
rules for complex systems. Furthermore it is not clear which
kind of measurements would exhibit a non-local effect and how a
separation of the system into (at least) two parts could be
achieved. It is a matter of fact that even in quantum physics
only very specific systems and very specific observables (e.g.
spin measurements on a singlet system) show a difference between
the classical and the quantum theoretical expectation values.

From the phenomenological model discussed above it seem
plausible that the "internal correlations" or "coherent states"
of an organizationally closed system may be good candidates to
show non-local correlations. For this purpose it would be
necessary to prepare an organizationally closed system by means
of suitable pragmatic information which enables the system to
become self-referential but on the other hand allows a

separation of two measurements on the system in order to ensure
that the correlation is in fact non-local. In the
EPR-experiments the separation is achieved by the spacelike
separation of the two spin-measurement devices. This setup
guarantees that no signal transfer from one measurement device
to the other one is possible at the moment of the measurement.

In the case of macroscopic complex systems the separation can be
introduced in two different ways: 1. by a sufficient shielding
against any physical influence and/or 2. by a temporal and thus
causal displacement of the two independent measurements. Since
it is very difficult experimentally to realize these conditions
it is necessary to test independently whether the alleged
non-local correlation includes an underlying signal. Thus we can
define a non-local correlation in a complex macroscopic system
as a correlation between two independent measurements which are
causally separated in the sense that no signal transfer is able
to produce the correlation.

The distinction between signals and correlations is of outmost
importance in this respect. Only in classical systems any
(meaningful) correlation is based on a signal transfer. This
does not mean that there exists no cause for the quantum
mechanical non-local correlation but here the cause is not a
functional one like a signal transfer but a structural one like
a conservation rule or a symmetry principle of the system. In
classical systems any correlation for which no signal transfer
can be found is considered as "pseudo correlations". The case of
"partial correlations" where a common factor produces a
correlation between two independent measurements cannot be
regarded as a counterexample because in this case there is still
a signal transfer from the common factor involved. In non-local
correlations such a signal transfer cannot be found. This is
obvious in the EPR-experiment otherwise it would be possible to
construct a "superluminal" telegraph with an EPR-apparatus. If
in fact the correlation of the two spin measurements would be
caused by an "instantaneous" signal transfer at the moment of a

measurement on one particle a signal could be coded by a
sequence of measurements and detect this sequence independently
at the spacelike separated measurement device. If on the other
hand such a (coded) signal could be found this would indicate
that the measured correlation is a local one.

5. Experiments to prove non-locality in complex systems

The first experiment which could be interpreted as a search for
macroscopic non-local effects was designed to test the so-called
"subjectivistic interpretation of quantum physics". This
interpretation assumes that the "conscious observer" of a
quantum mechanical measurement "collapses the state vector" (see
Wigner 1967).

In general, physicists do not take seriously the "subjectivistic
interpretation of quantum physics". They simply conceive it as
an unrealistic metaphor to illustrate the measurement problem.
Therefore it is not astonishing to find hardly no experimental
investigation along these lines although it would be rather
simple to accomplish such an experiment. The only experiment
which was published in this context led to a negative result
which means that no difference was found between a "normal"
quantum physical measurement and a measurement in which a
previous observer could have "collapsed" the statevector before
the measurement (see Hall, Kim, McElroy & Shimony 1977). In
literature this experiment is normally cited as the definite
proof that the "subjectivistic" interpretation of quantum
physics must be wrong (see for instance Selleri 1983). A closer
examination of the experiment however shows that the conclusion
the authors drew is not conclusive in relation to the
measurement problem. They only tested if a second observer could
"feel" whether a quantum event has already been observed by a
previous observer or not. However this is an additional
psychological assumption which might not be valid. Nevertheless,
the idea that the observation of a quantum-mechanical event
might include some psychological correlates to the observer

seems to be rather fruitful.

Wigner's proposal that the consciousness of the observer finishes the measurement could easily be tested if the observer would consciously try to "influence" the outcome of a measurement. However, this "influence" should not be considered as a normal physical influence or a signal because this would contradict the notion of the superposition of wavefunction of the observer and the observed. As in the EPR-case a signal transfer via the wavefunction is not possible. Therefore it is not expected that the expectation values of the quantum mechanical prediction could be arbitrarily changed by the observer in a way that would not be "allowed" by the physical process itself. Nevertheless it is still imaginable that a non-local correlation between the physical and the psychological measurement could exist.

From our system-theoretical point of view the superposition of the quantum mechanical state of the observer and the state of the observed process creates an organizationally closed unit if and only if the observer attributes some "meaning" to the process. In this case the process produces pragmatic information which is "understood" by the observer. Moreover, this view allows a "non-subjectivistic" interpretation of the measurement process because the observation produces real physical changes in the "observing subsystem" which again contributes to the state of the whole system . However, if the observed process is "meaningless" for the observer no pragmatic information and thus no change of the system is produced and there is no reason why observer and observed should be regarded as a joint system. It is merely a matter of semantics whether such an observer should still be called "observer".

The purpose of the following experiment was to find out whether the observation of "meaningful" quantum-physical events may lead to non-local correlations inside the joint system.

Naturally, the main experimental problem was how to establish an organizational closure between the observer and the quantum system and to make a prediction about the non-local correlations which might occur in the system due to the alleged organizational closure. Another problem was how to guarantee that no other physical influences of the observer could be responsible for the possible correlations in the system except the assumed "non-local observer effect". To solve the latter problem we separated the quantum process spatially from the observer and used additional electromagnetic, thermal, and mechanical shielding (as it was also used in Hall's experiment).

Since it would exceed the scope of this paper we will report here only the essential parts of the experiment. A detailed report on this experiment has been recently published (see Lucadou 1986).

The quantum physical process used in the experiment was the radioactive decay (e-) of a SR-90 source of about 0.045 mCi centered in a circle of five Geiger-Mueller-tubes. The observers of the process were 299 unselected subjects (mainly students). In one part of the experiment (which we describe here) they could observe the decay process "rather directly" in such a way that changes in the momentary decay rate could be observed. For this purpose a counter sampled a fixed number of events (e.g. 112) from one Geiger-Mueller tube. The resulting sampling time was used as a measure for the momentary decay rate. This momentary decay rate (sampling time Z_i) was then compared with the previous one. A decrease of the momentary decay rate was indicated as a "1" and an increase as a "0". If no change of the decay rate occurred the event was omitted.

$$Z_{i-1} > Z_i = "0"$$
$$Z_{i-1} < Z_i = "1" \tag{9}$$
$$Z_{i-1} = Z_i = "no\ result"$$

However, the latter case did not occur often due to the

sufficient variance of the sampling time.

It can be shown that the result of this procedure is a
Markow-chain which is completely determined by the following
transition matrix:

$$
pij = \begin{vmatrix} p00 & p01 \\ \\ p10 & p11 \end{vmatrix} = \begin{vmatrix} 1/3 & 2/3 \\ \\ 2/3 & 1/3 \end{vmatrix} \tag{10}
$$

It can further be shown that this matrix holds in general and
does not depend on the distribution of the random process if the
measured single events are stochastically independent.

The resulting Markow-chain was displayed to the observer on a
vertical string of 16 red lights (LEDs) in such a way that,
beginning with a string of the 8 lower lights switched on, a "1"
was indicated by the next light at the top of the line being
switched on and a "0" was indicated by switching off the light
at the top of the line. Since the generation rate of the
Markow-sequence was about 10 per second the observers got the
impression of a fast randomly fluctuating "thermometer column".

This technique to measure the momentary change in the decay rate
by the Markow-chain was used for all five counters around the
radioactive source. However, only one of them was displayed to
the observer but all five sequences were stored in a
computerfile. After the experiment the five parallel
Markow-chains could be compared with each other by means of
their cross-correlation function. The idea behind this procedure
was to find out whether possible changes in the decay rate could
be due to possible common local influences on the radioactive
source or the Geiger-Mueller tubes or the common high voltage
supply or the subsequent electronics.

Furthermore the stored Markow-chains were objected to several
detailed analysis procedures such as auto-correlation techniques

and statistical tests in order to detect any deviation from the "normal" behaviour which may artificially produce an "observer effect". Such artefacts, however, are rather unlikely and could only be produced if electrostatic influences, electrical transients, thermal effects or mechanical influences could overcome the shielding. In such a case, however, they should be detected by these analytical methods.

The main problem of introducing an organizational closure between the quantum-physical system and the observer was solved in the following way: The observers were instructed (in a standardized way) to observe the fluctuating light column at the display and to "push" it to the top of the row as often as possible or to keep it there as long as possible. This should be done only by "concentration", "mental powers" or "psychokinesis" or what so ever. It is known by other studies that many persons in fact believe to possess such "powers" and that the attitudes concerning these beliefs are emotionally loaded very much (another part of the study was to find out more detailed information about such belief systems). It was assumed that the personal interest or emotional involvement of the self-selected(!) subjects in such questions would generate a motivational factor that should be sufficient to produce an organizational closure. However, since the subjects may consciously or unconsciously react quite different to such an "odd" instruction it cannot be assumed that they would consistently "produce" an identical observer-effect.

Moreover, quite different observer-effects could be expected depending on the specific structure of each self-organizing system. This means that it is plausible to assume that the observer effect depends on personality characteristics of the observer. From this point of view it was decided to use personality questionnaires as independent psychological variables. Of course, the subjects had to fill out the questionnaires before the experimental session took place to avoid that the feedback information on the random process might influence the way the subjects filled out their questionnaires. Thus the questionnaires can be regarded as an independent measurement on the "observer sub-system" which is causally separated from the physical random process.

The following psychological variables of the subjects were measured: Believes about "psychic powers"(SG), locus of control (IPC) (Krampen 1981), personality traits (FPI) (Fahrenberg et al. 1977), present state (mood) (EWL) (Jahnke & Debus 1978) and a confidence score (PRIOR) of the subject's success in the experiment.

Each subject did 4 runs with a Markow-chain of 600 trials under two conditions, namely with and without feedback on the quantum physical process. (In fact the experiment also contained two further conditions with a different random generator which, for brevity, are not reported here. The final results and conclusions under these conditions, however, are similar but less distinct).

To summarize the situation we can hypothetically describe the whole system under feedback condition as a quantum physical self-referential system with two parts which we call the "observer" and the "observed". The organizational closure is introduced by the instruction and by the feedback to the observer. The meaning of the instruction is only defined in the context of the given feedback information. Two independent (locally separated!) sets of variables are measured: The physical variables of the Markow-chains (the observed) and the psychological variables of the observers. If the underlying structure of this system would in fact be isomorphic to the axiomatic structure of quantum theory we would expect significant correlations between the psychological and the physical variables indicating the existence of non-local correlations within the system. In the non-feedback condition, however, the system is totally separated because there is no feedback loop which may establish the organizational closure.

There is no "meaning" of the process for the subject because he or she cannot observe it. As a consequence we would not expect significant correlations between the psychological and physical variables.

The relevance of the feedback information is especially taken into account by a specific physical variable DIM which measures the "pragmatic information" of the feedback given to the observer by the display. As we have already discussed, the pragmatic information can only be defined in the context of the system. In this case the relevant context is defined by the instruction and the special form of the display. Since the instruction says that the observer had to "push" the column to the top of the row the meaning of a "hit" ("1") depends on the actual position of the light column on the display. If, for instance, the column has already reached the top the observer could not even detect a further hit. If the column has reached the bottom a further "miss" ("0") could not be detected and of course the challenge for the observer is quite different. The variable DIM weights the hits depending on it's actual position on the display. If pragmatic information would in fact be the relevant descriptor of organizational closure we would expect this variable to show consistently the highest correlations to the psychological variables, but only under the feedback condition. All the other physical variables which are not discussed here in detail, measure special statistical features of the Markow-chains and were used to find out whether (local) signals might have overcome the shielding and may be superimposed on the Markow-chains. To a certain extent they may also contribute to non-local correlations since they may also exhibit certain features of pragmatic information (for further details see Lucadou 1986).

Finally there should be made a remark concerning the characteristics of a possible observer effect on the physical variables. One of the usual misunderstandings in this respect is the expectation that a specific signal could be found in the physical variables alone that indicates the existence of an observer effect without correlating them with the psychological variables. However, this would be in variance to the assumed non locality of the state function. Any real signal would include instantaneous signal transfer which would contradict to the requirement of Lorentz invariance. This does not mean that it would be impossible to predict the size and direction of the correlation. However, any experimental manipulation to use this correlation as a mean to transfer information would change the system in such a way that the correlation would vanish. In a recent paper (Lucadou 1989c,d) it is shown that this property of non-local correlations leads to a new understanding of the problem of repeatability of psychological experiments and that the ususal methodological requirement of double blindness is not sufficient to prevent an experimenter effect (see Rosenthal 1969).

From this point of view we would not expect a specific physical process or variable indicating an "influence of the observer" on the physical process but merely correlations between pairs of the independent, randomly distributed physical and psychological variables. The notion of "an influence of the observer" which is often used in the context of the so-called "subjectivistic" interpretations of quantum theory (and which we have used as a psychological trick to attribute "meaning" to the random process) is indeed inconsistent with quantum theory because it would imply a non-local signal transfer. Thus we cannot distinguish the observed physical process from its "normal" random behaviour without making reference to the psychological variables. The experimental distribution of all physical variables of the random process shows indeed no deviation from the expectation values under the null hypothesis.

On a more qualitative basis and from some previous investigations we were able to make some predictions about the psychological content of the non-local correlations. Also from the point of view of the required "flexibility" of the observer to "understand" the "meaning" of the feedback information, we would expect that those observers who can be classified as non-anxious, non-neurotic, non-depressive, non-inhibited and extraverted would be more successful in establishing an organizational closure than those who can be classified along the corresponding antagonistic categories. These hypotheses were formulated before the experiment to avoid any post-hoc attribution.

The results of the experiment are shown in two correlation matrices of figure 1, one for the non-feedback case and the other for the feedback condition. Each line of the matrices represents a psychological variable (factor) and its correlation to the physical variables which are represented by the rows. A significant correlation is indicated by a hatching of the crossing. A light hatching means a correlation at a significance level of $p = .05$ and a dark hatching $p = .01$. If a line shows up many correlations this means that the corresponding psychological variable is a relevant one for the establishment of the organizational closure. If many correlations are indicated in a row this means that this physical variable is very sensitive in relation to the non-local correlation. Figure 2 gives the names of the variables which roughly describe their meaning (for details see Lucadou 1986). A rough estimation shows that the number of the significant correlations under the non-feedback condition does not exceed the number of random correlations which would be expected at this significance level. This means that these correlations can be considered as accidental correlations without meaning. Thus we can conclude that according to our expectations we could not find a non-local correlation under the non-feedback condition.

Under the feedback conditions, however, the number of correlations is increased significantly. Furthermore the tendency of most of the correlations is in agreement with the expectations of our hypotheses. This is indicated in the last two rows. The row labeled H1/T shows the expected tendency of

the correlation ("+" for positive correlations and "-" for
negative ones, and "?" if no hypothesis was formulated
beforehand) and the row TNDZ shows the measured direction of the
correlation (if no significant correlation was found this is
indicated by "?").

Finally, a direct support of the relevance of the feedback and
also a support of the phenomenological model of pragmatic
information can be found in the fact that indeed the variable
measuring the pragmatic information of the feedback DIM showed
not only the highest number of significant correlations to the
psychological variables but also consistently the strongest
correlation coefficients.

However, we still have to ask whether this result could have
been produced by some unexpected local physical influences which
may have overcome the shielding. For this purpose several
correlation techniques (as mentioned above) were used to find
physical signals which may modulate the random process. Without
going into the details here, it can be said that within the
accuracy of the methods applied, neither a common influence on
the five Geiger-Mueller tubes could be found nor does any other
variable exhibit a distribution indicating a signal different
from stochastical behaviour.

These results can be interpreted in the following way: "Allowed"
stochastical fluctuations of the psycho-physical system (process
plus observer) can be "triggered" or "selected" by the observer
according to his or her psychological structure if and only if
an organizational closure has been established. However, this
does not imply that the observer could "influence" or control
the observed as could be done by local signal transfer. Feedback
is an essential condition for this process. The "observer
effect" can only be measured as non-local correlation between
psychological and physical variables. This non-local correlation
mirrors the organizational closure of the system which is caused
by the pragmatic information of the feedback and the

self-reference of the observer.

Furthermore the result is suggestive for the assumption that
indeed macroscopic, self-referential complex systems may exhibit
an underlying structure which is isomorphic to the structure of
quantum theory. This may also be regarded as a further argument
for the assumption that the axiomatic structure of quantum
theory is system theoretical in nature. A recent article of Jahn
& Dunne (1986, 1987) presents a first approach, how such a
quantum mechanical structure could be implemented in
psychological models.

In the meantime the results of this study had been corroborated
by independent data and independent methods. The data stem from
different experiments which where performed to measure an
"observer effect" on quantum physical stochastical processes
(see Nelson & Radin 1989). Unfortunately in most of these
experiments the psychological variables are only described
qualitatively. Thus the impression arises that the observer
could in fact exert an influence on the physical process. A
closer analysis, however, shows that the data do not support the
notion of an "influence" because they do not hit the criteria
for signals (for details see May, Hubbard & Humphrey 1985).
Since most researchers still adhere to the classical paradigm
the discussion about "influence" versus "non-local correlation"
is still rather controversial (see Lucadou 1989c).

6. Possible consequences for dealing with complex systems

If further research should in fact corroborate the idea that
complex systems can be better described by a theoretical
structure which is isomorphic to quantum theory than by
classical models this would include several consequences which
are not only of theoretical but also of methodological and
practical relevance.

From our phenomenological point of view the existence of

complementarity, uncertainty relations and the non-local
correlations in complex systems imply not only quantitative but
primarily qualitative differences to classical models. As long
as the model is not elaborated more in detail it is not useful
to discuss the former ones. But the latter are interesting
enough: If structure and function would turn out to be
complementary observables for which equation 2 holds this would
imply that any measurement of a sufficient complex system would
lead to a "preparation" of the system which either puts more
emphasis on the structural or on the functional aspect. This
means that the systems may show a rather specific behaviour
depending on the way we "look" at it. From equation 5 follows
that a preparation of a system which leads to a high degree of
autonomy of the system will make the system less reliable and
produces a lot of novelty while the same system could produce
much confirmation if it is prepared in a mode of reliability.
This means that a "close examination" of a complex system could
effectively limit the reactions of the system (for examples see
Lucadou 1989a).

Another limitation of the behaviour of a complex system deals
with possible internal non-local correlations. If for instance
two measurements on the system include such non-local
correlations. A replication of this measurement is only possible
if there is no actual possibility to use the non-local
correlation as a signal transfer (see Lucadou 1989d). This leads
to a "second order experimenter effect": If the experimenter
knows the experimental finding of the previous experiment and
the experimental setup would allow him to use this knowledge to
introduce a signal transfer via the non-local correlation this
would cause it at least to decline or vanish. There are many
examples in psychology where later replications of an initial
experiment show such an "inter-experimenter decline effect" (see
Rosenthal & Rubin 1978).

However, the main practical relevance of our phenomenological
model is the possibility that quantum-mechanical fluctuations of

a physical system may depend on psychological conditions of an observer. Even if this "observer effect" cannot be applied in a purposeful manner (because this would include a signal transfer) it is still possible that certain fluctuations occur in an unpredictable way if certain psychological conditions are met. The author knows several anecdotical or jet unpublished reports indicating that in the neighbourhood of persons under stress the error rate of technical devices is increased. In all these cases a psychological relevance of the incidents was attributed by the persons who reported the events. Under the classical paradigm this could only be interpreted as a post hoc attribution of the reporters but from our point of view it might well be that the psychological conditions indeed created an organizational closure which made it possible that the random fluctuations coincided with the psychological "need" for it.

Even though such ideas seem to be rather speculative there is no reason not to consider them seriously because it is relatively simple to investigate them experimentally and also because they may have some relevance for our understanding of complex systems.

Address of the author:
Dipl.-Phys. Dr.rer.nat. Dr.phil. Walter v. Lucadou,
Hildastraße 64, D-7800 Freiburg i. Br.

References

Aspect, A., Dalibard, J., Roger, G. (1982). Experimental Test of Bell's Inequalities Using Time-Varying Analyzers. Physical Review Letters, 25, 1804.

Bell, J.S. (1966). On the problem of hidden variables in quantum mechanics. Review of Modern Physics, 38, 447.

Büchel, W. (1975). Naturphilosophische Überlegungen zum Leib-Seele-Problem. Philosophia Naturalis, 15, 308-343.

Bünning, E. (1943). Quantenmechanik und Biologie. Naturwissenschaften 31, 194–197.

Clausner, J.F., Shimony, A. (1978). Bell's theorem: experimental tests and implications. Reports on Progress in Physics, 41, 1881–1927.

Clough, S. (1988). The wrong side of the quantum tracks. New Scientist, 117 (No. 1605) 24 March: 37–40.

Devereux, G. (1972). Ethnopsychoanalyse. Die komplementaristische Methode in den Wissenschaften vom Menschen. Frankfurt: Suhrkamp 1984

Einstein, A., Podolsky, B., Rosen, N. (1935). Can quantum mechanical description of physical reality be considered complete? Physical Review, 47, 777–780.

Fahrenberg, J. (1979). Das Komplementaritätsprinzip in der psychophysiologischen Forschung und psychosomatischen Medizin. Zeitschrift für Klinische Psychologie und Psychotherapie 27: 151–167.

Fahrenberg, J., Selg, H., Hampel, R. (1978). Das Freiburger Persönlichkeitsinventar FPI. Göttingen: Hogrefe (3.Aufl.)

Finkelstein, D., Finkelstein, S.R. (1983). Computational Complementarity. International Journal of Theoretical Physics, 22, (8), 753–779.

Geert, P.van (1989). Growth, Non-Linearity, and the Limits of Prediction. Paper presented at the ESSCS, 7th Workshop, Saint-Maximin-La-Sainte-Baume, Provence, France, June 1989.

Gernert, G. (1985). Measurement of pragmatic information. Cognitive Systems, 1, (3), 169–176.

Hall, J., Kim, C., McElroy, B., Shimony, A. (1977). Wave-Packet Reduction as a Medium of Communication. Foundation of Physics, 7, 759–767.

Heisenberg, W. (1976). Was ist ein Elementarteilchen? Naturwissenschaften, 63, 1–7.

Hill, O.W. (1986). Further Implications of Anomalous Observations for Scientific Psychology. American Psychologist, 1170–1172.

Jahn, R.G. (1981). The Persistent Paradox of Psychic Phenomena: An Engineering Perspective. Proceedings of the IEEE, 70, (2), 136–170.

Jahn, R.G., Dunne, B.J. (1986). On the Quantum Mechanics of Consciousness, with Application to Anomalous Phenomena. Foundations of Physics, 16, (8), 721–772.

Jahn, R.G., Dunne, B.J. (1987). Margins of Reality. The Role of Consciousness in the Physical World. San Diego, New York, London: Harcourt Brace Jovanovich.

Jahnke, W., Debus, G. (1978). Die Eigenschaftswörterliste EWL. Eine mehrdimensionale Methode zur Beschreibung von Aspekten des Befindens. Göttingen: Hogrefe.

Jammer, M. (1874). The Philosophy of Quantum Mechanics. New York, London, Sydney, Toronto: John Wiley & Sons.

Jordan, P. (1947). Verdrängung und Komplementarität. Hamburg: Storm.

Kawamoto, H. (1987). Macroscopic quantum tunneling in SQUID. Japanese Journal of Applied Physics Supplement, 26, suppl. 26-3, pt. 2, 1389–1390.

Kornwachs, K. (1987). Cognition and complementarity. In: Carvallo, M.E. (ed.): Nature, Cognition and Systems I. (pp. 95-127). Amsterdam: Kluwer.

Kornwachs, K., Lucadou, W.v. (1975). Beitrag zum Begriff der Komplexität. Grundlagenstudien aus Kybernetik und Geisteswissenschaft, 16, 51-60.

Kornwachs, K., Lucadou, W.v. (1976). Beschreibung und Entscheidbarkeit. Grundlagenstudien aus Kybernetik und Geisteswissenschaft, 17, 79-86.

Kornwachs, K., Lucadou, W.v. (1978). Funktionelle Komplexität und Lernprozesse. Grundlagenstudien aus Kybernetik und Geisteswissenschaft, 19, 1-10.

Kornwachs, K., Lucadou, W.v. (1982). Pragmatic Information and Nonclassical Systems. In R. Trappl (ed.), Cybernetics and Systems Research (pp. 191-197). Amsterdam, New York: North Holland.

Kornwachs, K., Lucadou, W.v. (1984). Komplexe Systeme. In: K. Kornwachs (Hrg.): Offenheit-Zeitlichkeit-Komplexität. Zur Theorie der Offenen Systeme (pp. 110-166). Frankfurt: Campus.

Kornwachs, K., Lucadou, W.v. (1985). Pragmatic Information as a Nonclassical Concept to Describe Cognitive Processes. Cognitive Systems, 1, (1), 79-94.

Kornwachs, K., Lucadou, W.v. (1989). Open systems and complexity. In: G.J. Dalenoort (ed.): The Paradigm of Self-Organization. (pp. 123-145). New York, London, Paris: Gordon and Breach.

Krampen, G. (1981). IPC-Fragebogen zu Kontrollüberzeugungen ("Locus of Control"). Göttingen: Hogrefe.

Lucadou, W.v. (1986). Experimentelle Untersuchungen zur Beeinflußbarkeit von stochastischen quantenphysikalischen Systemen durch den Beobachter. Frankfurt: H.-A. Herchen Verlag.

Lucadou, W.v. (1989a). Psyche und Chaos. Freiburg: Aurum Verlag.

Lucadou, W.v. (1989b). Non-locality in complex systems. In: Rossel, E. Heylighen, F., De Meyere, F. (eds.): Self-steering and Cognition in Complex Systems. Studies in Cybernetics Vol II. New York, London, Paris: Gordon and Breach (in print).

Lucadou, W.v. (1989c). Non-locality in Complex Systems - a Way out of Isolation? Paper for the Sugarloaf Meeting on "The Relationship Between Mind and Matter", May 4th to 7th, Philadelphia, (to be published).

Lucadou, W.v. (1989d). The problem of repeatability in experiments with complex systems. Paper presented at the 7th Workshop of the European Society for the Study of Cognitive Systems, St. Maximin-La-Sainte-Baume, Provence, France, 19-22 June 1989.

Lucadou, W.v., Kornwachs, K. (1983). The Problem of Reductionism from a System-Theoretical Viewpoint. Zeitschrift für Allgemeine Wissenschaftstheorie, 14, (2), 338-349.

May, E.C., Hubbard, G.S., Humphrey, B.S. (1985). New Evidence for Interaction Between Quantum Systems and Human Observers. Paper given at the Fourth Annual Meeting of the Society for Scientific Exploration at Charlottesville.

Morgenstern, R., Niehaus, A., Thielmann, U. (1976). Interference of autoionizing transitions in time-dependent fields. Journal of Physics B: Atomic and Molecular Physics, 10, (6), 1039-1058.

Nelson, R.D. & Radin, D.I. (1989). Statistically robust anomalous effects: Replication in random event generators. Foundations of Physics, 20 (in print).

Neumann, J.v. (1966). The Theory of Self-Reproducing Automata (pp. 74-80). Urbana IL: University of Illinois Press.

Pattee, H.H. (1987). The measurement Problem in Physics, Computation and Cognition. (To be published).

Paty, M. (1977). The recent attempts to verify quantum mechanics. In: J.L. Lopes, M. Paty (eds.). Quantum Mechanics, a Half Century Later (pp.261-289). Dordrecht: D.Reidel.

Ristik, M.L. (1981). Makroskopische Quanteneffekte. In: J. Nitsch, J. Pfarr, E.-W. Stachow (Hrg.): Grundlagenprobleme der modernen Physik (pp.243-255). Mannheim, Wien, Zürich: Bibliographisches Institut.

Rosenthal, R. (1969). Interpersonal expectancies: the effects of the experimenter's hypothesis. In: R. Rosenthal and R. Rosnow (eds), Artifact in Behavioral Research (pp. 181-277). New York: Academic Press.

Rosenthal, R., Rubin, D.B. (1978) Interpersonal expectancy effects: The first 345 studies. Behavioral and Brain Science, 3, 377-415.

Schuster, H.G. (1984). Deterministic Chaos: An Introduction. Weinheim: Physik Verlag.

Sellerie, F. (1983). Die Debatte um die Quantentheorie. Braunschweig, Wiesbaden: Vieweg.

Varela, F.J. (1981). Autonomy and autopoiesis. In: G. Roth, H. Schwengler (eds.): Self-organizing systems (pp. 14-23). Frankfurt/New York: Campus.

Walker, E.H., May E.C., Spottiswoode, S.J.P. & Piantanida, T. (1987). Testing Schrödinger's Paradox with a Michelson Interferometer". International Workshop on Matter Wave Interferometry. Technische Universität Wien, Atominstitut der Österreichischen Universitäten, 14-16 September.

Weizsäcker, E.v. (1974). Erstmaligkeit und Bestätigung als Komponenten der pragmatischen Information. In: E. v. Weizsäcker (Hrg.): Offene Systeme I (pp. 83-113). Stuttgart: Klett.

Wigner, E.P. (1967) Symmetries and Reflections. Bloomington: Indiana University Press.

Versuchsbedingung: Nondisplay Markoff

Physikalische Variablen

Psychologische Variablen	F1	T	ZC	ZD	DIM	F2	ETG	BEC	PIG	F3	ETS	BES	PIS	F4	D	K	ZE	F5	V	R	F6	ZTA	ZB	H1/T	INDZ
PRIOR1																	▢							?	?
SGFAK1														▢	▨									?	?
SGFAK2																								?	?
SGFAK3														▢	▢									?	?
IPCFAK1																								?	?
IPCFAK2																								?	?
IPCFAK3																								?	?
FPISK1														▨	▢	▢								?	?
FPISK2		▢																			▨	▢		?	?
FPISK3																								?	?
FPISK4																	▢							?	?
FPISK5																								?	?
FPISK6						▢																		?	?
FPISK7				▢	▢																			?	?
FPISK8	▢			▢	▢																			?	?
FPISK9																								?	?
FPISK10								▢																?	?
FPISK11														▢			▨							?	?
FPISK12																								?	?
EWLFAK1														▢	▢									?	?
EWLFAK2														▢										?	?
EWLFAK3														▢	▢	▢								?	?
EWLFAK4				▢																				?	?
EWLFAK5											▢				▢	▢						▢		?	?
Anzahl n	180	299	180	180	180	180	180	180	180	180	180	180	180	180	180	180	180	180	180	180	180	180	180		

Versuchsbedingung: Display Markoff

Physikalische Variablen

Psychologische Variablen	F1	T	ZC	ZD	DIM	F2	ETG	BEC	PIG	F3	ETS	BES	PIS	F4	D	K	ZE	F5	V	R	F6	ZTA	ZB	H1/T	INDZ
PRIOR1					▢																			+	+
SGFAK1										▢														+	?
SGFAK2																					▨	▢		+	?
SGFAK3																								+	?
IPCFAK1					▢						▢													–	?
IPCFAK2																								–	?
IPCFAK3															▢									+	?
FPISK1	▢	▨		▢	▢						▢	▨	▢											–	–
FPISK2																								?	?
FPISK3	▢	▢	▢	▢	▨						▢						▢							–	–
FPISK4	▢	▨	▢	▨		▢		▢						▢			▢	▢						–	–
FPISK5					▢																			+	+
FPISK6		▢	▢								▢													+	+
FPISK7											▢													?	?
FPISK8		▢	▢		▨						▢		▢		▢		▢							–	–
FPISK9																								?	?
FPISK10																								+	+
FPISK11											▢		▢											?	?
FPISK12	▨	▨									▨									▨	▨			+	+
EWLFAK1				▢	▢		▢													▢				–	–
EWLFAK2					▨																			–	–
EWLFAK3																								+	?
EWLFAK4						▢					▢										▢			–	?
EWLFAK5		▢	▢	▢							▢								▨	▨				+	+
Anzahl n	299	299	299	299	299	299	299	299	299	299	299	299	299	299	299	299	299	299	299	299	299	299	299		

Figure 1: Correlation matrices for the non-feedback condition
and for the feedback condition

Figure 2:

List of the variable names

psychological variables		physical variables	
symbol	name	sym	Name
PRIOR1	confidence-score	T	number of "hits"
SGFAK1	personal belief	V	variance
SGFAK2	belief in own powers	R	runscore variance
SGFAK3	motivation	D	Kolmogorow-Smirnow
IPCFAK1	externality	K	auto-correlation
IPCFAK2	internality	K'	cross-correlation
IPCFAK3	fatalism	ZTA	generation time
FPISK1	nervousness	ZB	mean or variance
FPISK2	spont. agressivity	ZC	joint fluctuations
FPISK3	depressivity	ZD	time fluctuation
FPISK4	excitability	ZE	chaotic influence
FPISK5	sociability	ETS	positive novelty
FPISK6	calmness	BES	duration of ETS
FPISK7	dominance	PIS	product of ETS,BES
FPISK8	inhibition	ETG	negative novelty
FPISK9	openness	BEG	duration of ETG
FPISK10	extraversion	PIG	product of ETG,BEG
FPISK11	lability	DIM	prag. inf. feedback
FPISK12	masculinity		
EWLFAK1	non active mood	experimental conditions	
EWLFAK2	depressed mood		
EWLFAK3	high spirits	Markoff-RNG feedback	
EWLFAK4	nervous mood		
EWLFAK5	frustration	Markoff-RNG no feedback	
number of runs / condition: 4		number of subjects: 299	

Names of the factors of the physical variables

factor names	physical variables
F1 hits	T,ZC,ZD, ETS,ETG, DIM
F2 negative fluctuation	ETG,BEG, PIG
F3 positive fluctuation	ETS,BES, PIS
F4 distribution	D,K,ZE
F5 variance	V,R
F6 time deviation	ZTA

SYSTEMS AND UNCERTAINTY

Eduardo R. Caianiello

Dipartimento di Fisica Teorica - Università di Salerno
Via S. Allende, 84081 Baronissi (SALERNO)

ITALY

I. Perspectives from systems theory

I propose to report on my search for a common paradigm to sciences which have to this day been regarded as unrelated: a kind of "unification", which from only fundamental premises of Systems and Information theories (in the differential form: inference theory) yields, through the Cramér-Rao inequalities, uncertainty relations among pairs of conjugate quantities such as are considered typical only of quantum mechanics. Geometrical structures arise from some kind of entropy. I have considered in some detail two specific cases as far apart as I could guess, classical thermodynamics and (pre)quantum geometry: I shall show some of the results that follow for the first, after only a brief mention of the second (for which ampler accounts can be found in the references).

In systems theory we have a clear indication that "uncertainty relations" are a general feature, to be expected whenever we deal with conjugate, or correlated, variables. The deep issue is, which concepts should be taken as "primitive" in the construction of a theory. Considering the physical world as a system, and therefore quantum physics as a study of phenomena, leads to take as "fundamental" concept the measure of that uncertainty. Hence, I have taken h, the Planck constant, as fundamental; together, of course, with c, the speed of light (thus, x and t are "derived" concepts: how many "think" this way?). I shall return later to this point, which is essential to the present approach. I must first tell a few things about systems theory, in the light cast upon it by R. E. Kalman, and can do no better than quote his own words:

"It seems to be an imperative fact ... that human nature cannot tolerate uncertainty. Heretofore in modeling and in the soft sciences the classical solution to this deeply felt "need" has been the **elimination of uncertainty by the application of prejudice.**"; where (ib.) **prejudice** = "assumptions unrelated to data, independenmt of data, which cannot be (or simply, are not) checked against data."

For a system-theoretic approach we have to study, by degrees:

□ **probability**, which is about dice and the like;

□ **statistics**, which studies the fit of data to some given distribution (sample spaces, measures and all that);

□ **statistical inference**, which purports to guess the distribution that fits (a scatter of) data. It includes "estimation theory". It is called "parametric" when a solution is sought within a class fixed by some parameters (prejudice again!).

All this is prerequisite "mathematics". For "systems" we have:

A   "system plus noisy environment"

B   **"noisy behavioral data"**

C   "model"

We have only B: A is postulated to account for **B**; **C** determines systems which could generate **B**.

"Noise" is what we know nothing about, and handle with prejudices, such as the assumption that there "exists" a **true**, a **unique** value measurable in principle with **unlimited** accuracy, or that it is generated by some simple probabilistic mechanism.

I have stepped in at this point, and **replaced** "noise" with "error" + "uncertainty"; "error" can be ideally suppressed by improving measurement, "uncertainty" **has** to be incorporated in the model. Kalman (by studying a linear two parameter model by Gini, 1921) finds radical differences in the parameters' estimation under the assumptions:

a) these parameters "exist", or

b) their values "depend on measurements".

I require that good models only allow for "noise". "Uncertainty" is to be expected in all systems where an **extensive** quantity is conjugate to an **intensive** quantity (e.g. controls the first); in general, when variables are correlated. Infinite precision **in both** is not attainable.

Examples: "value of the dollar" (the cost of an apparatus to measure it with arbitrarily high accuracy would obviously alter it correpondingly); "diffusion theory", a source of well known uncertainty relations; "classical thermodynamics" (see section **II**); "Quantum Mechanics".

My arguments are drawn from **differential estimation theory** and the well known **Camér–Rao Inequality**, which I can only mention here.

"Shannon entropy" goes wrong in too many respects (the name is misleading for a physicist; it has many drawbacks in the continuum); I shall use instead (out of the infinitely many entropies defined or definable) the "Kullback–Leibler Entropy" = Cross Entropy = Information = (directed) distance between two distributions =

$$H_C\,(x_1,\,x_2)\ =\ \int\ dz\ \ \rho\,(x_1|z)\ \ lg\ \ \frac{\rho\,(x_1|z)}{\rho\,(x_2|z)}$$

where x = parameters; z = random variables.

Symmetrized, it is called "Jeffreys divergence" J; its first non-vanishing differential is

$$dJ\ =\ 2\ dH_C\ =\ \Sigma\ g_{hk}\,(x)\ dx^h\ dx^k\ \sim\ dS$$

where $\quad g_{hk}\ =\ g_{kh}\ =\ \int\ dz\ \rho\ \partial_h\ lg\ \rho\ \ \partial_k\ lg\ \rho$

$$=\ \text{Fisher metric}\ =\ <\ \partial_h\ lg\ \rho\ |\ \partial_k\ lg\ \rho\ >$$

which defines the Fisher metric.

I remark here that

$$g_{hk} \equiv 4 \int dz\ \partial_h \sqrt{\rho}\ \partial_k \sqrt{\rho}$$

hence, first **suggestions** (for a study of pre-quantization):

a) $\qquad \sqrt{\rho} \sim \Psi$

b) $\qquad$ rimannian $\rightarrow$ hermitian

(note that the most different entropies acquire this same form when their differential is taken, no matter how they differ globally).

The most general connection compatible with Fisher metric is (Chentzov):

$$\overset{\alpha}{\Gamma}_{ijk} = [ijk] - (\alpha/2) \int dz\ \partial_i \lg \rho\ \partial_j \lg \rho\ \partial_k \lg \rho$$

Example: Take the Gaussian ($x_1 = \mu$, $x_2 = \sigma$); here, we get for the metric G and the curvature tensor R

$$G = \begin{pmatrix} \sigma^2 & -2\mu\sigma^2 \\ -2\mu\sigma^2 & 4\mu^2\sigma^4 + 2\sigma^4 \end{pmatrix}$$

$$\overset{\alpha}{R}_{1212} = (1-\alpha^2)\ \sigma^6\ ,$$

but only $\alpha=0$ is compatible with metricity, thus $R = \sigma^6$ measures dispersion = uncertainty which shows up as curvature, **another suggestion!**

I recall next that E. T. Jaynes's principle of **maximum entropy** (or "honesty"; J. N. Kapur) extremizes

a) information (=cross−entropy) plus

b) **constraints** (Lagrange multipliers ...), **both** over **all** parameter space.

I step in again, by changing a) and b) as follows:

1) keep a), but on some element $\omega \subset \Omega$ (we get the same equations and metric);

2) replace b) with the correct choice of the **connection** Γ and all that.

The Cramér-Rao inequalties, which always hold and which are by now a classical tool of estimation theory, free from many objections to which some Q.M. uncertainty relations are subjected, are my starting point. It will suffice here to consider them in one dimension, where they read, setting:

$$\Delta x^2 := \mathrm{var}_x (\tilde{x}) , \; \tilde{x} \text{ "estimator"},$$

$$\Delta x^2 \; g_{11} \geq 1$$

where $g^{11} = < (\partial_x \lg \rho)^2 >$ is called "Fisher information".

I recall now the previous suggestions and set:

$$\sqrt{\rho} = \Psi ; \quad p_x = -i\hbar \, \partial/\partial x ;$$

$$\Delta x^2 \; 4 \int dz \; (\partial_x \sqrt{\rho})^2 \geq 1$$

to get

$$\Delta x \; \Delta p_x \geq \hbar/2 .$$

We can do much better, but it becomes a bit technical. I just mention our result (extension is not unique): Instead of hermitian straight Fisher metric

$$\gamma_{ij} = \overline{\gamma}_{ij} = < \partial_i \Psi_m \mid \partial_j \Psi_m > ; \; m \equiv (q, \, p)$$

(Ψ_m parametric family of normalized Hilbert space vectors), take

$$1/4 \; \tilde{\gamma}_{ij} (x) = < \partial_i \Psi_m \mid \partial_j \Psi_m > - < \partial_i \Psi_m \mid \Psi_m > < \Psi_m \mid \partial_j \Psi_m >$$

Then both the real part , with $\rho_m = |\Psi_m> <\Psi_m|$,

$$\text{Re } \tilde{\gamma}_{ij} = 2 < \partial_i \sqrt{\rho_m} \mid \partial_j \sqrt{\rho_m} >$$

and the **imaginary part** yield uncertainty relations for (q,p) and (t,E), and a **chain** of inequalities:

$$\text{var } F \cdot \gamma (X,X) \geq \text{var } F \cdot \tilde{\gamma} (X,X)$$

$$\geq \text{var } F \cdot g_{Fisher} (X,X) \geq | < dF | X > |^2$$

which have to do with osmotic velocity in Nelson stochastic processes, etc.

Example: By taking n = 2 ,

$$\Psi_m = \exp \left((i/\hbar) (qP + pQ) \right) \Psi_m ; \quad m = q , p ;$$

we get $\quad \text{var } Q \cdot \text{var } P \geq 1/4 \, \hbar^2 + \text{cov } (Q,P)$

(Schrödinger, 1930), with

$$\text{cov}_m (A,B) = <m| 1/2 [AB+BA] |m> - <m|A|m> <m|B|m> .$$

From these premises one can obtain all (first) quantization, as well as additional results on mass spectra, wave and geodesic equations, etc., as reported in the references.

II. Application to classical thermodynmics

1. From affine to metric space
It is instructive to illustrate our systems-theoretic approach with an example from unsuspected quarters. Says L. Brillouin about thermodynamics:

"Dans cet espace P, V, T, il nous est impossible de définir la distance de deux points! Entre le point p répresentant un état du gaz et un point voisin, quelle est la distance? Un telle question n'a aucun sense physique."

Contrary to this opinion, I shall introduce a metric into thermodynamical space, utilizing Landau's fluctuation theory. Our metric, besides yielding his results in a more direct manner, leads naturally to uncertainties and other results. We note first that classic mechanics and thermodynamics have long been known to exhibut the same formal structure.

A thermodynamical system can be specified by its entropy $S(x_1, \ldots ,x_n)$ which determines its equilibrium properties: its differential can be expressed as

$$dS = \Sigma \ R_i \ dx^i$$

where R_i are the extensive variables conjugate (with respect to S) to the differential intensive variables x_i. This can be compared to

$$d\Phi = \Sigma \ p_i \ dx^i$$

of a function Φ which describes a dynamical system in the Hamilton-Jacobi theory. Both differentials play the same role in the description of the respective systems; the same can be said for the Legendre transformation. These formal coincidences give a differential equation for S analogous to the Hamilton-Jacobi equation. In the case of closed systems, taking the entropy as a funtion of volume V and energy E, the conjugate variables are respectively P/T and 1/T; the form of the latter term is of interest, because of its well known correspondence to time t in the standard association of statistical mechanics with Euclidean field theory.

We us the notation which is standard in this field. We shall assume the entropy to be maximum at point x=0 of parameter space, and denote with x_i the "small" fluctuation of the corresponding parameter around 0 ($x_i \equiv \Delta x_i$ or dx_i in other parts of this paper) to use Landau's own symbols, as well as his notation for indices (all in lower position). Following him, from Einstein's probability distribution for fluctuations

$$W(x) \sim \exp \left(S(x) \ / \ k \right)$$

we get, to second order,

$$S(x) - S(0) = - (1/2) \ \beta_{ij} \ x_i \ x_j$$

where $\quad \beta_{ij} = \left(\partial^2 S / \partial x_i \, \partial x_j \right) \Big|_{x=0}$

We introduce with Landau the conjugate variables

$$X_i = - \partial S / \partial x_i = \beta_{ij} \, x_j$$

which have the correlations

$$\overline{x_i \, x_j} = k \, (\beta_{ij})^{-1} \; ; \quad \overline{x_i \, X_j} = k \, \delta_{ij} \; ; \quad \overline{X_i \, X_j} = k \, \beta_{ij}$$

We consider for simplicity the case of $n = 2$, $x = (x_1 , x_2)$. These coincide with the second moments of the fluctuations

$$(x_i, x_j) = \int x_i \, x_j \; W(x) \; \mu \, (\, dx_1, \, dx_2 \,)$$

with invariant measure defined as

$$\mu \, (\, dx_1, \, dx_2 \,) = \left(\sqrt{\det \beta} / 2\pi \right) dx_1 \, dx_2$$

so that $\quad (x_i, x_j) = \overline{x_i \, x_j}$

The uncertainty relations follow from Schwartz's inequality (the X_i depend linearly in the x_j). Of particular interest is the one independent of the metric, which is formally identical with Heisenberg's, if Planck's constant replaces Boltzmann's:

$$|X_i| \, |x_j| \geq k \, \delta_{ij}$$

2. Some consequences

As an example, consider a closed system: in Landau's notation

$$x_1 = E' - E = \Delta E \, , \quad x_2 = V' - V = \Delta V \, , \quad X_1 = 1/T \, , \quad X_2 = P/T$$

as the fluctuations around the equilibrium values P, T, V, E. The quadratic approximation to S is

$$S \, (x) - S \, (0) = (1/2k) \left(\, \Delta E \, \Delta(1/T) + \Delta V \, \Delta(P/T) \, \right) .$$

For dielectrics, e.g., the same holds if we replace

$$P \rightarrow -\varepsilon \quad \text{(external electric field)}$$
$$V \rightarrow \rho \quad \text{(polarization)}$$

For all such systems we have

$$\left(\Delta(1/T) , \Delta(P/T) \right) = 0 ; \quad \left(\Delta E , \Delta V \right) = 0$$

(no correlations) and, what is most interesting for us, the uncertainty relations:

$$|\Delta E| \, |\Delta(1/T)| \geq k ; \quad |\Delta V| \, |\Delta(P/T)| \geq k$$

which are more general than Landau's.

All classical results of thermodynamics: Maxwell's relations, relations among response functions, the expressions of the first in terms of the second, follow straightforwardly from this approach: they amount, essentially, to stating that "a cosine cannot be grater than 1".

When β reduces to

$$\beta = \begin{pmatrix} 1/(C_V \, T^2) & 0 \\ 0 & 1/(T \, V \, \beta_S) - C_V/\Gamma_V^2 \end{pmatrix}$$

one has

$$|\Delta(1/T)|^2 = k / (C_V \, T^2) ; \quad |\Delta E|^2 = k \, C_V \, T^2 ;$$

$$|\Delta(P/T)|^2 = k \left(1/(VT\beta_S) - C_V/\Gamma_V^2 \right) ;$$

$$|\Delta V|^2 = k / \left(1/(VT\beta_S) - C_V/\Gamma_V^2 \right) .$$

Since $C_V > 0$, also $|\Delta(1/T)|^2$ and $|\Delta E|^2$ are positive. Imposing furthermore that $|\Delta(P/T)|^2$ and $|\Delta E|^2$ are positive one has

$$\Gamma_V^2 \geq V \, \beta_S \, C_V \, T$$

Such inequalities can also be derived from general principles.

The magnetic analogue of the last is

$$C_H \geq (T/\chi_M) \, (\partial M/\partial T)_H^2$$

where $C_H = T \, (\partial S/\partial T)_H$ is the heat capacity at fixed H, M is the magnetization, and $\chi_M = (\partial M/\partial H)_T$ the susceptibility. Let us apply it near T_C, for a substance which looses its ferromagnetic properties above T_C to become a paramagnet. Near and below T_C we have

$$M = A \, t^\beta \; ; \quad \chi_M = B \, t^{-\gamma'} \; ; \quad C_H = D \, t^{-\alpha'}$$

where $t = 1 - T/T_C$ and A, B, D are factors independent of T. Thus

$$t^{-\alpha'} \geq F(A, B, D) \, t^{2\beta + \gamma' - 2}$$

and hence

$$\alpha' + 2\beta + \gamma' \geq 2$$

the well known Rushbrooke-Coppersmith inequality. As a last consideration, let us compare the micro- and macro-canonical ensembles: in the first case, energy is bounded within $E \pm \Delta/2$, $\Delta << E$; in the second, it has no bound. We can compute from the previous formulae the relative fluctuation in energy:

$$\Delta E \, / \, E = \sqrt{k \, C_V \, T^2} \, / \, E$$

which is $O \, (N^{-1/2})$ and hence negligible in the thermodynamical limit. This is the standard result of ensemble theory: another consequence of metrization.

References

E. R. Caianicllo, "Hermitian metrics and the Weyl–London approach to the "Quantum Theory" ", Lett.N.Cim. 25, 225 (1979); "Some remarks on quantum mechanics and relativity", ib. 27, 89 (1980); "Is there a maximal acceleration?" ib. 32, 65 (1981); "Geometry from quantum mechanics", Il N.Cim. 59B, (1980); "Geodesics of free particles in QM phase space: free mass as a quantum effect", Lett.N.Cim. 35, 381 (1982); "Geometrical identification of quantum and information theories", ib. 38, 539 (1983); "Maximal acceleration as a consequence of Heisenberg's uncertainty relations", ib. 41, 371 (1984); Spineurs simples, Urfelder and factorizations of Dirac equations and spinors", Phys. Scripta 37, 197 (1988);

(with G. Vilasi:) "Extended particles and their spectra in curved phase space", Lett.N.Cim. 30, 469 (1981); (with S. de Filippo and G. Vilasi:) ib. 33, 55 (1982); (with W. Guz:) The Dirac-like equation as a phenomenological model for families of spin 1/2 baryons", ib. 43, 1 (1985); (with S. de Filippo, G. Marmo, G. Vilasi:) "Remarks on the maximal acceleration hypothesis", ib. 34, 112 (1982); (with G. Marmo and G. Scarpetta:) "Quantum geometry and quantum force: wave and geodesic equations", ib. 36, 487 (1983); (with G. Marmi and G. Scarpetta:) Geodesic and hamiltonian equations in quantum geometry", ib. 37, 361 (1983); (with G. Marmo and G. Scarpetta:) "Pre-quantum geometry", Il N.Cim. 86A, 337 (1985); (with G. Landi:) "Maximal acceleration and Sakharov's limiting temperature", Lett.N.Cim. 42, 70 (1985); (with G. Di Genova:) "Some consequences of phase space geometry". In: F. Mancini (Ed.), "Quantum Field Theory", North–Holland (1986); (with C. Noce and W. Guz:) "Quantum Fisher metric and uncertainty relations", Phys. Lett. A 126, 223 (1988).

R. E. Kalman: "Identification from real data", in: M. Hazelwinkel and A. H. Rinnoy Kan (eds.): "Current developments in the interface: Economics, Econometrics, Mathematics". Reidel 1982.
R. E. Kalman: "Identification of noisy systems", 50th Anniv. Symposium, Steklov Inst. of Mathematics, Moscow 1984.
E. T. Jaynes in: Brandeis Theor. Phys. Lectures on Statistical Physics, vol. 3 (New York).
J. N. Kapur, Journ. Math. Phys. Sci. 17, 103 (1983).
N. N. Chentzov, "Statistical decision rules and optimal conclusions" (in Russian), Moscow 1972.

The Construction of Space and the Logics of Quantum Mechanics

Dr Piergiorgio Quadranti
57 Bd du pont d'Arve
CH-1205 Genève

Theoretical terms and Interpretatio

The main statement in ontology of this paper is the introduction of interpretatio. This means that we can assert the existence of new entities in our construction of knowledge. The main task is therefore to recognize the rules of some activity.

The interpretatio is a universal activity we practice in all domains, in physics as in psychology or in our communication.

The interpretatio involves something to be interpreted (the interpretandum) and the new entities, introduced as "meaning" (the interpretata). The interpretata seem to be theoretical terms. If this assertion is right, then the difference between actual researches, as made by Sneed, Moulines, Gähde, Balzer (W. Stegmüller Band II Teil F-G) is that I don't try to recognize theoretical terms in a theory, but I construct directly a theory with them.

The theory of interpretatio isn't already written. Only a construction is suggested, as preparation for this theory. But the construction covers many domains.

Quantum mechanics will be approached with something like a three-valued propositional calculus. Orthogonality can be defined with a particular connective. Notions as superposition and separation receive a "natural" interpretation. In particular, the not separation becomes the normal case, and the separation has to be justified. The classical case needs a unique axiom and the separation case doesn't need axioms. The quantum case needs the habitual three axioms, atomicity, weak modularity and covering law.

We will present here the theory of time and the first steps of the construction of space. This notions allow us to introduce the foundation of quantum mechanics with the notion of question, presented in the theoretical frame of interpretatio. Here we define only the concepts necessary to express our comprehension of E.P.R.'s paradox.

1 Interpretatio

Let be a world. The introduction of new entities (introduction of meaning) suppose a partition of this world (or a partition of a set, describing an aspect of all this world).

Let V be the world and {M} = V its partition.

The new entities to introduce will be described by a set S of sets I_M, with a one-to-one map f

$$f: V \longrightarrow S$$

$$M \longmapsto I_M$$

M are the interpretanda, I_M the interpretata and f the interpretatio.

The map f is of a particular nature. We must introduce S and f together. We can't suppose S and then define f as a set included in the cartesian product of V and S.

The structure of I_M can be asserted only as a function of the properties of elements of M. But we can't limit ourselves to the intrinsic properties (properties between only elements of M). This aspect appears in the theory of question.

We meet two groups of problems.

1) To give the conditions for the partition of V

2) To establish the rules for structuring the I_M (interpretative rules).

It was impossible to me to find out immediately such conditions and rules. Therefore, I made a long detour through a construction provided with provisional rules.

1.2 The draft of the conditions for the partition and of the interpretative rules

a) Conditions for the partition, or criteria to accept definitions of M.

1) The definition of M has to be the same for all M (universality of the definition).

2) The definition of M can't be contradictory.

3) The definition can't be tautological (Popper).

4) (provisional formulation) All properties, satisfied at the given level of construction, must be considered.

5) The stability. The definition has to allow the stability of M. The notion of stability will be defined later.

6) Separation. An M must be separated in two sets if at the level of construction there are separated elements of M. The separation will be defined later.

7) We will perhaps need a higher criterion of choice between possible definitions (?).

8) Priority to the higher context. A more complex world allows a change of definitions.

b) Interpretative rules

1) The rules must be universal.

2) The initial rules must consist of one-to-one relations.

3) They can't introduce a contradictory interpretatum.

4) They can't lead to a trivial interpretatum (A unique interpretatum).

5) The interpretatum has to be stable and not divergent or always shifted in time by interpretations of higer order (if we apply again the rules).

6) Priority to the higher context.

The mains steps that the construction will cross are represented in a table (see appendix).

2) The simplest world and the first interpretatio: sensations, memory and time

At the beginning, idealism: the first world and the subject S which knows it, coincide.

Sensations.

The subject S is described by a set of pairs of states:

$$A_i = \{ a_i, b_i \}$$

The elements of a set A_i can't be satisfied together.

The subject S has a first knowledge on these states (sensations). This knowledge is described by a new set of states:

$$D := \prod_i A_i, \quad d \in D.$$

In the paper (Piaget)[2], I give more details and theoretical considerations about this point.

Memory.

Every state we introduced to describe the sensation has a copy in the memory.

$$A_i = \{a_i, b_i\} \longrightarrow A_i^m = \{a_i^m, b_i^m\}$$

$$D = \{d\} \longrightarrow D^m = \{d^m\}$$

A observer, external to the subject S and equiped with time, sees évery d replaced by another, whereas every d^m remains.

Interpretatio: time

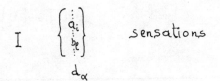

I   sensations

d_α

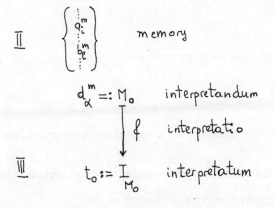

II   memory

$d_\alpha^m =: M_o$ interpretandum

$\downarrow f$ interpretatio

III $t_o := I_{M_o}$ interpretatum

We can consider the successive step to generalize.

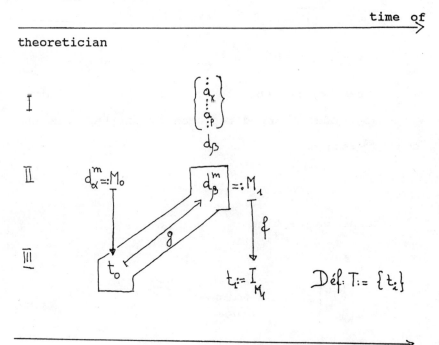

Definition of the map g:

g : T ⟶ D^m

Remark: the set T changes with the construction.

Let d_μ^m be an element of memory, for which there exists in the sensation the element d_μ.

Let t_{max} be the greatest t (see later the definition of the order relation on T).

1) If $g(t_{max})$ =/= d_μ^m and there exists t such that

{ $g(t) = d_\mu^m$ <u>and</u> $f(t,g(t)) = t_{max}$}

then the subject S does nothing.

2) If t_{max} is such that $g(t_{max})$ =/= d_μ^m and there doesn't exist t such that

$g(t) = d_\mu^m$ <u>and</u> $f(t, g(t)) = t_{max}$

then the subject S sets

$g(t_{max}) = d_\mu^m$ and $M = \{(t_{max}, d_\mu^m)\}$.

3) If $g(t_{max}) = d_\mu^m$ then the subject sets

$M = \{(t_{max}, d_\mu^m)\}$.

Definition of the order relation on T.

1) The "near" relation.

Near(t,t') iff $f(t,g(t)) = t'$ <u>or</u> $f(t',g(t)) = t$

2) Set E_p of neighbours

A sub-set of T is an E_p iff every element of E_p is exactly near to two elements, except at the most two elements which are near to a single element.

3) The past H(t) of t.

H(t) is an E_p containing t_0 and exactly one element near to t. Only t_0 is such that $f^{-1}(t_0) \in D^m$.

4) Order relation on T

t < t' iff H(t) $\subset$ H(t')

3 Space

A spatial structure can be constructed on the time structure: this is the thesis.

3.1 Points and facts (events)

Points

Def: Every set A_i^m is a point p_i, and $E := \{p_i\}$ is the space of these points.

Facts (events).

By construction of time, we can define a map Ch_i from T on every p_i, $p_i = \{ a_i^m, b_i^m \}$.

Indeed $f^{-1}(t) = (t'; d^m)$ and by projection π_D on D^m we have d^m. Finally by projection π_{pi} on p_i we obtain an element of p_i.

$$Ch_i := \pi_{pi} \circ \pi_D \circ f^{-1}.$$

A fact means that t and t', t and t' near, have different images by Ch_i. The set Φ^i of facts of p_i is a subset of the cartesian product $Ch_i \times Ch_i$. An element f^i_t of Φ^i can by defined as following:

Def: fact f^i_t:

$f^i_t := [(t'; Ch_i(t')) ; (t; Ch_i(t))]$ such that $Ch_i(t) =/= Ch_i(t')$ and $t' < t$ and (t near to t').

We can define the following sets:

$$\Phi_t := \{f^i_t\}_{i= 1 \ldots n} ; \Phi_{tot} := U_i \Phi^i = U_t \Phi_t.$$

$$\Phi := \{\Phi_t\}_{t \in T}.$$

Φ is totally ordered by T (time).

Remark: $\Phi_{tot} =/= \Phi$.

We have to define a near relation on E and sets of neighbours which can be ordered. This means that we have to define segments.

We will begin with the most simple case, defined by the hypothesis that there is only a fact for each time t: $|\bar{\Phi}_t| = 1$ for each t.

With this hypothesis, we define:

Near relation on E

Def: p_i near to p_j at t iff $\bar{\Phi}_t = \{f^i{}_t\}$ and $\bar{\Phi}_{t'} = \{f^i{}_{t'}\}$, where t near to t' and t' < t.

The restriction "at t" is necessary: the near relation can change in time (what isn't a motion!).

Path

We can define segments, containing at least two points and having at most two common points with another segments. Such segments will be called path and can be defined as following.

We introduce a partition

$K = \{c^\wedge{}_n\}$ n=1... of $\bar{\Phi}$. For this, we remark that

$\bar{\Phi}_t < \bar{\Phi}_{t'} <==> t < t'$. Then we set

i) $\bar{\Phi}_{t0} \in c^\wedge{}_0$ and

ii) for $x \in \bar{\Phi}$, $x \in c^\wedge{}_n <==>$

(there exists $\bar{\Phi}_{ta} \in c^\wedge{}_n$

and $x \geq \bar{\Phi}_{ta}$ and there don't exist $\bar{\Phi}_{t'}$ & $\bar{\Phi}_{t''}$ in $c^\wedge{}_n$ before x such that there exists p_i with $(\bar{\Phi}^i \cap \bar{\Phi}_{t'} =/= \phi$ et $\bar{\Phi}^i \cap \bar{\Phi}_{t''} =/= \phi))$

iii) $\underline{\phi}_{ta}$ has to exist. We must say if a new $c^\wedge_{n'}$ has an element. Hence, let be $c^\wedge_n =/= \phi$. There exists at least $c^\wedge_0 =/= \phi$. We define then

post$(c^\wedge_n)$:= $\{\phi_t|$ there exists $\phi_{t'''}; \phi_{t''}; \phi_{t'} \in c^\wedge_n; \phi_{t''} > \phi_{t'}$ and $\phi_t > \phi_{t''}$, and [there exists $p_i \in E$ such that $(\phi^i \cap \phi_{t'} =/= \phi$ and $\phi^i \cap \phi_t =/= \phi)]\}$. This means that we take the ϕ_t following a repetition of an element of $c^\wedge_n$. From ii) we have post$(c^\wedge_n) \cap c^\wedge_n = \phi$. ϕ is ordered and we can define

$\overline{\phi}_{ta}$:= Inf(post$(c^\wedge_n)$) and set

$\overline{\phi}_{ta} \in c^\wedge_{n'}$.

Every $c^\wedge_{n'}$ with its extremities, define a unique path.

Triangulation

We need criteria for accepting the near relation, defined above.

1) Two paths with common extremities merge into a unique path without near relation.

2) Set H of hedgehogs h.

h is a hedgehog iff h $\subset$ E and h $= \cup_i c_i$, where the paths c_i are such that there exists $e_h = \cap_i c_i$.

e_h will be called the heart of the hedgehog h.
The paths c_i will be called the aculei.

We can meet different configurations.

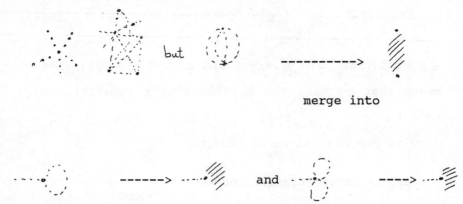

merge into

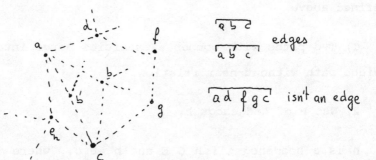

3) Edge b of hedgehog h.

b is an edge of a hedgehog h iff b C E and b is a minimal set of paths without aculei and joining two ends of aculei of h.

Two ends oh h can be joined by several edges.

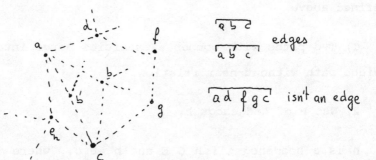

4) Umbrella p_{hi}.

p_{hi} is an umbrella iff

$p_{hi} := h_i \cup B_{hi}$ where $B_{hi} := \{ b_i \}$, b_i is an edge of hedgehog h_i with $h_i \subset h$, and every end of h_i belongs exactly to two b_i.

5) Change in near relation.

Let p_h and p_h' be two umbrellas at time t

i) with a single edge {a, b} in common

ii) such that this edge and respective aculei of both hedgehogs contain only two points.

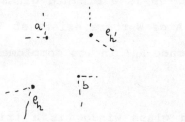

iii) If at t', t' > t, we have "e_h near to e_h'", then we set " $\neg$(a near to b)".

iv) If we have umbrellas and two facts $f^i{}_t$ et $f^j{}_{t'}$, t' near to t, such that p_i and p_j belong two different umbrellas not in the situation of i) to iii) then these facts don't imply p_i near to p_j.

Remark: the foregoing rules don't define motion. Indeed we have only a change in the structure we attribute to the points. Motion supposes such spatial structure and can be defined only later.

6) Chalice Cal_h.

An umbrella p_{hi} is a chalice Cl_h iff h_i = h. A chalice is an umbrella containing all aculei of a hedgehog h.

7) Rosace R_h.

A chalice Cal_h is a rosace R_h iff

i) It is unique for the hedgehog h

ii) It doesn't contains umbrellas.

8) Stained glass window W.

A set of rosaces is a stained glass window W iff for each subset A of W there exists at least a rosace R_h in A and a rosace R_h' in its complement such that $R_h \cap R_{h'}$ =/= $\emptyset$.

9) A stained glass window is a triangulation of its points.

Examples

To draw a point out of a rosace.

Lines represent the near relation.

1)

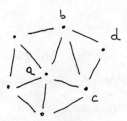

2)

(a near d) => ¬(b near c)

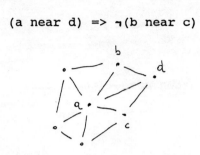

And so on. We obtain

6)

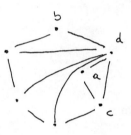

Then 'dac' and 'dc' merge, with loss of near relation.

7)

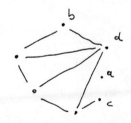

8) Now, we can have

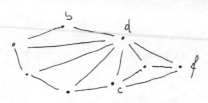

The point has be drawn out of the rosace.

Introduction to the general case.

Now we don't suppose that we have a unique fact (event) at t.

Simple case.

We suppose the following:

M_0 is the set of facts at t_0 .

M_n is the first set such that $M_0 \subset M_n$, or $M_0 \cap M_n \neq \emptyset$

We define $M_{m\ b} := M_m \setminus M_0$ and $M_{n\ b} := M_n \setminus M_{n-m}$.

For $i < m$, $M_{i\ b} := M_i$.

Then we can consider the set A of sets $M_{m\ b}$. We ca define a near relation on A.

If we have a new set B of such sets $N_{p\ b}$, we can have that $N_{p\ b} \cap M_{m\ b}$ is a singleton for $i < m\ r$. Then we can infer a near relation on $N_{p\ b}$.

So we obtain again a triangulation.

Two problems appear at least 1) to search the spatial structures which are possible by the foregoing definitions (by a transcription in a computer programm) and 2) to find out the conditions which lead to the " every day" space. The first thesis is: the process allows to obtain the every day euclidian space if and only it allows to define objetcs.

We reserve it for a forthcoming work.

Introduction to the theory of objects.

The objects are interpretata of a set of configurations of points. This configurations are identified by facts and the near relation. We meet something akin to the 'pattern recognition'. But we don't suppose a three dimensional space. The three dimensions can be introduced interpretatively with the objects.

Now, by rotation, it is possible to define a set of aligned points. This is the fundamental notion for axiomatic geometry.

The theory of objects is a particular case of the theory of question.

4) Introduction to the theory of questions: quantum mechanics

We give here only some definitions and suggestions. A complete discussion will appear in Quantum Ontology.

I suppose a partition of V, V $\mapsto$ (M), $f(M) = I_M$

We have time and space.

$$\text{time} \quad \text{------------------>}$$

$M := \quad . \quad . \quad . \quad . \quad . \quad . \quad . \quad . \quad . \quad . $

$$m \quad m'$$

$m, m' \in M$

Situations := the set A of intrinsic properties of the elements of the M's. This means that a situation is a set of properties connecting two eéelements of a M.

This definition needs some restrictions. We must 1) consider the constructible properties, 2) restrict ourselves to the level we have reached and 3) we restrict ourselves to a finite theory.

Conditions :- the set B of extrinsic properties
of the elements of the M's. They are the properties
connecting elements of M and the external world.

Answer := Set of subsets of C, where C is the set
of properties connecting two successive elements m,
m' of a M. r $\in \mathcal{P}(c)$.

Questions: The set $\mathcal{Q}$:= B $\times \mathcal{P}(c) \times$ [0; 1].

A question q, q $\in \mathcal{Q}$ identifies one element $\underline{\alpha}$ of
U I_M: f^1 (q) = $\underline{\alpha}$. The q's are the interpretanda and
the $\underline{\alpha}$ are the interpretata.

Frequency. The frequency $\Omega(r)$ of r for a
situation a and for elements of M is defined by

$\Omega(r)$ = [N (number of realization of the
condition)] / [n (number of the answer r)]

We can decompose f^1

$$f^1(q) = (f^1_\alpha (b; r); f^1_\Omega (\Omega_r)) = (\alpha; n(\Omega)) = \underline{\alpha}$$

We will call α a determination.

Connectives on { α }

We consider V $\times$ A. The domains of values of α are
subset of V $\times$ A. We consider the evaluations of α at
the situations. A measure involves three moments

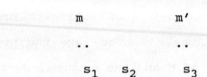

$$s_1 \qquad s_2 \qquad s_3$$

s_1 = the end of situation, s_3 = the end of answer, s_2 begins before s_3 and ends after s_1.

Evaluations at s_1

Frequencies		Values
$\Omega(\alpha) = 1$	<------------------>	c (actual)
$\Omega(\alpha) = 0$	<------------------>	f (false)
$0 < \Omega(\alpha) < 1$	<------------------>	π (potential)

We indicate the domains of the different values as following: E^c_α (for c), E^f_α (for f) and E^π_α (for π).

Connectives are defined as subset of following table

Table

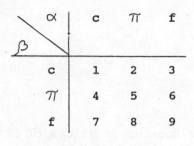

$\beta \backslash \alpha$	c	π	f
c	1	2	3
π	4	5	6
f	7	8	9

Structure on $\{ \alpha \}$

$\alpha < \beta$ iff $E^c_\alpha \subset E^c_\beta$

$\alpha \equiv \beta$ iff $E^c_\alpha = E^c_\beta$

a := [α] (equivalence class). "a" will called a property.

$E^c_a = E^c_\alpha$ and $E^f_a = \bigcap E^f_\alpha$

The orthogonality between a and b is defined by the emptiness of the cases 1, 2 and 4. The case 5 can be not empty. This fact allows to introduce the notion of superposition.

Infimum

$a \wedge b := a \text{ and } b$

Supremum

$\bigvee a_i := \bigwedge_{a_i < b} b$ where each b isn't defined as [or α_{ij}]

Classical case

It is defined by the condition: $E^{\pi}_{\alpha} = \phi$ for each α. The orthocomplement a' of a, is defined by $a' = \neg a$

Quantum case

In this case we can have $E^{\pi}_{\alpha} =/= \phi$. For each a, we can define α, $\alpha \in a$, such that $E^{c}_{\alpha} = E^{c}_{a}$ and $E^{f}_{\alpha} = \cup E^{c}_{p}$ for $a \perp p$.

Then we can define $a' := [\neg\alpha]$

It is possible now to define the notions of superposition and super selection rules. But we will only introduce some considerations about the E.P.R. paradox.

If we admit that at time s_1 we have only a single fact and, on the contrary, if we admit two facts as an answer at time s_3 we have a measure considered by E.P.R. Now, if both answers belong the same set M, then we have the same interpretatum, the same system. Logically, even if not intuitively, we have the conception proposed by the quantum theory.

The correlation is the normal case. The separation has to be justified.

Such a justification can be constructed. We have to decompose the question with double answer into two questions, with a single answer, related by a connective. Then we have the separation iff the case 5 doesn't give more information: this means that all combinations definable in this case are given.

In such a construction, both new notions of superposition and correlation are defined using the case 5. If we use three values, then both notions can be defined, and have to be defined "spontaneously" in the construction, even if not "intuitively". That is the main result of our choice of initial notions.

Conclusion: 1) interpretatio allows us to distinguish between interpretanda (what we observe) and interpretata (the physical system). If we identify both levels, paradoxs arise. 2) Using a three evalued propositional calculus, notions as superposition and correlation can be spontaneously defined. Indeed, case 5 gives new information on the physical system. Information that we can't define by connectives of first level, as we have introduced with the subsets of the foregoing table. The reason being that the domain of each value of such first level connective can only contain the totality of each case. To define the information carried by the case 5 we must consider subsets of this case.

TABLE-1

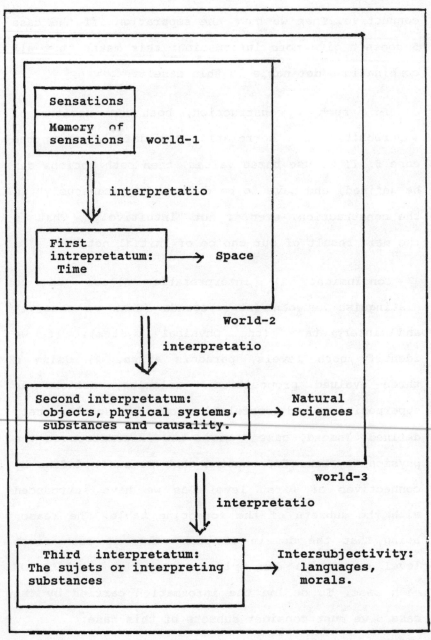

BIBLIOGRAPHY

Aerts,Dirk (1981). The One and the Many. Bruxelles: Vrije Universiteit Brussel Faculteit der Wetenschappen.

- (1983). In Les fondements de la mécanique quanti-que. 25e Cours de perfectionnement de l'Association Vaudoise des Chercheurs en Physique. Montana, du 6 au 12 mars 1983. C. Gruber, C. Piron, Trân Minh Tàm, R. Weill (eds) A.V.C.P., Case postale 101, CH-1015 Lausanne

Van Fraassen,Bas C. (1980). The scientific image N.Y. Claredon Press Oxford

Howard,Don (1985). Einstein on locality and separability. In Stud. Hist. Phil. Sc. Vol 16,3 pp171.201 (1985)

Ludwig,Günther (1978). Die Grundstrukturen einer physicalischen Theorie Berlin Springer-Verlag

- (1985). An Axiomatic Basis for Quantum Mechanics Berlin Springer-Verlag

Piron, Constantin (1976). Foundation of Quantum Physics. London: W.A.Benjamin.

- (1983). In Les fondements de la mécanique quanti-que. 25e Cours de perfectionnement de l'Association Vaudoise des Chercheurs en Physique. Montana, du 6 au 12 mars 1983. C. Gruber, C. Piron, Trân Minh Tàm, R. Weill (eds) A.V.C.P., Case postale 101, CH-1015 Lausanne

Quadranti, Piergiorgio (1984). Interpretation and Time. In WOPPLOT 83. Parallel Processing: Logic, Organisation, and Technology. Proceeding, Neubiberg (Germany) 1983 (= Lecture Notes in Physics 196). J. Becker and I. Eisele (eds), 69-79. Berlin-Heidelberg: Springer.

- (1988) Kant, Piaget et UNITYP. Cahiers Ferdinand de Saussure 42(1988), 65-95

- (1988) Objet et causalité. In L'objectivité dans les différentes sciences, Evandro Agazzi (ed),169-175. Université de Fribourg. (This edition contains many misprints)

- (Piaget)^2 à paraître.

- Quantum Ontology. A paraître

CELLULAR AUTOMATA AND THE CONCEPT OF SPACE

Dieter GERNERT, München (FRG)

ABSTRACT: After some introductory remarks on space in mathematics and phsyics, an overview of existing applications of cellular automata in physics is given. There are several arguments, partially based on mathematical physics, which impose restrictions on the possible number of dimensions of physical space. One of them claims that our space must have at least three dimensions in order to avoid disturbances between processes of signal transmission. This argument is refuted by means of a cellular-automata model.

1. Introductory remarks on space in mathematics and physics

In the central part of this article the proposal will be advanced to develop models of physical space based on cellular automata. For a better understanding of the specific features of such models, but also of the difficulties involved, it will be helpful to begin with some general remarks on the concept of space in mathematics and physics.

Physics in its historical development started from everyday experience. Two traditional branches, optics and acoustics, are named after human senses, and similarly mechanics has its roots in the sense of touch.

Mathematics and physics jointly evolved in a positive feedback loop. It is well known that open problems in physics gave rise to new branches of mathematics or to extensions of existing ones. Still more important, but less noticed, is the inverse phenomenon: if some branch of mathematics had attained a high standard - or at least was fashionable - there was an increased chance for a transfer to physics.

Similar observations can be made with respect to other disciplines, and it seems to be justified to name this the "phenomenon of the earlier tool": an intellectual tool which comes later has less chances of being accepted, no matter what its true qualities may be; and there must be very strong arguments - combined with personal engagement - in order to give a later tool its fair chance.

An example is given by the way how mathematical tools were selected for theoretical economics and related fields. In earlier stages the time-honoured differential calculus was transferred to economics and social sciences where their real utility is rather limited; and it took much time and labour to advance the use of discrete mathematics in these fields.

Originally, "higher mathematics for physicists" mainly consisted of differential and integral calculus and the theory of finite-dimensional real vector spaces; later on group theory was adopted. In spite of all achievements it should be kept in mind that these branches of mathematics are only subsets, but not at all the totality of mathematics. Whenever problems in physics persist we should at least have enough courage to test alternative types of models.

The term "space" as such is not defined in mathematics. Generally, mathematical dictionaries only give references to more specific terms which are formed by combining the word "space" with an adjective or a proper name, e.g. "affine space", "Banach space", etc. A mathematical dictionary (NAAS/SCHMID 1961, vol. II, p. 455) presents about 4o such references.

An "encyclopedic dictionary of mathematics" gives a circumscription which can be hardly considered a definition: "The term space is used in mathematics for any set when certain types of properties are to be discussed or when it is intended to use some sort of geometrical terminology." (SNEDDON 1976, p. 616)

The concept of space in physics started with the three-dimensional space from everyday experience; even the later adoption of non-euclidean metrics cannot obscure the fact that a finite-dimensional real vector space is still the basic underlying structure. Meanwhile the number of dimensions has been considerably increased; now there is a theory based on 26-dimensional space (see e.g. MANIN 1989), and it cannot be granted that this really is the record. (In this moment only such dimension numbers are focussed which are immediately ascribed to physical space; therefore the dimension numbers of phase spaces and other mathematical structures cannot be included in a comparison.)

2. Cellular automata

2.1 Definition

A cellular automaton (also called cellular space, cellular network, polyautomata network etc.) consists of many automata each of which is connected exactly with those automata which belong to a certain "neighbourhood". A definition of neighbourhood must be given in the beginning; no concept of locality or connectedness must be observed. An initial configuration must be defined, too.

According to WOLFRAM (1984a, page vii), cellular automata have five fundamental defining characteristics:

1. They consist of a discrete lattice of sites (where each "site" can be identified with a simple automaton).

2. They evolve in discrete time steps.

3. Each site takes on a finite set of values (possible states).

4. The values (states) of each site evolve according to the same deterministic rules (transition function).

5. The evolution of a site depends only on its momentaneous state and on the states in its neighbourhood.

The most famous example of a cellular automaton is CONWAY's "Game of Life" (see e.g. BERLEKAMP et al. 1984). Of course it is comfortable to start from an orthogonal net or another regular net; but on the other hand, there is a risk of wrong generalization if someone might impute features met in regular nets to cellular automata which are based on a quite different definition.

2.2 Existing applications to physics

Before proceeding to physics, it should be briefly mentioned that cellular automata have already found successful applications in various fields, including general systems theory (WOLFRAM 1984b,c), computer science (HÄNDLER et al. 1985), chemistry, biology (HOGEWEG 1980, PRESTON/DUFF 1984), and even in urban planning and ecology (HOGEWEG 1988).

At present, the idea of modelling physical processes in terms of cellular automata has been fairly accepted, although it is not yet integrated in the mainstream of theoretical physics.

First of all, the contributions by Konrad ZUSE must be highlighted, who formulated this concept in a special monograph and several articles (ZUSE 1969, 1975, 1982). He firmly advocates the idea that a discrete space structure should be taken as the basis for modelling fundamental processes in physics. Among other examples, he proposes the "digital particle": the propagation of a particle can be modelled as the migration of a disturbance in a grid.

Other physical processes or problems which have already been attacked include diffusion and equilibrium, fluid dynamics, the theory of collective phenomena (TOFFOLI/MARGOLUS 1987), chaos thoery (GRASSBERGER 1984), wave-particle dualism and problems of measurement in quantum mechanics (GERNERT 1986, 1988). (For further references see the proceedings volumes edited by DEMONGEOT et al. 1985 and by FARMER et al. 1984.)

3. Cellular automata and physical space

3.1 Empty space as a carrier of physical properties

The following remarks on the concept of space start with a short summary of a more detailed analysis given by BÜCHEL (1965, p. 114-12o).

Space is something "extended". On the other hand, space is conceptually distinguished from the real three-dimensional things. Anything that is in space, or moves through space, cannot be identical with space. Therefore, space cannot be regarded as any kind of material reality; rather, space must be a symbolic construction. The "hole in the wall" is used as a metaphor: the hole is not material, only the bricks surrounding it are; the hole is a symbolic or intellectual construction, which is suitable to describe the real situation. The same holds for "space" in physics.

Space has a psychological aspect, too: Our perception and imagination will not permit that perceived or imagined objects stand isolated one at the side of another. Therefore, things are projected onto a joint frame of reference.

"Physical space" is that concept of space which underlies the language of physicists. The usage of this term is not uniform. Often the idea of a coordinate system is implicitly attached to it (BÜCHEL 1964, p. 119).

As soon as the idea of _field_ becomes essential in physics, the question of a "carrier" or "transmitter" of forces arises. If empty space acts as a transmitter of forces or energies, it must be capable of assuming different physical states at different locations. Then empty space is no longer an intellectual construction, but a real medium which is responsible for the transmission of forces and energies. Consequently, space takes over a role which goes far beyond its merely geometrical definition as a totality of all locations; space becomes a "carrier of physical properties", and thus also becomes a "subject of physical research, which has the task to discover its physical properties". (WESTPHAL 1950, p. 60-61)

This problem formulation, which was taken from a renowned textbook (and which is still valid although it has been abridged in later editions of the same book), is still far away from a systematic treatment. But exactly here an excellent starting-point for future research can be fixed.

If empty space really is a carrier of physical properties and occasionally takes on different states at different locations, then it must be highly promising to model space itself in terms of cellular automata. Of course, this has to be done in such a way that known physical phenomena can be derived from the model by projection and specialization, and - hopefully - the new models yield some new predictions, too.

3.2 How stringent are arguments in favour of three-dimensional space?

Although it is an empirical fact that our space has three dimensions, philosophers like to ask the question whether the number of dimensions must necessarily take on exactly that values. Several arguments have been formulated which impose restrictions upon the possible number of dimensions (BÜCHEL 1965, p. 151-156); only two of them can be summed up and commented on here.

The first such argument claims that optical and acoustical signal transmission is possible only if both reverberation and distortion are excluded. Reverberation means that a spoken word reaches the listener's ear repeatedly, and that the listener simultaneously hears several words which were spoken at different times. Distortion means that different pitches are transmitted at unequal intensities.

As a result of mathematical physics (COURANT/HILBERT 1962, vol. 2, p. 695-698), signal transmission without reverberation is possible only in a space with an odd number of dimensions (together with other arguments this finally leads to three-dimensional space).

Unfortunately, this argument is not compulsory in its present form, but will require additional support. In the beginning it cannot be excluded that reverberation does occur in our physical space and that the second and all following arrivals of the same signal are filtered out by our perception. There would be no problem to build "artificial creatures" in the style of BRAITENBERG (1984) which, despite of an artificial environment with reverberation, would have a functioning capability of orientation and communication. Only experience with visible speech and similar techniques can have the unintentional side effect to refute this speculation.

Another argument states that our space must have at least three dimensions because in a two-dimensional space organic life - in the form familiar to us, with an advanced nervous system - would be impossible. Nerves would cross each other like streets in a city, and at the crossing-points the cables would disturb one another; there would be no chance to avoid crossings by bridges or tunnels.

But a deeper analysis will show that this argument cannot be maintained in its present form. In cellular automata, information crossing is possible without an underlying three-dimensional structure.

In the following models at least three different states - a "ground state" and two or more "activated states" a,b,... - are needed. A process of signal transmission is represented by a chunk of activated states which travel jointly.

As a first step, an ordinary rectangular grid is presupposed and a signal transmission is defined by the transition rules in Figure 1. Here, the frames are drawn for better orientation, empty boxes stand for the ground state, and the two rules must be sup-

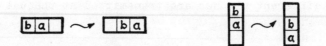

Figure 1

plemented symmetrically for right-to-left and bottom-up motions.

Next, a conflictless crossing of two signals can be followed in

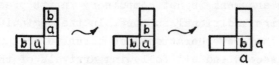

Figure 2

Figure 2. Here, a general rule can be studied, the "<u>convoy principle</u>":

> A correct transaction is possible only if 1. each of the involved single processes is represented by a chunk of activated states which remains essentially unchanged at least for a nontrivial time interval ("convoy"), and if 2. the transition rules are adjusted to these chunks and their interactions.

A similar definition is possible for a hexagonal grid in the plane. In the general case of an irregular grid some technical complications will arise, but a conflictless signal crossing is possible, too. Here the convoy principle must be observed in a more severe version (a minimal configuration with just two elements and two states will no more suffice), as indicated by Figure 3.

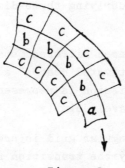

Figure 3

4. Concluding remarks

There is no doubt that cellular-automata models in physics have a serious disadvantage: numerical calculations are neither easy nor accurate. Nevertheless, it should not be forgotten that the principal purpose of mathematics in physics is not to provide numerical calculations of single physical processes, but to contribute to a better qualitative understanding. Numerical precision is not the same as insight.

It is just in this sense that cellular-automata models will offer a chance. There is no use to ask whether they are "better" than traditional models. For a great variety of problems differential equation will remain indispensable. A broader acceptance of cellular automata in physics would have nothing to do with a "paradigm change" (this term has already been pushed up too frequently in a wrong context); it will simply be a supplementation of the tool-kit, just as it was the case when group theory was adopted.

It seems that the chances of the new type of models lie in "really fundamental" problems, as e.g. the "properties of empty space" (see Section 3.1), the offspring of a field, or the mechanism underlying the influence of a field upon a particle. Some problems of physics, like wave-particle dualism, will be better understood, and the question of the number of dimensions will lose its significance as soon as physical processes are described as projections of coordinated state changes of underlying structures; and finally, one or another open question will be unmasked as a pseudo-problem.

References

BERLEKAMP, E.R., CONWAY, J.H., GUY, R.K., Winning ways for jour mathematical plays, vol. 2. Academic Press, London 1982

BRAITENBERG, V., Vehicles. Experiments in synthetic psychology. MIT Press, Cambridge/Mass. 1984

BÜCHEL, W., Philosophische Probleme der Physik. Herder, Freiburg 1965

COURANT, R., HILBERT, D., Methods of mathematical physics, vol. 2. Interscience, New York 1962

DEMONGEOT, J., GOLÈS, E., TCHUENTE, M. (eds.), Dynamical systems and cellular automata. Acad. Press, London 1985

FARMER, D., TOFFOLI, T., WOLFRAM, S. (eds.), Cellular automata
    (Proc. Workshop Los Alamos, March 1983). North-Holland,
    Amsterdam 1984 (= vol. 1o (1984) of Physica D)

GERNERT, D., A cellular-space model for studying wave-particle
    dualism. In: R. TRAPPL (ed.), Cybernetics and systems '86,
    Reidel, Dordrecht 1986, p. 117-122

GERNERT, D., Cellular-space models for processes of measurement.
    In: R. TRAPPL (ed.), Cybernetics and systems '88, part 1.
    Kluwer, Dordrecht 1988, p. 1o9-115

HÄNDLER, W., LEGENDI, T., WOLF, G. (eds.) Parcella '84 (Proc.
    Workshop Berlin, Sept. 1984). Akademie-Verlag, Berlin 1985

HOGEWEG, P., Locally synchronised developmental systems. Int.J.
    General Systems 6 (198o) 57-73

HOGEWEG, P., Cellular automata as a paradigm for ecological mod-
    elling. Applied Mathematics and Computation 27 (1988)
    81-1oo

MANIN, Yu. I., Strings. Math. Intelligencer 11 (1989) 59-65

NAAS, J., SCHMID, H.L., Mathematisches Wörterbuch. Akademie-Ver-
    lag, Berlin 1961

PRESTON, K., DUFF, M. J. B., Modern cellular automata. Plenum
    Press, New York 1984

SNEDDON, I.N. (ed.), Encyclopedic dictionary of mathematics for
    engineers and applied scientists. Pergamon, Oxford 1976

TOFFOLI, T., MARGOLUS, N., Cellular automata machines. MIT Press,
    Cambridge/Mass. 1987

WESTPHAL, W. H., Physik. Springer, Berlin, 14. Aufl. 195o

WOLFRAM, S., Preface (to the proceedings volume listed under
    FARMER). Physica D 1o (1984) vii-xii (1984a)

WOLFRAM, S., Cellular automata as models of complexity. Nature
    311 (4 Oct. 1984) 419-424 (1984b)

WOLFRAM, S., Some recent results and questions about cellular
    automata. In: DEMONGEOT et al. (eds.) 1984, p. 153-167
    (1984c)

ZUSE, K., Rechnender Raum. Vieweg, Braunschweig 1969

ZUSE, K., Ansätze einer Theorie des Netzautomaten. Nova Acta
    Leopoldina, N.F., Nr. 22o, Bd. 43 (1975) 5 - 46

ZUSE, K., The computing universe. Int. J. of Theoret. Physics
    21 (1982) 589-6oo

Address: Dieter GERNERT,
         Hardenbergstr. 24, D-8ooo München 5o

MULTIPURPOSE AND SPECIAL PURPOSE COMPUTERS

Gerhard Fritsch
Universität Erlangen-Nürnberg
Department of Computer Science (IMMD)
D-8520 Erlangen, F.R. Germany

Abstract:

Very different computational requirements of user problems have given rise to a broad spectrum of computers ranging from general purpose via specialized multipurpose to special architectures. For the choice of the appropriate computer system, the user has to take into account both the computational demands of the target application and involved economic considerations. After a short characterization of scientific and engineering problem classes, main hardware and software means for specialization are described. Because of the versatility of multiprocessors, such systems are especially well suited for adapting the architecture to the user problem. This issue is discussed for architectural concepts such as hierarchical bus systems and shared memory systems.

1. INTRODUCTION

The conjunction "and" in the title simply means existence of both types of computers. From a user's point of view the conjunction should rather be "or", as the user has to decide, in view of the target application which of the two types is the most appropriate one. However, for certain heterogeneous applications, a computer system which comprises both types may be more suitable.

The classical von Neumann computer is known as a universal computer. Since then, the computational performance has been increased by several orders of magnitude due to technological advance in hardware components, specialization and parallel processing.

There has been an enormous gain of speed by very large scale integration technology, by fast special components and units (floating point arithmetic units, address calculation units etc.), by hierarchically organized memories and so on. Nevertheless, limits of technically possible performance of the uniprocessor are showing up. Users demand more and more computing power and storage capacity. On these grounds, the interest in parallel processing systems has strongly been growing within the last three or five years. With parallel systems also the structure of the interconnection system can be adapted to multipurpose or special purpose use. This possibility does not exist with uniprocessor systems.

Whether to use a general purpose or a special purpose machine is primarily an economic problem. A user needs a general purpose machine if it is aimed to a broad spectrum of application problems. On the contrary, he wants a highly efficient special purpose machine if the application is well defined and featured by a small variation of its computational parameters. Anyway, the user will be interested in a system which provides for the needed computational power and storage capacity at a minimum price/performance ratio.

The next chapter presents the characterization of large problem classes from natural and engineering sciences. The third chapter is dedicated to computer-architectural means of specialization, in particular with parallel computers. Finally, some conclusions on the issue are drawn.

2. SCIENTIFIC AND ENGINEERING APPLICATIONS

Multiple processing systems have additional overhead compared to uniprocessor systems. This is due to unbalanced load in the nodes of a multiple processing system and due to interprocessor communication.

Despite these drawbacks there are two reasons to build multiple processing systems:

- To offer higher performance for computing large user problems than can be obtained with uniprocessors at comparable price/performance ratio, and

- to realize a fault-tolerant computing system, for instance in case of safety requirements. Obviously large multiple processing systems need some fault recognition and recovery mechanisms.

We only refer to the high performance issue. User problems with high computational and memory capacity requirements mostly belong to scientific or engineering application fields. Principal problem classes are:

(1) Numerical grid problems:

Approximate solution of sets of partial differential equations (PDEs) by discretization of space and time. This problem class plays an essential role in scientific supercomputing as it comprises many application fields in science and engineering (e.g. numerical simulation of physical phenomena in material science, aerodynamics etc.).

(2) Particle and field models:

Such models are used for a broad spectrum of applications in plasma physics, solid state physics, aerodynamics, chemistry and others. For the mathematical description Molecular Dynamics and Monte-Carlo methods are applied.

(3) Simulation of very large scale integrated (VLSI) electronic circuits (design and verification): According to the simulation level to be considered different methods and mathematical procedures are applied:

- System level: Functional specification.
- Logic and switch level: Logic operations.
- Electric circuit and timing level: Sets of coupled, nonlinear ordinary differential equations, Fourier transform and others.
- Device and process level: Modeling of physical phenomena in electronic devices and technological processes, e.g. mathematical description by sets of partial differential equations (PDEs).

(4) Ab initio quantum mechanical calculations of chemical systems for the theoretical prediction of the properties of new materials or for the interpretation of new physical phenomena (e.g. high temperature superconduction). Such problems comprise the calculation of 1- and 2-electron-integrals, operations of large matrices and the determination of eigen values.

These four problem classes have in common a strong demand for fast numerical operation units and large memory capacity. The degree of inherent parallelism generally is high. In the case of VLSI simulation, this is only true for device and process simulation where highly parallel algorithms can be used. A quite different situation is found in the following problem class:

(5) Artificial intelligence and automation. This problem class comprises application areas as for instance image processing, pattern recognition, computer vision, speech understanding, intelligent robotics, expert systems and others. Depending on the particular problem to be solved, there are requirements for massively parallel operations (e.g. image preprocessing) up to complex computations with low inherent parallelism (e.g. knowledge engineering, decision making).

This coarse classification of user problems in science and engineering shows the heterogeneity of computational structures involved. Moreover, given one type of a problem, there can be a broad variation of computational parameters even between similar problems. As an example, the number of floating point operations to be executed per data access can vary by orders of magnitude. For performance optimization, the designer has to balance processor performance and data supply rate (e.g. memory accesses per second). Maximum performance can be achieved for only one case, that is, for one set of parameter values.

Due to the heterogeneity and broad variation of the computational structure of a great deal of user problems each processor node must work under its own control unit in order to allow for independent processing of subtasks and flexible interprocessor communication.

Appropriate parallel computers belong to the category of multiple-instruction-multiple-data (MIMD) systems. Such systems are realized as multiprocessor or multicomputer machines based on various architectural concepts. Such systems can be used with great versatility as the nodes can execute the same or different instructions and operate on the same or different data.

The computational requirements vary considerably from one user problem to another. The specific requirements of a given user problem entail special design properties of the computer system. Conversely, a given computer architecture can efficiently be used only for suitably structured applications. The interaction between computer architecture and the problem structure can be considered at different specification levels, as the coarse description of the problem class, the mathematical model, the used calculation methods and the computational, algorithmic level (Tab.1).

107

USER PROBLEM ⟷ COMPUTER	
APPLICATION Science and Engineering • Definition of the problem • Theory/Modelling	PERFORMANCE ESTIMATION • CPU Performance • Memory capacity • Peripherals
MATHEMATICAL MODELL • Set of PDE • Statistical Models, etc.	
COMPUTATIONAL METHODS • Numerical grid • Matrix operations • Numerical integration • Monte Carlo methods etc.	SPECIAL HW and SW • Fast arithmetic units • Fast logic units • Random number generator • Look-ahead table • Nearest-Neighbor communication • Main & Secondary Memory
PROGRAMMING • Subtask decomposition • "Grain size" • Process - process communication • No. of operations per memory access • Macro-instructions etc.	PROCESS/PROCESSOR ASSIGNMENT NODE INDIVIDUALIZATION MULTIPROCESSOR CONFIGURATION • Balanced processor-memory data transfer • Inter-processor communication, etc.

Tab. 1: Interaction between computer architecture and application

3. SPECIALIZED MULTIPURPOSE SYSTEMS

3.1 General

Currently, most successful parallel computers are **vector computers** which use different functional units to perform different stages of a computational task, one after the other. Upon a sequence of input data, fast vector-processors can deliver one result at the output of the last functional unit - the end of the "pipeline" - every clock cycle. This paper does not deal with "pipeline" parallelism, but with "real" parallelism, also called concurrency: Parallel execution of complete operations at the instruction or subroutine level. Such parallel systems are subdivided into two classes: **Array processors** and **multiprocessors** /Volk88/. The former typically contain many processing elements, e.g. multiple arithmetic units, but only one control unit. In the following we concentrate on the highly flexible architecture of multiprocessors. We start from a rather general purpose base of parallel architecture and look at hardware and software means suitable for specialization.

Multiprocessor systems can be specialized in a twofold way:

(i)   By special hardware or software components of the individual computer, that is, the processor-memory node of a multiprocessor system. The design of the node has to take into consideration the computational requirements, as for instance fast floating point operations, fast logical operations, generation of random numbers.

(ii)   By topological means, that is, by designing a suitable interconnection network. The appropriate network topology depends on the subtask decomposition of the target user problem. In particular, the computational structure of the subtasks and their communication dependencies must be considered.

3.2 Individualization of the node

Parallel computers can achieve higher performance, if designed with special architectural features. However, this curtails the useful application range. On the other hand, if the system design is aimed at a large problem class, the resulting performance is likely to be lower.

Frequently, specializing features of the node have been realized by including special processors, supplementary to a general purpose standard processor which fulfills the master function. The basic node structure is depicted in Fig. 1. Typically the special processor may contain the following units /BFHH85/:

- Floating point arithmetic unit(s),
- fixed point unit,
- fast logic unit,
- address calculation units serving the special processor with operands from the node's main memory and delivering the results,
- microprogram unit for fast computation of parts of algorithms (e.g. relaxation on a numerical grid) through corresponding microprograms (firmware).

Especially with the microprogramming facility high versatility of the special processor can be obtained. The instruction set can be dynamically adapted to a particular application. Moreover, this microprogramming approach can be extended to store various microprograms which represent different operation modes of the processor. The concept of a Standard-Processor which unifies different operation modes was proposed by W. Händler /Haen85/. This approach allows for dynamic processing such as to change from one operation mode to another. Possible operation modes are:

- General purpose processing,
- higher level language processing (level of big objects instead of single machine words),
- reduction automation (demand-driven),
- data-flow processing (data-driven),
- associative parallel processing and others.

This concept essentially aims at more efficient and flexible use of the processor hardware. The operation mode can be chosen and changed by the programmer. There is a unique operating system for all operation modes. A change of the operation mode does not necessarily entail a complete exchange of the corresponding instruction set (set of microprograms).

As an example, Fig. 2 represents a block scheme of the Erlangen MEMSY-node (Modular Expandable Multiprocessor System) /Hess89/, /BFHH85/, /FKLV83/. The special processor fetches the data from the dual-port memory (SM), special problem-adapted instructions are stored in the microprogram memory (MPM). The main processor (P), in general a commercially available microprocessor, performs inter-node-

and I/O-communication, effects data transfer to and from the special processor memory (SM), computes minor subtasks of the user problem etc. In the local memory (LM) standard and special software (and data) are stored. The design allows for adapting to technological advance of high performance components. Thus, recently announced high performance processors, as for instance the Intel i860 (80 MFLOPS + 40 MOPS integer, peak) or the Motorola 88100 may be used as a special processor (replacing SP and MPM). Furthermore, the special processor unit can be adjusted to the application's needs both by multiplying the same unit and/or by connecting different special units. For instance with highly compute-intensive user problems, the special processor can be composed by two or three numerical "co-processors" working in parallel on data transferred to their respective dual-port memories (SM) by the main processor. On the other hand, for rather heterogeneous user problems, the special processor can be constituted by various functionally different co-processors specialized for, say, numerical, logical, graphical tasks. As an alternative design, the node could contain two identical high performance processors instead of two different ones, for the main processor (P) and for the special processor (SP). Then the node structure is symmetric with respect to each high performance processor, but each processor plays a different role: The "administrative" role and the "number crunching" role. Moreover, this design concept provides for a fault-tolerant feature of the node, in so far as one processor can assume the task of the other.

The SUPRENUM-node /BGMu86/, /Gilo88/ is another example for the special processor approach (SUPRENUM-Project financed by the German Federal Ministry of Research and Technology). The SUPRENUM-multiprocessor is aimed at compute-intensive numerical applications. Therefore, the special processor is provided with a fast pipelined floating point unit (Weitek WTL 2264/2265). The main processor used for administrative tasks is a Motorola MC 68020.

Artificial intelligence (AI) problems belong to another, quite different class of applications. Symbolic programming is widely used and requires special hardware and software components. However, task and data partitioning in this field is often less straightforward than in the numerical field and the parallelization degree is often rather small, e.g. the degree of OR-parallelization of PROLOG programs is smaller than ten /HöWi87/. In several cases multiprocessor systems with about 100 processors may be used with acceptable efficiency. PROLOG machines need special units for efficient computation, as for instance microprogramming units, unification unit (for unifying two data objects), address calculation unit(s), a special unit for supporting backtracking. Parallel systems need, in addition to these special units, an appropriate

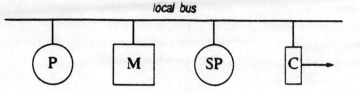

Fig. 1: Basic multiprocessor node: Main processor P, memory M, special processor
SP, communication unit C for communication with other nodes and peripherals.

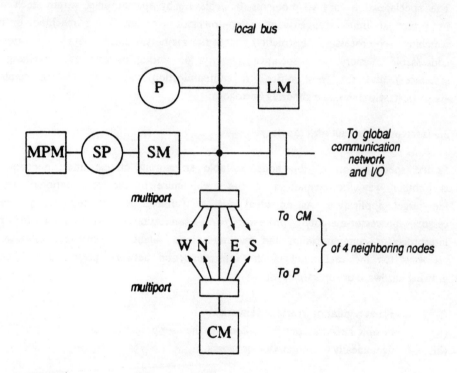

Fig.2: MEMSY-node:
Processsor-Module: Main processor P, local memory LM, special processor
SP, dual port memory SM, micro-program memory MPM.
Communication Memory CM.
Internode Connections:
Connections to 4 neighbor-nodes (W, N, E, S) in a
Nearest-Neighbor (NN) network

interconnection network in order to allow for efficient communication. This strongly depends upon the memory structure (global shared memory, distributed local memory or distributed shared memory with local communication paths). A more thorough discussion of the architectural requirements can be found in /Kobe88/.

An example of an AI multiprocessor system is the Fairchild AI Machine (FAIM-1) /Ande87/. It is designed as a highly concurrent, "general purpose" system for AI problem computation. The processing elements ("hectogons") are connected in a hexagonal mesh allowing for direct interprocessor communication with six neighbors. The specialization for AI problems is achieved by special units within each hectogon. Its prinioipal units are: Evaluation co-processor (for evaluating machine instructions), switching co-processor, instruction stream memory, local data memory, pattern addressable memory (an associative memory for storing patterns and matching complex structures) and the "post office", a communication co-processor which communicates via six ports with the six neighboring hectogons.

3.3 Interconnection network topology

Beside specialization of the node, suitable structuring of the interconnection network is another way for adaptation of the architecture to the computational structure of the target application. As an ideal goal, a suitable network topology (processor-pro-cessor-, processor-memory-, processor-I/0 - connections) should allow for minimum interprocessor communication and maximum load balancing. There are three architectu-ral ways for the realization of the interconnection network, presented with respect to growing hardware or software costs:

(i)   Fixed regular or irregular networks,
(ii)   flexible networks, possibly tailored to the user problems, and
(iii)   dynamically reconfigurable networks.

The last case (iii) is the most interesting one as to optimization of the communication structure. However, its realization is difficult. There are architectures based on switching-networks which are sometimes supplied with active (programmable) nodes so as to provide for appropriate communication paths at runtime (e.g. Denelcor HEP, the multi-stage Delta-network). Flexible, reconfigurable interconnection systems can be tailored according to the subtask - dependency - structure of the user problem. In order to match suitably the problem structure irregular configurations must often be accepted. This can produce heavy difficulties for the programmer. Therefore, there is a practical size limitation of about 20 nodes for irregular multiprocessor configurations. With the DIRMU-25 system (Distributed Reconfigurable Multiprocessor system) a great deal of experience was gained from implementations on different configurations (ring, pyramid, binary tree, rectangular arrays and others) /HaMW85/, /FBMV88/. For many user problems the ring configuration has proven appropriate. Moreover, the ring configuration can be easily made fault-tolerant by additional connections between every processor and its two second neighbors /MaMW86/.

High performance multiprocessor systems need a fixed and regular inter-node communication network for the purpose of programmability. Furthermore, such large systems should be scalable to allow for enhancement of performance. An important architectural feature for **scalability** /Haen77/ is **constant local connection complexity**, (i.e. constant number of communication paths to direct neighbor-nodes) so that with increasing size the global complexity grows linearly with the number of nodes /Haen80/. A counter-example is the hypercube as its topology is size-dependent. The number of communication paths of each node grows with the system size, that is the dimension of the hypercube. Programming data transfer between nodes has to take into account the cube dimension. This entails compatibility problems for programs to be run on hypercubes with different size.

The demand for regularity of the communication network restricts the topological adaptation possibilities. But fortunately the large class of compute-intensive problems of all kinds of simulation of physical phenomena have similar basic features: A gridlike computational structure and a predominantly local nearest-neighbor (NN) communication which arise from discretization of the 2- or 3-dimensional physical space. Fundamentally, this is a consequence of the fact that physical action takes place through an immediate medium (local interaction). To a smaller extent, also global communication facilities are required, as for instance for collection of partial results from the single nodes.

To cope with these communication demands a few architectural principles have been used in the design of high performance multiprocessor systems. Three main principles (Fig. 3 and 4) are discussed in the following.

114

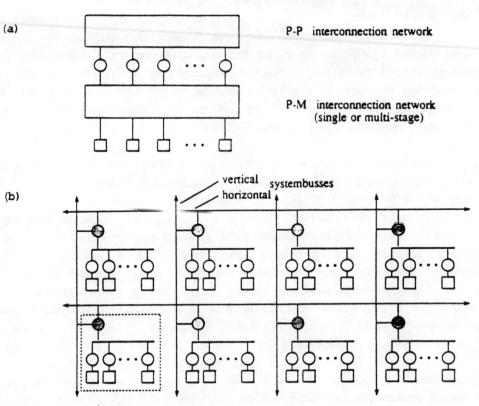

Fig. 3:   Multiprocessor architectures (◯ processor, ☐ memory):

(a) Switching networks (general scheme):
- Processor/processor communication (message - passing)
- Processor/memory (single or multi-stage) interconnection network with (programmable) switching nodes (global shared memory)

(b)   Distributed local memory, for instance SUPRENUM: 4 x 4 clusters (16 processors each):
- Lower level: Cluster bus,
- Upper level: Horizontal and vertical system busses.

115

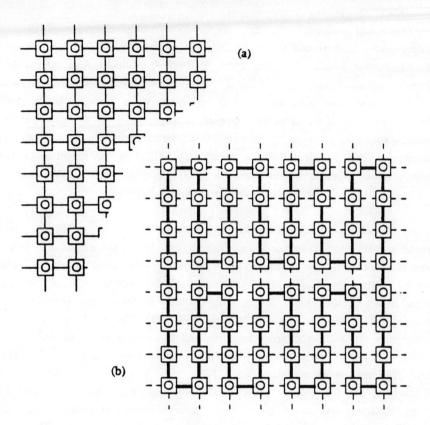

(a)

(b)

 Processor-Memory-Module

Fig. 4: Distributed Shared Memory Architecture (DSM):
MEMSY (Modular Expandable Multiprocessor System).

(a) Orthogonal processor array. Nearest-neighbor connection via memory coupling. Each Processor-Memory-Module (PMM) is coupled to 4 neighboring nodes via 5 communication memories CM ("own" CM and 4 neighbor CMs).

(b) Ring configuration by suitable activation (by software) of hardware configuration (a). Fault-Tolerant configuration.

The **clustering** principle was first realized with the Carnegie Mellon University multiprocessors C.mmp and Cm* /JonS80/. Presently, the SUPRENUM system based on a similar cluster structure is being built /BGMu86/, /Gilo88/. The inter-node communication is achieved via a hierarchical two-level bus system (cluster bus and system bus). Each node can communicate with any other by a message-passing protocol. This universal communication property, however, is detrimental to a highly parallel nearest-neighbor communication as demanded by numerical grid problems. This architectural disadvantage can be evaded by a virtual SUPRENUM NN-machine (a logical NN-communication structure on the real machine) which however implies some additional overhead. Scalability is limited due to the hierarchical bus system.

Another widely used interconnection principle is based on **switching networks**. In particular multi-stage interconnection systems can be versatile and can adapt well to different communication structures of applications even with a strongly heterogeneous subtask system. Beside the more classical switching networks (e.g. banyan network, omega network) the R256, a research parallel processor for scientific computation, represents an interesting new development /FKTI89/. The interprocessor communication of the 16 x 16 processor array is performed by a distributed parallel network which consists of 2 x 16 switching units placed in every row and column of the processor array. The network is controlled in a hierarchical fashion: The switching units of the i-th row and the i-th column are controlled by one local control unit, and all 16 local control units are controlled by a main control unit. The network was designed to resolve heavy communication load.

A third interconnection principle uses **memory-coupling** for inter-processor communication. Communication between two processors occurs through common variables stored in a memory shared by both processors. The communication mechanism is fast, as only about twice the memory access time is needed for one data exchange. Multiprocessor systems with one global shared memory represent a real general purpose architecture, but scaling up is limited due to contention on the memory bus. This drawback can be avoided if the shared memory is distributed all over the processor array in such a way that each partial shared memory is accessed by a small and constant number of "neighbored" processors. Such **distributed shared memory (DSM) systems** can be efficiently used for all kinds of numerical grid problems which allow for "natural" data partitioning and mapping onto the processor array. Due to inherent constant local complexity (i.e. with respect to every partial shared memory) such systems are - in principle - arbitrarily scalable. As an example for DSM systems, the worker processor array of MEMSY is organized as a orthogonal mesh of processors: Every processor embodies a communication memory which can be accessed by the processor itself and its four neighbored processors /FKLV83/, /BFHH85/, /FrVo88/, /Hess89/. MEMSY is hierarchically organized: The large worker processor array (level A, with e.g. 256 processors) is provided with data and controlled by the B-level processor array (with e.g. 64 processors). The overall control and the communication with the host is performed by the supervisor at level C. The pyramidal organization of MEMSY is essentially functional /Frit86/: Computation of the user problem at level A, operating system (higher functions) and I/0 to mass storage at level B, and overall control at level C. Regarding pyramidal architectures it may be mentioned here that this organisational principle has also been used for parallel image processing, e.g. with the PAPIA systems /Cant86/. PAPIA is a multi-SIMD (single-instruction-multiple-data) machine organized in a pyramidal structure such as to process an image in different stages (preprocessing, primitives extraction, description, interpretation) when the data is being transferred and reduced from the lowest level to the highest one.

3.4. Special Systems

In the preceding sections, we discussed specializing architectural means which can be applied to basically general purpose architectures of parallel computers, in particular multiprocessor systems. If the user wants more specialization to increase performance and lower the cost/performance ratio, then the design must closely match the computational structure of the user problem. Special parallel computers can be built using special components and/or special topologies for the interconnection network.

Many special computers have been built for special user problems. For instance highly efficient Monte-Carlo- and Molecular Dynamics (MD) processors were built for the simulation of many-particle ensembles. Highest speed can be achieved if most of the algorithm is "cast" in hardware. This was realized with the Delft Molecular Dynamics processor /BaBr88/. The different parts of the MD-algorithm are executed by dedicated hardware operators: Momentum update unit, position-update unit, the I/0-section for handling the interface between the MD-processor and the host (general purpose) computer, and the control unit. In addition, there is a "particle memory" where the current momenta and positions of the particles are kept. As the momentum update unit (the positions and momenta of two interacting particles are input and the new momenta of both particles are output) is the most compute-intensive part of the algorithm, a parallel and pipelined processor architecture was designed to speed up the computation.

An extensive problem class should here be mentioned for which very fast parallel computers have been realized: Image processing. The corresponding algorithms vary considerably from the first to the last stage. Thus, the algorithms for image description and in interpretation can be very different for one particular problem or other. This complex application can only be treated by a powerful and versatile computer system. Typically, a special and fast SIMD-system is used for preprocessing to cope with huge data amounts. Proceeding to the "higher" stages, more and more general purpose properties of the computer system are required. Complex image processing problems can only be computed by means of computer systems involving special purpose and general purpose processors.

4. CONCLUSIONS

Efficient computation can be achieved if the used computer is well adjusted to the computational needs of the user problem. As there is a broad spectrum of very different applications corresponding suitable computer architectures vary considerably. The wide range between general purpose and special purpose computer systems can be covered by systems which are adapted to the target application by some specializing architecture. This can be realized by special hardware units, microprogrammability, appropriate operating system functions (e.g. different operation modes), and in case of parallel computers by topological means. In particular with flexible multiprocessor systems there are several design possibilities for adaptation to the computational structure of the target application. Especially the interconnection network and the interprocessor communication mechanisms can be adapted to common features of problem classes. Moreover, in large multiprocessor systems **space sharing** can be used /Haen80/. This means that the total processor ensemble can be (flexibly) subdivided into sub-ensembles which are assigned to different tasks. The special case of functional "space sharing" was discussed in the preceding chapter for a distributed shared memory architecture (MEMSY) with hierarchical (pyramidal) organization: The lowest level is formed by the worker processor array (e.g. 256 processors), at the next higher level operating system functions and I/0 are implemented on a smaller array (e.g. 64 processors) and at the top level the supervisor processor is located with a connection to the host computer.

Because of wide variations of the computational needs of current and possibly future less known user problems, general purpose properties of modern computer architectures will be necessary. However, for the sake of efficiency special components, depending on the application, must be provided. As discussed above, multiprocessor systems represent a suitable architectural answer to the general purpose/special purpose issue. Both aspects can be suitably combined in multiprocessor systems. Furthermore, small multiprocessor systems (about up to 10 nodes) may be composed heterogeneously by nodes with different specialization (for floating point operations, logical operations, graphics and others), thus forming a "team of specialists". Such a system can be used as a workstation which can optionally be operated in a general purpose mode or a few special purpose modes.

5. REFERENCES

/Ande87/   Andersson, J.M. et al.: The architecture of FAIM-1, Computer, Jan. 1987, 55-68

/BaBr88/   Bakker, A.F. and C. Brain: Design and Implementation fo the Delft Molecular-Dynamics processor. In: Special Purpose Computers (B.J. Alder ed.), Academic Press Inc. Boston 1988

/BFHH85/   Bode, A.; Fritsch, G.; Händler, W.; Henning, W.; Hofmann, F.; Volkert, J.: Multi-grid oriented computer architecture. Proc. 1985 Int. Conf. Parallel Processing, St. Charles, 81-95. IEEE Comp. Soc. 1985

/BGMu86/   Behr, P.; Giloi, W.K.; Mühlenbein, R.: SUPRENUM - The German Supercomputer Architecture - Rationale and Concepts, Proc. 1986 Internat. Conf. on Parallel Processing, IEEE catalog no. 86CH2355-6, 483-486

/Cant86/   Cantoni, V.: I.P. Hierarchical systems: Architectural features. In: Pyramidal Systems for Computer Vision (V. Cantoni, St. Levialdi, ed.). NATO ASI Series F, Comp. and Syst. Sciences Vol. 25, Springer Verlag Berlin Heidelberg 1986

/FBMV88/   Finnemann, H.; Brehm, J.; Michel, E. and J. Volkert: Solution of the neutron diffusion equation through multigrid methods implemented on a memory-coupled 25-processor system. Parallel Computing 8 (1988), 391-398

/FKLV 83/   Fritsch, G .; Kleinöder, W.; Linster, C.U.;; Volkert, J.: EMSY 85 - The Erlangen Multiprocessor System for a broad spectrum of applications. Proc. 1983 Int. Conf. on Parallel Processing, 325-330 and in: Supercomputers: Design and Applications (K. Hwang, ed.), IEEE Comp. Soc. 1984

/Frit86/   Fritsch, G.: Numerical simulation of physical phenomena by parallel computing. WOPPLOT 86. Lect. Notes in Computer Science, No. 253 (J.D. Becker, I. Eisele, eds.). Springer Verlag Berlin Heidelberg 1987

/FKTI89/   Fukazawa, T. et al.: R 256: A research parallel processor for scientific computation. Proc. 16th Annual Int. Symp. on Computer Architecture, May 28 - June 1, 1989, Jerusalem. IEEE Comp. Soc. Press 1989 (ACM 0884 - 7495/89/0000/0344)

/FrVo88/   Fritsch, G. and J. Volkert: Multiprocessor systems for large numerical applications. Proc. PARCELLA 88, Berlin (DDR), Oct.17-21, 1988. Mathematical Research Vol. 48, Akademie Verlag Berlin 1588

/Gilo88/   Giloi, W.K.: The SUPRENUM Architecture, Proc. CONPAR 88, Part A, British Computer Society 1988, 1-8

/Haen77/   Händler, W.: Aspects of parallelism in computer architecture. In: Parallel Computers - Parallel Mathematics (M. Feilmeier, ed.). Int. Association for Mathematics and Computers in Simulation 1977

/Haen80/   Händler, W.: Thesen und Anmerkungen zur künftigen Rechnerentwicklung. Aus: G. Regenspurg (Hrsg.): GMD-Rechnerstruktur-Workshop, GMD St. Augustin/F.R. Germany, Bericht Nr. 128. R. Oldenbourg München Wien 1980

/Haen85/   Händler, W.: Dynamic computer structures for manifold utilization. Parallel Computing 2 (1985), 15-32

/HaMW85/   Händler, W.; Maehle, E.; Wirl, K.: DIRMU Multiprocessor Configurations, Proc. 1985 Int. Conf. on Parallel Processing, St. Charles 1985, 652-656. IEEE Comp. Soc. 1985

/Hess89/   Hessenauer, H. et al.: Internal Report. DFG-Sonderforschungsbereich 182: Multiprozessor- und Netzwerkkonfigurationen. IMMD III, Universität Erlangen-Nürnberg 1989

/HöWi87/   Höpfl, F.; Wirl, K.: Oder-Parallelisierung von PROLOG-Programmen auf dem DIRMU Multiprozessorsystem. 3.GI/PARS-Workshop, Jülich 1987

/JonS80/   Jones, A.K. and P. Schwarz: Experience using multiprocessor systems - a status report, ACM Computing Surveys, Vol. 12, No. 2, June 1980, 121-165

/Kobe88/   Kober, R. (Hrsg.): Parallelrechner-Architekturen. Springer Verlag Berlin Heidelberg 1988

MaMW86/   Maehle, E.; Moritzen, K. and K. Wirl: Fault-tolerant hardware configuration management on the multiprocesssor system DIRMU25. Proc. CONPAR 86, Aachen 1986. Lect. Notes in Computer Science, No. 237, Springer Verlag Berlin Heidelberg 1986

/Volk88/   Volkert, J.: Architectures, programming and performance of supercomputers. Kerntechnik 52 (1988), No. 2, 112-129

Implementation of Divide-and-Conquer Algorithms
on Multiprocessors

Renate Knecht

Zentralinstitut für Angewandte Mathematik

Kernforschungsanlage Jülich GmbH

5170 Jülich, Fed. Rep. Germany

Abstract: Algorithms with a divide-and-conquer structure are suitable candidates for parallelization. The idea of the divide-and-conquer paradigm is to fragment a problem into subproblems of the same kind, to solve the subproblems recursively, and, finally, to combine the solutions of the subproblems into a solution of the original problem.

Two algorithms of this structure namely an "approximation" algorithm for the Euclidean Traveling Salesman Problem and an algorithm to determine the convex hull of a two-dimensional point set have been implemented in FORTRAN on a CRAY X-MP using the CRAY multitasking facilities. For the parallel implementation of algorithms with a divide-and-conquer structure two methods are discussed. The goal was to find an implementation strategy which is independent of the available shared memory multiprocessor system and additionally independent of the number of processors which can be used to find the problem's solution.

Keywords: Multitasking, macrotasking, microtasking, CRAY X-MP, divide-and-conquer, parallel algorithm, Euclidean Traveling Salesman Problem, convex hull.

1. Introduction

Divide-and-conquer represents an important strategy successfully used on a wide range of problems. Efficient divide-and-conquer algorithms have been developed in sorting (Quicksort, Mergesort [1, 13]), searching (binary search tree [1, 13]), geometry (closest-point problems [2], convex hull [11]), graph theory (Traveling Salesman Problem [9]), Fast Fourier Transform [13], and for the approximation of NP-hard problems. Divide-and-conquer is also a useful strategy in hardware design.

The general idea of the divide-and-conquer strategy is to fragment a problem into subproblems of the same kind. This partitioning continues recursively until the problem size has been reduced to a predefined value, called the *maximum partition size*. Then, these subproblems can be solved independently. Finally, to get a solution of the original problem, the solutions of the subproblems have to be combined.

The divide-and-conquer strategy has the property that the subdivision in smaller problems forms a tree, which is usually a binary tree if the problems are subdivided in two subproblems.

In this case, each subproblem corresponds to a partition number i. The original problem has the partition number 1 whereas the children of partition i have the partition numbers $2*i$ and $2*i+1$.

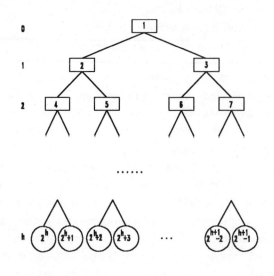

Fig. 1. Binary tree resulting from subdivision in smaller problems

Fig. 1 shows the binary tree resulting from the subdivision of the original problem in smaller problems. The nodes correspond to the partition numbers of the subproblems. The leaves which are indicated by circles specify those subproblems which are no longer subdivided and can be solved. The numbers on the left mark the level of the nodes in the tree of height h. The resulting tree is not necessarily a full binary tree, but the levels of the leaves differ at most by one.

2. Parallelizing divide-and-conquer algorithms

For the parallel implementation of divide-and-conquer algorithms two methods are discussed. The first possibility to parallelize an algorithm with a divide-and-conquer structure is to divide the problem in two subproblems. This can be done ' one processor. Now, a second task is created and both tasks subdivide their subproblems into smaller subproblems. Then, two additional tasks can be created. If the maximum number of available CPUs is reached, all tasks work on their subproblems asynchronously until they get a solution of their subproblems. Finally, the solutions of the subproblems have to be combined to get a solution of the original problem.

The idea behind the second method is to solve the problem in two steps:

1. Partitioning into subproblems or solution of the subproblems which are no longer subdivided, respectively.
2. Combination of the subproblems' solutions until a solution of the original problem is found.

To keep in mind which subproblems can be treated next, a circular queue is chosen as the data structure which contains the work still left. At the beginning only the original problem can be subdivided and this is the only problem which is queued. This element is removed from the queue and the problem is subdivided into two smaller subproblems. After this partitioning the two resulting subproblems are queued and so on.

Concerning the second step of the algorithm the combination of the solutions of two subproblems is done under the assumption that the processing of both subproblems is finished.

Considering the parallel execution of this second method, each task tries to get an element of the queue and, if the queue is not empty, each successful task can work on this partition.

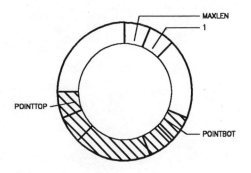

Fig. 2. Circular queue

The circular queue is graphically shown in Fig. 2. *MAXLEN* is defined to be the length of the array for the implementation of the queue. The pointer *POINTBOT* indicates the first element in the queue and the pointer *POINTTOP* denotes the last element in the queue. The queue has a first-in first-out behaviour. Elements can be added to the queue after the element pointed to by *POINTTOP* and they can be removed at position *POINTBOT*.

Comparing the two methods to parallelize divide-and-conquer algorithms it is obvious that the number of CPUs which can be used in the first method has to be a power of two. After the first partitioning a new task is created and both tasks create an additional task after the next subdivision and so on. The structure of the program has to reflect explicitly the number of CPUs. That means, the program has to be modified if the number of CPUs varies.

Contrary to the first method, the second implementation allows any number of CPUs to work efficiently on these kind of problems; the number is not restricted to a power of two. For this implementation the number of tasks has not to be defined explicitly, any number of tasks can work on the queue and, if the queue is not empty, an element can be removed. To allow a different number of tasks to work concurrently, only the constant defining the number of CPUs has to be changed.

The circular queue contains two sorts of information: The partition number of those subproblems that can be treated next and the phase for these subproblems. The phase of a subproblem indicates whether the subproblem is to be divided further or is to be solved, respectively (phase = 1), or if two subproblems are to be combined (phase = 2).

```
SUBROUTINE ENQUEUE(COUNT,POINTTOP,NUMBER,PHASE,OK)
INTEGER MAXLEN
PARAMETER (MAXLEN = 100000)
INTEGER QUEUE(MAXLEN,2),COUNT,POINTTOP,NUMBER,PHASE
LOGICAL OK
COMMON/SPACE/QUEUE
IF (COUNT.LT.MAXLEN) THEN
   POINTTOP = MOD(POINTTOP,MAXLEN) + 1
   QUEUE(POINTTOP,1) = NUMBER
   QUEUE(POINTTOP,2) = PHASE
   COUNT = COUNT + 1
   OK = .TRUE.
ELSE
   OK = .FALSE.
ENDIF
RETURN
END

SUBROUTINE DEQUEUE(COUNT,POINTBOT,NUMBER,PHASE,OK)
INTEGER MAXLEN
PARAMETER (MAXLEN = 100000)
INTEGER QUEUE(MAXLEN,2),COUNT,POINTBOT,NUMBER,PHASE
LOGICAL OK
COMMON/SPACE/QUEUE
IF (COUNT.GT.0) THEN
   NUMBER = QUEUE(POINTBOT,1)
   PHASE = QUEUE(POINTBOT,2)
   POINTBOT = MOD(POINTBOT,MAXLEN) + 1
   COUNT = COUNT - 1
   OK = .TRUE.
ELSE
   OK = .FALSE.
ENDIF
RETURN
END
```

Fig. 3. Subroutines to handle the circular queue

To handle the queue two procedures are implemented: *ENQUEUE* and *DEQUEUE* (Fig. 3). ENQUEUE puts an element in the queue and DEQUEUE gets an element from the queue. Fig. 3 further shows the implementation of the circular queue as the two-dimensional array *QUEUE* which contains the number of the subproblem (*NUMBER*) and whether the problem is to be treated in the first or in the second step of the algorithm (*PHASE*). The variable *COUNT* specifies the number of elements which are in the queue at a time and *OK* signals if the ENQUEUE or DEQUEUE operation succeeded.

It has to be noted that at one time only one processor may operate on the queue. Therefore, operations on the queue must be embedded in critical regions.

3. Two divide-and-conquer algorithms

To show the results of implementations of the parallelization strategies described above, two different algorithms are considered. The first one is Karp's partitioning algorithm to find a suboptimal solution for the Euclidean Traveling Salesman Problem (ETSP) [9]. The second one is a divide-and-conquer algorithm to compute the convex hull of a two-dimensional point set [11].

Divide-and-conquer has a large number of variations, depending on the work needed to create the subproblems and the work needed to combine the solutions of the subproblems. Considering the two given examples, in the ETSP algorithm the combination is the simpler part whereas in the convex hull algorithm the subdivision and the solution are simpler in comparison to the combination of two subproblems which is more time consuming.

3.1 Euclidean Traveling Salesman Problem

Given n vertices and the distance between each pair of vertices, the Traveling Salesman Problem is to find the shortest tour in a labelled graph where every vertex is visited exactly once. For this problem, which seems to be very simple, the time to find the exact solution increases exponentially with n for the algorithms known today. For this reason, approximate algorithms - algorithms which find a suboptimal solution in polynomial time - are attractive. Many of these algorithms find tours which are usually only a few percent longer than the shortest tour, yet have a low-order polynomial time complexity [10]. One of the many approximate algorithms available for the ETSP is Karp's partitioning algorithm which has a divide-and-conquer structure.

Karp's partitioning algorithm is a recursive method to solve the ETSP. In the ETSP each vertex has a location in the Euclidean space, and thus the distances are Euclidean. Since the successor vertex to any vertex in the tour is usually a nearby vertex, the problem can be "geographically" subdivided into two subproblems which then can be solved independently. Combining the resulting subtours into a single tour yields an approximate solution to the ETSP [14].

A set of vertices is divided into two subsets. Partitioning can be done either by x-coordinates or by y-coordinates, depending on whether the distribution of the vertices is greater in the x- or y-direction, respectively. The only vertex which is included in both subsets, is the vertex with the medium x- or y-coordinate. The vertex, common in both subsets, is called the *source* of these subtours. Thus, the extremal x- and y-coordinates of the set of vertices in any partition must be calculated in order to determine the partitioning direction. Once the direction is chosen, the proper median point must be found, so that the set can be evenly divided into two subsets. A subset contains at most one more element than the other subset.

This divide-and-conquer process recursively continues until a set of vertices is left whose cardinality is smaller than a predefined value, the *maximum partition size*. The resulting sub-problems are solved using the farthest-insertion heuristic [4]. The farthest-insertion heuristic springs from the intuition that if the rough outline of the tour can be constructed from the widely-separated vertices, then the finer details of the tour resulting from the inclusion of nearer vertices can be filled in without increasing its total length significantly. The algorithm adds the farthest vertex from any vertex in the partially-completed tour in such a way that the increase in the length of the partially-completed tour is minimized (Fig. 4).

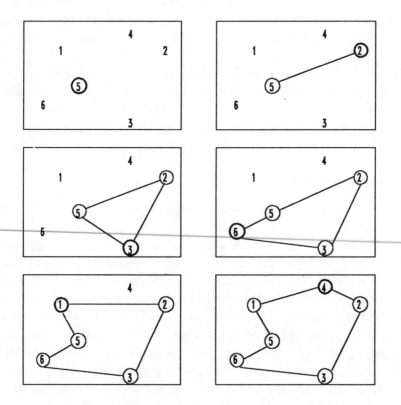

Fig. 4. **Illustration of the farthest-insertion heuristic**

For the combination of two subtours there are two possibilities (Fig. 5):

1. The edge in partition i directed to the source and the edge in partition j directed from the source are deleted. Then the predecessor of the source in partition i and the successor of the source in partition j are joined.
2. The edge in partition j directed to the source and the edge in partition i directed from the source are deleted. Then the predecessor of the source in partition j and the successor of the source in partition i are joined.

Joining two subtours is done by deciding which of these two possible shortcuts shortens the tour by the greatest amount.

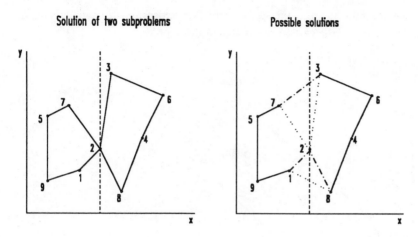

Fig. 5. Illustration of Karp's partitioning algorithm

An important factor concerning not only the resulting tour length but also the execution time is the value of the maximum partition size for solving a subproblem. On the one hand the execution time for the considered problems decreases when the maximum partition size increases and on the other hand the solution results in a shorter tour length. That means, the more combinations exist to find a solution, the longer the resulting tour length can be because of the simplicity of the combination, but it is more time consuming. For example, the sequential execution time for $n=420$ measured on a CRAY X-MP/416 is 65.31 ms for a maximum partition size of 14. The resulting tour length is 185.60. The maximum partition size of 3 for the same example leads to an execution time of 394.72 ms and to a tour length of 215.73.

A compromise for the considered examples is not to subdivide the problem to such a small partition size but also to guarantee several subdivisions. Therefore, a value of 14 turned out to be a good selection for the considered problems.

3.2 Convex Hull of n points in two dimensions

Given a point set S consisting of n points in two dimensions, the convex hull is the boundary of the smallest convex domain in E^2 containing the point set S [12].

Preparata and Hong [11] developed a divide-and-conquer algorithm to solve this problem. Based on this algorithm the implemented algorithm includes as a preliminary step the sorting of the elements in the point set S according to the x-coordinates. To sort the elements the Quicksort algorithm is used in its sequential form.

To compute the convex hull, the point set S is subdivided with respect to x into two subproblems which then can be solved independently. This divide-and-conquer process recursively continues until the subsets contain at most 3 points. The combination of the resulting convex hulls of the subsets yields the convex hull of the original problem. The algorithm presented finds the ordered convex hull of a set of two-dimensional points, i.e. not only the set of vertices of the convex polygon representing the hull but also the sequence of the vertices on the boundary of this polygon. The sequence contains the vertices in counterclockwise order.

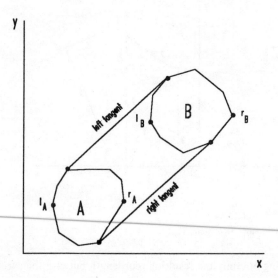

Fig. 6. Illustration of the planar merge procedure

Let A and B be the two convex polygons in the plane which are the subproblems that have to be combined. Considering Fig. 6 l_A is the leftmost point in A with the minimum x-coordinate and r_A is the rightmost point in A with the maximum x-coordinate; l_B and r_B are defined similarly. The convex polygon of A and B is obtained by determining the left and right tangent and by eliminating the points of A and B which become internal to the resulting polygon.

4. Multitasking on CRAY X-MP

On the multiprocessor vector-supercomputer CRAY X-MP parallelism can be exploited beyond vectorization on the programming language level by several multitasking strategies [3]. Using multitasking a program can be structured into two or more tasks that can execute concurrently. With *macrotasking*, multitasking is supported on the subroutine level. Task creation, synchronization, and communication are specified explicitly by the programmer using subroutine calls. With *microtasking*, multitasking is supported on the statement level. Subroutines, loops, or possibly sets of statements may be executed in parallel on multiple processors of a CRAY X-MP. Microtasking uses preprocessor directives inserted by the programmer. These directives are interpreted by a preprocessor which generates subroutine calls for the creation of parallel tasks and their synchronization. Within microtasking, tasks communicate by means of shared registers. The overhead of the communication and synchronization via registers is much smaller than the overhead caused by using shared variables in the main memory as it is realized with the macrotasking library routines.

A new and easy to use method of multitasking is *autotasking* which is not considered in this paper.

In order to use macrotasking, the application has to be partitioned into a fixed number of tasks, which have to be created explicitly by the user. The task management is done by the library scheduler.

The macrotasking library consists of three different parts:

1. Routines to manipulate tasks:

    *   TSKSTART(*tskarray,name* [,*list*])
        generates a new task with identification *tskarray* for the subroutine *name* and passes the parameters of *list*.
    *   TSKWAIT(*tskarray*)
        waits for the end of the task identified by *tskarray*.

2. Routines for the synchronization of tasks:

    *   EVASGN(*evid* [,*value*])
        declares the INTEGER variable *evid* as an event variable.
    *   EVWAIT(*evid*)
        waits for the event *evid*.
    *   EVPOST(*evid*)
        signals the event *evid* to the scheduler.

3. Routines to control critical regions:

    *   LOCKASGN(*lockid* [,*value*])
        declares the INTEGER variable *lockid* as a lock variable.

- LOCKON(*lockid*)

  signals the scheduler that this task will enter the critical region connected to the variable *lockid*.

- LOCKOFF(*lockid*)

  signals the scheduler that this task leaves the critical region connected to the variable *lockid*.

As mentioned above the microtasking features are provided by preprocessor directives which must be inserted in the program by the user.

The microtasking preprocessor directives can be classified according to their usage:

1. Demand and release of logical CPUs.

   - CMIC$ GETCPUS *n*

     declares the maximum number of CPUs the program may use.

   - CMIC$ RELCPUS

     releases the required CPUs back to the operation system.

2. Definition of control structures:

   - CMIC$ DO GLOBAL

     marks a DO loop to be executed in parallel, if more than one CPU is available.

   - CMIC$ PROCESS

     marks the beginning of a program part which must be executed by one processor.

   - CMIC$ ALSO PROCESS

     marks the beginning of a further program part which could be executed in parallel to other processes.

   - CMIC$ END PROCESS

     marks the end of a PROCESS part.

   - CMIC$ STOP ALL PROCESS

     provides a way to stop parallel execution before all computations are finished.

3. Safeguarding of critical regions:

   - CMIC$ GUARD *m*

     marks the beginning of the critical region named *m*.

   - CMIC$ END GUARD *m*

     marks the end of the critical region named *m*.

Using multitasking to implement the first method, namely defining the tasks explicitly, on a CRAY X-MP only macrotasking is suitable for parallelization. A new task can be created by calling the routine TSKSTART. If the maximum number of CPUs is reached, each task works on its partition until a solution is found. The necessary synchronization before combining the solutions of the subproblems can be done using the routine TSKWAIT. A disadvantage of this implementation is the large overhead of the macrotasking library routines. Microtasking can not be applied in the same way, because it is not possible to create two tasks at a time and later on two additional tasks and so on.

The second method based on the circular queue can be implemented with microtasking using the DO GLOBAL directive for parallel execution of DO-loops. The algorithm is based on a RE-PEAT loop realized with an IF and GOTO statement. Some special characteristics of the microtasking implementation can cause a lot of problems for a code of this type [8]. Because of the small granularity of these algorithms which have been implemented on the CRAY, it is a big advantage that the overhead using the microtasking directives is considerably smaller than the overhead for the macrotasking library routines.

5. Results

The algorithms to find an approximate solution for the ETSP and the algorithm to determine the convex hull of a two-dimensional point set were implemented in FORTRAN on a CRAY X-MP/416 at KFA Jülich. The CRAY X-MP/416 has 4 CPUs (8.5 ns cycle time) and a main memory of 16 megawords. The sequential programs have been compared with the parallel versions using the CRAY multitasking facilities, macrotasking and microtasking. The execution times were measured on a dedicated machine using the function IRTC to count the clock periods. As input the algorithms get two arrays containing the x- and y-coordinates. In our examples, the uniformly distributed values were produced using a random number generator.

Tab. 1 shows the CPU-times (in ms) on a CRAY X-MP/416 for the ETSP algorithm and the convex hull algorithm, respectively.

Problem Size	ETSP	Convex Hull
64	12.28	76.29
256	53.55	313.23
448	105.25	472.36
640	114.46	632.40
832	124.20	790.54
1024	220.82	1260.63

Tab. 1. CPU-times on CRAY X-MP/416 for ETSP and convex hull algorithm sequentially

Fig. 7 shows the speedup values achieved with macrotasking for Karp's partitioning algorithm where the first parallel implementation method is used [5, 6].
The speedup is defined as follows:

$$\text{Speedup} = \frac{\text{Time for serial execution}}{\text{Time for parallel execution}}$$

The speedup obtained with two processors for a problem size of 1000 nodes is about 1.8. However, if the problem size is small, i.e. less than 200 nodes, the speedup achieved with 4 CPUs is smaller than with 2 CPUs because the overhead becomes too large for such small problems, namely the overhead for the task creation. If the problem size is larger than 800, a speedup of 2.7 is reached. It has to be noted that the first three subdivisions and the last three combinations have to be done sequentially or at most by two processors. This reduces the program part which can be executed in parallel.

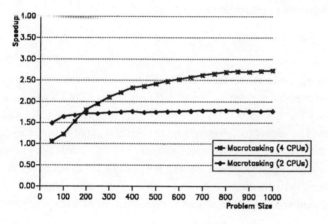

Fig. 7. Traveling Salesman Problem on CRAY X-MP/48 with macrotasking, maximum partition size = 14

The following figures show the results for the microtasking implementation which corresponds to the second parallel implementation method using a circular queue.

In Fig. 8 the speedup values for the ETSP algorithm using 2, 3, and 4 CPUs are shown. It can be observed that the resulting curves do not increase uniformly if the problem size increases. Looking at the speedup values achieved with 4 CPUs for a problem size of 416 and 448 respectively, one can notice a decreasing speedup. To explain these results the obtained speedup with 4 CPUs will be considered in the range from 416 to 448 in more detail.

In Fig. 9 the problem size increases by one. Comparing the results obtained for a maximum partition size of 14 with the results obtained for a maximum partition size of 3, it is obvious that the last curve represents a better speedup. This good performance is due to the fact that the execution time increases if the maximum partition size decreases and therefore the overhead caused by the microtasking routines has less importance. Another advantage is the better CPU-utilization because of the more balanced workload.

On the other side it has to be remarked that in most cases the resulting tour length becomes longer when the subdivision continues to a partition size of 3 elements, because the simplicity of the combination of two subtours can lead to a worse solution. The more combinations exist to find a solution the longer the resulting tour can be.

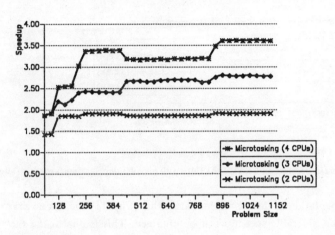

Fig. 8. Traveling Salesman Problem on CRAY X-MP/416 with microtasking, maximum partition size = 14

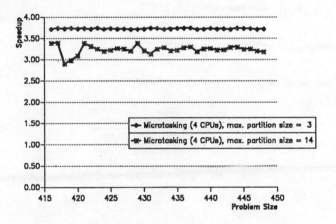

Fig. 9. Traveling Salesman Problem on CRAY X-MP/416 with microtasking

Fig. 10 shows the resulting speedup values obtained for larger problems where the maximum partition size is 14. Due to the increasing execution time the overhead caused by the microtasking routines has less importance and the speedup obtained is quite satisfactory.

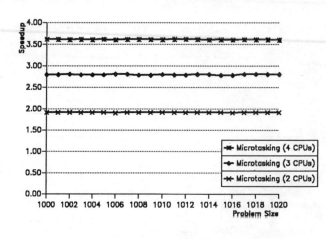

Fig. 10. Traveling Salesman Problem on CRAY X-MP/416 with microtasking, maximum partition size = 14

Concerning the implementation of the convex hull algorithm (Fig. 11) one can observe that also for small problems a considerable speedup can be obtained. This is due to the fact that the maximum partition size in this algorithm is 3. The speedup with 4 processors for a problem size of 1000 elements is about 3.8.

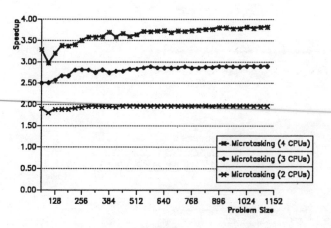

Fig. 11. Convex hull on CRAY X-MP/416 with microtasking

6. Conclusions

Two possibilities for parallelizing algorithms with a divide-and-conquer structure have been discussed. An algorithm to find a suboptimal solution for the ETSP and an algorithm to compute the convex hull of a two-dimensional point set are described and have been implemented on a CRAY X-MP using macrotasking and microtasking. The speedup which can be achieved for both algorithms using up to 4 CPUs is about 90-95% of the theoretically attainable speedup. It can be observed that microtasking leads to a better speedup than macrotasking, especially if the problem size is small.

The goal was to find an implementation which is independent of the shared memory multiprocessor system and of the number of available CPUs.

In the future, the algorithms will be implemented on a CRAY Y-MP using more than 4 processors. Additionally, autotasking can be used for the parallelization. Moreover, it could be interesting to implement this strategy on various supercomputers, for example on an IBM 3090 multiprocessor system.

Concerning the convex hull algorithm an implementation for three-dimensional point sets will be investigated.

Acknowledgment

The author wants to thank Dr. P. Weidner for his continuous encouragement. Several discussions with J.-Fr. Hake and W.E. Nagel are gratefully acknowledged.

References

[1]   A.V. Aho, J.E. Hopcroft, J.D. Ullman, *Data Structures and Algorithms* (Addison-Wesley, 1987).

[2]   J.L. Bentley, *Multidimensional Divide-and-Conquer*, Communications of the ACM 23 (1980) 214-229.

[3]   *CRAY Y-MP, CRAY X-MP EA and CRAY X-MP Multitasking Programmer's Reference Manual* (CRAY-Research Inc., SR-0222 F, 1989).

[4]   B. Golden, L. Bodin, T. Doyle, W. Stewart, jr., *Approximate Traveling Salesman Algorithms*, Operations Research 28 (1980) 694-711.

[5]   R. Gurke, *Graphenalgorithmen für MIMD-Rechner* (KFA Jülich, Jül-Spez-355, 1986).

[6]   R. Gurke, *The approximate solution of the Euclidean traveling salesman problem on a CRAY X-MP*, Parallel Computing 8 (1988) 177-183.

[7]   E. Horowitz, A. Zorat, *Divide-and-Conquer for Parallel Processing*, IEEE Transactions on Computers C-32 (1983) 582-585.

[8]   F. Hossfeld, R. Knecht, W.E. Nagel, *Multitasking: Experiences with Applications on a CRAY X-MP*, Parallel Computing, to appear.

[9]   R. Karp, *Probabilistic Analysis of Partitioning Algorithms for the Traveling Salesman Problem in the Plane*, Mathematics of Operations Research 2 (1977) 209-224.

[10] E.L. Lawler, J.K. Lenstra, A.H.G. Rinnooy Kan, D.B. Shmoys, *The Traveling Salesman Problem* (Wiley, 1985).

[11] F.P. Preparata, S.J. Hong, *Convex Hulls of Finite Sets of Points in Two and Three Dimensions*, Communications of the ACM 20 (1977) 87-93.

[12] F.P. Preparata, M.I. Shamos, *Computational Geometry - An Introduction* (Springer, 1985).

[13] R. Sedgewick, *Algorithms* (Addison-Wesley, 1988).

[14] M.J. Quinn, *The design and analysis of algorithms and data structures for the efficient solution of graph theoretic problems on MIMD computers*, Ph.D. Thesis, Computer Science Dept., Washington State Univ., Pullman, Wash. (1983).

Some remarks on synchronization problems

Roland Vollmar
Universität Karlsruhe
D-7500 Karlsruhe
Federal Republic of Germany

Abstract: In this paper some hints to synchronization problems and to their solutions in cellular automata are reviewed and two modifications are introduced.

0. Introduction

Cellular automata are understood as models for massively parallel working systems and also as very rough models for biological systems. Usually they are characterized by a homogeneous structure and by a synchronous working mode. Both properties do not seem appropriate for biological systems: Living systems are strongly influenced by failures, and although the existence of some clocks in such systems is known, normally a strict synchronous behaviour cannot be observed. In this paper we consider modifications of cellular automata which reflect these properties.

To investigate the relations of the modified models to the original one a usual method is the proof of the possibilities of mutual simulations. Here we renounce such a comparison, rather we restrict our considerations to the aspect of synchronization: On the basis of the conditions of the well-known "firing squad synchronization problem" (fssp) for cellular automata corresponding processes are investigated for the modifications where especially the needed time is of interest. This problem was chosen because it is a problem which is only meaningful for parallel structures and for which "good" solutions need (in some sense) a high degree of parallel working.

1. Cellular automata and the fssp

Intuitively a cellular space is given by countable infinite many deterministic
Moore automata of the same kind which are "attached" at the integer valued
points of a finite-dimensional Euclidean space. Each of these automata is connected to a finite number of automata according to a certain scheme. This function is called local transition function of the cellular space because it describes
the transition behaviour, i.e. the next state and the output of each automaton, in
a local way. The global transition function is determined by one simultaneous
application of the local transition function to each automaton in the whole space
and therefore it describes the global behaviour of the cellular space. To clarify the
function of a cellular space a global clock is assumed which causes the synchronous transitions of all automata.
Formally (see e.g. [13]), a cellular space

$$\alpha = (A, d, N, F)$$

is determined by

- the state alphabet A , i.e. the finite, nonempty set of states which an automaton can assume. A special state, called quiescent state, is distinguished by the
  convention that the quiescent state of an automaton remains unchanged if all
  directly connected automata are also in quiescent states
- the dimension d of the space
- the neighbourhood index N , i.e. an n-tuple of distinct elements of Z^d , which
  describes the relative positions of the automata directly connected to a given
  automaton
- the global transition function F ; it determines (under certain conditions) in a
  bijective manner the local transition function.

It should be remarked that we will consider cellular spaces which are homogeneous as well with respect to the assignment of automata as with respect to
their neighbourhood structure.
In the following we will restrict our considerations to so-called cellular automata: From the beginning only a finite part of a cellular space is considered and
moreover any pair of automata in it is connected (according to the neighbourhood
index). Formally this is obtained by the introduction of boundary or border automata which form a layer such that no information can pass these boundaries.
The basis of the following part is a one-dimensional cellular automaton with a
special neighbourhood index, the so-called von Neumann neighbourhood index
(fig.1). Such an index restricts the direct influences of a single automaton to
these automata which are neighboured in the usual meaning.

Fig.1: One-dimensional cellular automaton with von Neumann neighbourhood

Some further notions are needed: By a configuration we mean an assignment of a state to each automaton of the cellular space resp. cellular automaton, and in this way the global state is described, too. Starting with an initial configuration, i.e. the configuration at time t=0, by the repeated application of the global transition function F a unique sequence of configurations, called the propagation, is produced.

Now we are able to introduce a synchronization problem for cellular automata, the so-called firing squad synchronization problem.
At a first glance it may seem unnecessary to synchronize a cellular automaton because it is controlled by a central clock. However the assumption about the quiescent state prevents an easy global synchronization. Less formally one could argue by energy requirements; in real systems hierarchical structures could be used.
As Moore states, the problem to synchronize a finite (but arbitrary long) chain of finite automata was devised about 1957 by Myhill:
All cells except one at a border, the so-called general, of a one-dimensional cellular automata with von Neumann neighbourhood are in the quiescent state at time t=0. After a certain period all cells must assume simultaneously the same state, the firing state, which is not assumed before by any automaton.
We will only sketch a procedure (of Balzer [1] ; see also fig.2); his method is the basis of most of the solutions. The general (at the left border) sends out two signals, one with unit speed (each time step the signal proceeds for one automaton) and one with speed 1/3. The first signal is reflected at the right border and this reflected signal meets the second one in the middle of the chain. Depending on the length of the cellular automaton one automaton or two automata are situated there. "Middle automata" act in the same manner as the general at the beginning, and so the two quarter automata are found. This process is repeated until the firing condition is fulfilled: An automaton assumes the firing state if and only if it is marked and its neighbours are marked or one of them is a border automaton. (It is important that this condition is a local one but produces a global effect.)
The time for this solution is (for a cellular automaton of length n) <3n.

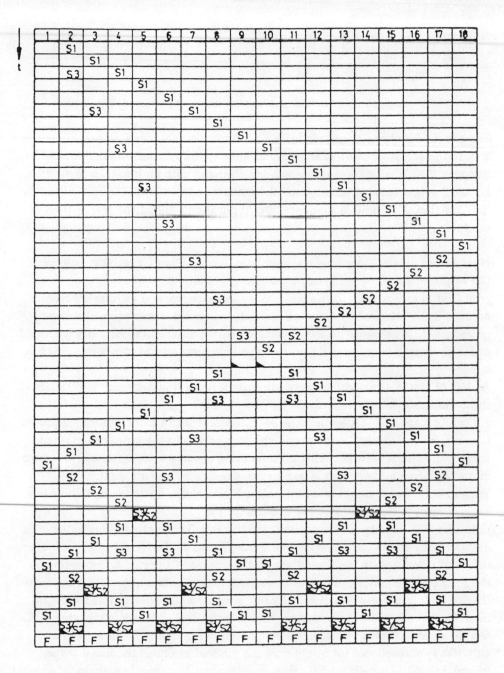

Fig.2: Solution of the firing squad synchronization problem
(Only the "most relevant" state transitions are shown)

The solution described is an example of a non-minimal time solution. It is well known that the minimal time for the solution of the fssp is 2n-2 , and there exist many algorithms working in this time.

To reach the optimal time, the accomplishment of the repeated bisection must be done with a higher degree of parallelism. In the algorithm of Waksman [15] e.g. additional waves are sent. With increasing length of the chain of automata more and more waves with different speeds have to be sent. By the finiteness condition for the state set this cannot be done in a naive way, but a trigger process is successful: From the wave moving with unit speed to the right, signals are sent to the left, which have been partly absorbed and partly cause a moving up of waves to the next automaton.

There are a lot of results concerning the fssp; a recent overview about some aspects can be found in Mazoyer [10]. During about 20 years the smallest known number of states for a minimal time solution was 8 (Balzer [1]), then improvements to 7 (Gerken [3]) and 6 (Mazoyer [9]) were found independently at almost the same time. (By the way, it is not known whether a 5-state solution exists.)

2. Modified cellular automata and the fssp

Possible connections between biological phenomena and synchronization processes are mentioned in [4].

Under the aspects of biological modelling cellular automata with symmetric transition functions may be of interest. For symmetric transition functions it cannot be distinguished which input comes from the right and which one comes from the left neighbour. Herman gives a solution to the fssp using a concept from biology (refractory periods). Szwerinski has improved this result with respect to the number of states and to time [11] (and he has shown the principal equivalence of non-symmetric and symmetric cellular automata [12]).

Herman et al. [5] treat the synchronization problem for growing cellular automata. Their solution works in all cases in which the cellular automata grow at the ends with a rational rate smaller than one per time step; the growth can take place at the ends with a different speed. The solution is also based on a division on parts of equal length.

In the theory of polyautomata the fssp and modifications have been considered more often as it can be motivated by its relevance for real world tasks. Reasons may be that the fssp provides the base for many algorithms for cellular automata and that algorithms for the solution of this problem show in an ideal manner in which way only local influences can produce a global effect. This "myopic" feature is typical for "nice" cellular algorithms. Therefore -as mentioned above- we will discuss the following modified cellular automata only under the aspects of synchronization.

Realistic systems -as well biological ones as technical ones- contain failures, and in general the probability that failures occur increases with the number of elements of a system. Massively parallel systems have to be investigated thoroughly under these aspects, but on the other side it can be expected to circumvent defects because the possibilities to communicate are well developed.

In the sequel we consider one-dimensional cellular automata which contain defective parts. The defects are restricted in that sense that the single automata are only able to transmit informations but not to process it.

At a first glance this restriction may be seen as a weak one, but if non-probabilistic models are considered, a stronger one, namely completely defective automata would cause interruptions (in the one-dimensional case) and so only disjoint parts could be studied.

The results about this topic were obtained in cooperation with M. Kutrib [6],[7].
In our model
- intact automata are able to recognize if neighboured (in the sense of the neighbourhood index) automata are defective
- defective automata are only able to transmit information
- the role of the quiescent state is the same as in "normal" cellular automata.

For the fssp in defective cellular automata we assume that at least the rightmost and leftmost automata neighboured to the borders are intact.

The initial configuration contains the general at the first place, the intact automata in the quiescent state, and the defective automata in a special state, corresponding to the quiescent state. A solution is reached when the intact automata assume simultaneously the firing state, whereas the contents of the defective automata at this time are arbitrary.

The synchronization time for a defective cellular automaton of length n with the initial configuration of fig.3 is
$$\geq \max (2n-2, \max_{1 \leq j \leq k} (n+1+2d_j)) .$$

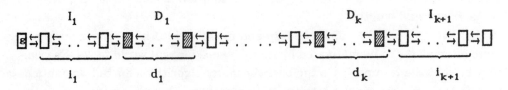

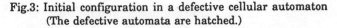

Fig.3: Initial configuration in a defective cellular automaton
(The defective automata are hatched.)

This minimal synchronization time depends on the length of the whole cellular automaton but also on the length of the longest defective part, whereas it does not depend e.g. on the positions of defective automata.

If the length of the largest defective part is $\leq (n-3)/2$ no additional steps are needed -compared with the classical fssp.

In general the minimal synchronization time cannot be obtained by a defective cellular automaton with a finite state set.

We will not discuss this result in detail, rather it only should be mentioned that a ingle automaton which is not influenced by its neighbours eventually reaches a periodic state sequence. Defective parts -which may be sufficiently large- cause such an isolation.

Now under the aspects of this result it may be interesting to discuss the conditions under which the minimal time can be obtained. Two cases have been investigated in detail:

1) Defective cellular automata are modified in the sense that the single automata are substituted by so-called memory augmented automata introduced by Dyer and Rosenfeld [2].

2) Additional assumptions about the lengths of the defective parts are made.

1) In memory augmented cellular automata the single automata consist of a finite number of counters with lengths depending on the size of the cellular automaton.

With such modified cellular automata it is possible to obtain a synchronization in minimal time.

The solution bases on a determination of the length of the cellular automaton and a decreasing of corresponding contents of the counters. To give an impression of the procedure in fig.4 a small example is sketched in a rough way. But with this naive approach about 5n steps are needed. To reach the minimal time a lot of improvements have to be added (for details cf. [7]).

2) The minimal time synchronization is also reachable for a "normal" defective cellular automaton if for the lengths of the intact parts the following demand is fulfilled:

$$\bigwedge_{1 \leq j \leq k} \mid I_j \mid \geq 2(\lfloor ld\ n \rfloor + 1) ,$$

i.e. that between defective parts are sufficiently large (dependent on the length of the whole cellular automaton) intact parts. In principle the solution works as the algorithm mentioned above but instead in counters the corresponding information is stored in some adjacent automata, i.e. chains of automata are considered as registers. It may be clear that a lot of problems emerge in such a concept; a solution is described in [6].

```
        (-,1)
          (-,2)
            (-,3)
              (-,3)
            (1,4)        (-,3)
                            x
            (2,5)       x
          x
            (3,5)
          (3,5)
              (3,5)
                (-,6)
                  (-,7)
                    (-,7)
                  (2,8)          x
                           x
              (2,8)
          (2,8)
                  (-,9)
                    9
                9    8
            7        7
        6    6        6
    6    5    5        5
6        4    4
    6
        3        3    3    3
    2    2        2    2    2
1    1    1        1    1    1
F    F    F        F    F    F
```

Fig.4: A simple solution of the fssp in defective cellular automata
(Only the "most relevant" information is shown.)

In the following the problem of a sort of asynchronous working of cellular automata is considered.

Lipton, Miller, and Snyder [8] introduced so-called delay bounded asynchronous parallel structures. Since we are only interested on some principles of their work we will only give an informal, somewhat restricted and altered explanation, and we only quote the result about synchronization.

Starting point is again a one-dimensional cellular automaton with von Neumann neighbourhood. The following informal definitions are cited from [8], but slightly modified: The time is measured in a relative fashion, with one step elapsing whenever some automaton changes its state. A given automaton is said to become active when it is first capable of a transition. Once active, the automaton can perform the state change at any step. However, no automaton can remain active, without changing state, for more than D steps. The delay D is an integer value which gives the number of steps any automaton is allowed to remain idle prior to completing a computational step. In contrast to [8], a global clock is assumed which produces signals periods of equal lengths. Hence, when D=0, no idle steps are allowed, and we have a classical cellular automaton. When D>0, the automata operate at worst case rate of once every D+1 steps. No assumption is made how long it takes for a given automaton to change state, except that it is bounded. for a given automaton the execution time may change for any reason whatsoever. Considering the fssp it would be foolish to expect that such a delay bounded cellular automaton with D>0 could solve the problem, since an automaton may or may not choose to "fire" at the appointed moment. But it seems reasonable to require that all automata should fire within an interval of length D or at least O(D).

Lipton, Miller, and Snyder have shown that even this simpler problem cannot be solved and indeed they proved a much stronger result:

The firing of all automata of a delay-bounded cellular automaton of length n can only be "guaranteed" in an interval of length O(n).

In which way can this process be speeded up or, with other words, which modifications can shorten the "synchronization interval"?

We propose to consider the following system:

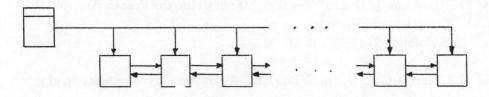

Fig.5: Delay-bounded cellular automaton with broadcasting

A global unidirectional bus and a memory-augmented automaton (in the sense of Dyer/Rosenfeld) are added. In addition to its state set it contains a finite number of counters. In one of the counters the length n of the delay-bounded cellular automaton is stored. The additional automaton works in a usual manner, i.e. without delays (- but this is not essential for the following). Along the unidirectional bus messages of the form "address+(finite)information" may be sent to automata. It is assumed that all automata know their own addresses.

A "synchronization" algorithm is very simple: The "broadcasting automaton" forces $n^{1/2}$ automata at (about) equal distances to assume the firing state successively. After that they send firing commands (on their local connection lines) to their neighbours. Since after the initial settings sectors of lengths $O(n^{1/2})$ remain to synchronize, the synchronization time is of this order.

This implies that the firing of all automata of this delay-bounded cellular automaton with bus can be guaranteed in an interval of length $O(n^{1/2})$.

Although this algorithm reduces the maximal distances between firing, it may seem too naive. But in effect it can be shown -along the same lines as in a proof in [14] - that the result is optimal (in the order of magnitude).

References

[1] R.M. Balzer   An 8-state minimal time solution to the firing squad synchronization problem
   Information and Control **10** (1967), 22-42

[2] C.D. Dyer, A. Rosenfeld   Parallel image processing by memory-augmented cellular automata
   IEEE Trans. on Pattern Analysis and Machine Intelligence PAMI-3 (1981), 29-41

[3] H.-D. Gerken   Über Synchronisationsprobleme bei Zellularautomaten
   Diploma Thesis, Braunschweig, 1987

[4] G.T. Herman, W.H. Liu   The daughter of Celia, the French flag, and the firing squad
   Simulation **21** (1973), 33-41

[5] G.T. Herman, W.H. Liu, S. Rowland, A. Walker   Synchronization of growing cellular arrays
   Information and Control **25** (1974), 103-121

[6] M. Kutrib, R. Vollmar   Minimal time synchronization in restricted defective cellular automata
submitted for publication

[7] M. Kutrib, R. Vollmar   The firing squad synchronization problem in defective cellular automata
submitted for publication

[8] R.J. Lipton, R.E. Miller, L. Snyder   Synchronization and computing capabilities of linear asynchronous structures
J. of Computer and System Sciences 14 (1977), 49-72

[9] J. Mazoyer   A six-state minimal time solution to the firing squad synchronization problem
Theoretical Computer Science 50 (1987), 183-238

[10] J. Mazoyer   An overview of the firing squad synchronization problem
In: C. Choffrut (Ed.) Automata Networks
Springer, Berlin, 1988, 82-94

[11] H. Szwerinski   Time optimal solution of the firing squad synchronization problem for n-dimensional rectangles with the general at an arbitrary position
Theoretical Computer Science 19 (1982), 305-320

[12] H. Szwerinski   Zellularautomaten mit symmetrischer lokaler Transformation
PhD Thesis, Braunschweig, 1982

[13] R. Vollmar   Algorithmen in Zellularautomaten
Teubner, Stuttgart, 1979

[14] R. Vollmar   FSSP for cellular automata with broadcasting
submitted for publication

[15] A. Waksman   An optimum solution to the firing squad synchronization problem
Information and Control 9 (1966), 66-78

EFFECTIVE IMPLEMENTATION OF DISTRIBUTED ARBITRATION IN MULTIPROCESSOR SYSTEMS

M. Makhaniok[1]), V. Cherniavsky[1]), R. Männer[2]), O. Stucky[2])

1) Institute of Technical Cybernetics, BSSR Academy of Sciences, Surganov Str. 6, 220605 Minsk, USSR
2) Physics Institute, University of Heidelberg, Philosophenweg 12, D-6900 Heidelberg, West Germany

ABSTRACT

Distributed arbitration schemes as implemented, e.g., in Futurebus and MultibusII, use a fixed arbitration time for the acquisition procedure defining new bus mastership. This time corresponds to the maximum settling delay required by a decentralized combinational logic of the arbitration scheme and depends on the set of all possible arbitration numbers which are assigned to processors requesting the bus. In this paper, a new method for reducing bus acquisition time is described which exploits the dependency of the delay time required on the arbitration numbers. The theorem that defines this dependency is derived. The application of the proposed method allows to reduce the delay time in each module and therefore the whole time of the arbitration process considerably. It is shown, that the average bus acquisition time can be shortened, e.g., by 30%. The proposed method is directly suitable for improving the performance of standard multiprocessor buses.

I. INTRODUCTION

Todays multiprocessor systems contain mostly a large number of processors. All of them are interconnected by buses to allow communication with other processors, memories, or with peripheral devices. Examples are MultibusII, Nubus, Futurebus, and Nanobus [1, 2].

In most multiprocessor applications, a global task is decomposed into rather independent processes which are dispatched to available processors. In a real-time environment, some tasks are time-critical, so that priorities are assigned to all processes. These priorities depend on several parameters, e.g., on execution deadlines. Usually, some kind of hardware priority of a processor is associated with the priority of the process it executes. This hardware priority can, e.g., be used to lock out interrupts or to access the bus system.

The process and the hardware priority of processor i are represented by a m-bit binary arbitration number $A_i = <a_{i1},...,a_{im}>$. When a group of processors competes for bus mastership at the same time, then the processor with the highest priority will be selected as the next bus master. This bus acquisition procedure of making the "won/lost" decision is known as arbitration.

The performance of multiprocessor systems depends to a large extend on the efficiency of the arbitration scheme that takes this decision. In most arbitration schemes, for each bus request processors have to wait a predefined time necessary for the arbitration procedure. Therefore one of the main problems in multiprocessor bus design is to determine the minimal value of this time to reduce an unnecessary loss of processing power.

All modern buses mentioned above use the same arbitration scheme, which is based on a sequential bit comparison of m-bit priority numbers A_i [3]. The distributed arbitration scheme on Futurebus uses m arbitration lines AB and three synchronization lines AP, AQ, and AR, which are all implemented in wired OR logic; and local combinational comparison circuits which are identical for all modules. One example of such a local circuit which is used currently, e.g., in MultibusII and Futurebus [4, 5], is shown in Fig. 1. For the discussion below, it is sufficient to use only a single synchronization line E instead of the three lines AP, AQ, and AR.

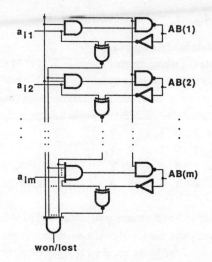

won/lost

Fig. 1: Local combinational comparison circuit

With such an arbitration scheme all processors can execute the bus acquisition procedure by using local information only. Particularly, no processor has to have any information about the technical characteristics of others. Accordingly, processor i has to assert its request (a logic 1) onto the line E for a time $t(i)$ that depends only on the delays of its own circuit and on values that are specified for the whole system [6]. The new bus mastership can be determined when all requestors remove their 1-signals from the line E. This will be after time

$$\text{Max } t(i),$$
$$i \in Z$$

where Z is the set of processors currently participating in the arbitration process. The arbitration process will be finished correctly if at least one processor holds its individual request signal on line E for a time $t(i)$ that exceeds time t_a of the real bus arbitration process.

II. STATEMENT OF THE TASK

For the arbitration scheme, a suitable expression for t_a was defined by Taub in 1984 [3] and 1987 [6] which we write in an abbreviated form as

$$t_a = 4t_p + t_s + \sum_{k=1}^{m} \text{Max}\{t_k\} + \text{Max}\{\tau_i\}, \qquad (2)$$

where

t_p is the end-to-end propagation delay along the bus,

t_s is the signal start skew,

t_{ik} is the processing time of the arbitration logic in processor i between two level changes on the AB(k-1) and AB(k) lines ($0 < t_{ki} < \infty$),

τ_i is the delay of processor i for making its won/lost decision ($0 < \tau_i < \infty$).

Taub [3] also proved that the individual arbitration time

$$t(i) = 4t_p + t_s + (m+1)Max\{t_{ik},\tau_i\} \qquad (3)$$

of processor i gives one of the solutions of the equation (2).

As can be seen from formula (3), taking the coefficients in front of **Max**{ } equal to $(m+1)$ for every i gives usually an unnecessary high value for the bus acquisition time t_a, because the real arbitration process finishes earlier in most cases. If we take, e.g., $m=6$, then the maximum value $(m+1)$ of the coefficient is needed only if the following four types of arbitration numbers take part in the competition

$$
\begin{array}{lll}
A_{42} & = 1\ 0\ 1\ 0\ 1\ 0, & \\
A_{41} & = 1\ 0\ 1\ 0\ 0\ 1, & (4) \\
A_i & = 1\ 0\ 0\ 1\ X\ X, & i \in \{36,37,38,39\}, \\
A_j & = 0\ 1\ X\ X\ X\ X, & j \in \{16,17,...,31\},
\end{array}
$$

where arbitration number A_i corresponds to the binary representation of its index value i, and X to an arbitrary value of the binary digit. It is easy to prove that for all other cases the coefficient can be lowered. Therefore, the minimum value of the coefficient is defined by the actual set Z of processors which request the system bus. To define new individual coefficients n_i instead of $(m+1)$, we introduce the following condition that guarantees a correct operation in the arbitration process:

$$\underset{i \in Z}{Max}\ t(i) \geq 4t_p + t_s + t(Z),\ \text{see (1)},$$

where $t(Z)$ is the total switching time of all combinational circuits in the arbitration scheme.

A reduction of the arbitration time can therefore be achieved if values of the coefficients $n_i \leq (m+1)$ are found, according to

$$t(i) = 4t_p + t_s + n_i\,Max\{t_{ik},\tau_i\}.$$

This task is solved if at least one relation of the following system is true for all possible $0 < t_{ik} < \infty$, $0 < \tau_i < \infty$:

$$n_i\,Max\{t_{ik},\tau_i\} \geq t(Z),\quad i \in Z,$$

where $t(Z)$ is the sum of the individual module delays t_{ik} and τ_i for the actual set of the arbitration numbers. This means that

$$t(Z) \leq \sum_{i \in Z} C_i(Z)Max\{t_{ik},\tau_i\},$$

where $C_i(Z) \in \{0,1,2,...\}$. From the last two relations we can conclude that for a correct operation of the arbitration process it is sufficient to solve the following task:

TASK It is required to find such minimal coefficients n_i for every set Z and different delays $t_{ik},\tau_i \in (0,\infty)$ so that the following relation holds true at least for one i

$$n_i Max\{t_k, \tau_i\} \geq \sum_{i \in Z} C_i(Z) Max\{t_k, \tau_i\},$$

$$Z^* = \{i, i \in Z, C_i(Z) \neq 0\}. \tag{5}$$

REMARK. It is possible to define "minimal" in the stated task in different respects. The function **F** of the coefficients n_i that has to be minimized depends on several parameters including the bus load. According to [7] and [8] we will take as the function **F** for simplicity

$$F = \frac{\sum_{i \in Z} n_i}{|Z|},$$

the average number of delay steps in the arbitration process of all modules. In this case, (5) assigns all possible boundary conditions to the task

$$F \rightarrow Min. \tag{6}$$

III. SOLUTION OF THE TASK

The task (6) with conditions (5) may seem to be simple to solve. But it is necessary to remember that each processor has no information about the delay times of other processors, which appear on the right side of the relations (5). The theorem given below establishes relations between the coefficients n_i and $C_i(Z)$ without using any unknown values.

THEOREM. It is a necessary and sufficient condition for the coefficients n_i and $C_i(Z)$ that

$$\sum_{i \in Z} \frac{C_i(Z)}{n_i} \leq 1 \tag{7}$$

if at least one of the relations (5) is true.

We will prove this theorem according to [8].

Sufficient: It has to be shown that, if condition (7) is true, there exists $i^* \in Z^*$, so that

$$n_{i^*} Max\{t_{i^*k}, \tau_{i^*}\} \geq \sum_{i \in Z^*} C_i(Z) Max\{t_k, \tau_i\}.$$

Let us assume the contrary. Then for every $i^* \in Z^*$ we have

$$n_{i^*} Max\{t_{i^*k}, \tau_{i^*}\} < \sum_{i \in Z^*} C_i(Z) Max\{t_k, \tau_i\}.$$

From this inequality it follows that

$$\frac{C_{i^*}(Z) Max\{t_{i^*k}, \tau_{i^*}\}}{\sum_{i \in Z^*} C_i(Z) Max\{t_k, \tau_i\}} < \frac{C_{i^*}(Z)}{n_{i^*}}.$$

If we sum both sides over $i^* \in Z^*$, we will get the following contradiction

$$\sum_{i^* \in Z^*} \frac{C_{i^*}(Z) \text{Max}\{t_{i^*k}, \tau_{i^*}\}}{\sum_{i \in Z^*} C_i(Z) \text{Max}\{t_{ik}, \tau_i\}} = 1 < \sum_{i^* \in Z^*} \frac{C_{i^*}(Z)}{n_{i^*}} = \sum_{i \in Z^*} \frac{C_i(Z)}{n_i} \ . \ \text{QED.}$$

Necessary: We will also disproof the contrary. Let us assume that (7) is false and that for every $0 < t_{ik} < \infty$, $0 < \tau_i < \infty$ at least one of the relations (5) is true. This means that the system

$$\begin{cases} n_i \text{Max}\{t_{ik}, \tau_i\} - \sum_{j \in Z^*} C_j(Z) \text{Max}\{t_{jk}, \tau_j\} < 0, \text{ for every } l \in Z^*, \\ \\ - \text{Max}\{t_{ik}, \tau_i\} < 0, \text{ for every } i \in Z \end{cases}$$

has no solution with respect to $\text{Max}\{t_{ik}, \tau_i\}$.

It is known from the Voronoi criterion [9] that in this case there exists always a solution $x_l > 0$, $y_l > 0$ for the following system of linear equations

$$C_i(Z) \sum_{j \in Z^*} x_j - n_i x_i + y_i = 0, \ \ i \in Z^*.$$

If we divide all elements of this system by n_i and sum over all $i^* \in Z^*$ then

$$(1 - \sum_{i \in Z^*} \frac{C_i(Z)}{n_i}) \sum_{j \in Z^*} x_j = \sum_{i \in Z^*} \frac{y_i}{n_i} \ .$$

If we take now, as assumed

$$\sum_{i \in Z^*} \frac{C_i(Z)}{n_i} > 1$$

we will have

$$\sum_{j \in Z^*} x_j < 0 \qquad \text{or} \qquad \sum_{i \in Z^*} \frac{y_i}{n_i} < 0$$

which requires that at least one x_i or one y_i is less than zero. This contradicts the Voronoi criterion. QED.

EXAMPLE. Let us consider the most critical case (4) that was given above. In this case only processors with four types of arbitration numbers compete for the bus. They are A_{42}, A_{41}, A_i, A_j, where $i \in \{36,37,38,39\}$, $j \in \{16,17,...,31\}$. Here we get the following set of relations, where all t_{ik} and τ_i may vary from 0 to ∞,

$$n_{42} \text{Max}\{t_{42k}, \tau_{42}\} \geq 3 \ \text{Max}\{t_{42k}, \tau_{42}\} + \text{Max}\{t_{41k}, \tau_{41}\} +$$
$$+ \text{Max}\{t_{ik}, \tau_i\} + \text{Max}\{t_{jk}, \tau_j\},$$
$$n_{41} \text{Max}\{t_{41k}, \tau_{41}\} \geq 3 \ \text{Max}\{t_{42k}, \tau_{42}\} + \text{Max}\{t_{41k}, \tau_{41}\} +$$

$$+ \; \text{Max}\{t_{ik}, \tau_i\} \quad + \; \text{Max}\{t_{jk}, \tau_j\},$$

$$n_i \, \text{Max}\{t_{ik}, \tau_i\} \quad \geq \; 3 \; \text{Max}\{t_{42k}, \tau_{42}\} + \; \text{Max}\{t_{41k}, \tau_{41}\} +$$
$$+ \; \text{Max}\{t_{ik}, \tau_i\} \quad + \; \text{Max}\{t_{jk}, \tau_j\},$$

$$n_j \, \text{Max}\{t_{jk}, \tau_j\} \quad \geq \; 3 \; \text{Max}\{t_{42k}, \tau_{42}\} + \; \text{Max}\{t_{41k}, \tau_{41}\} +$$
$$+ \; \text{Max}\{t_{ik}, \tau_i\} \quad + \; \text{Max}\{t_{jk}, \tau_j\},$$

at least one of which must be satisfied. By applying the given theorem instead of this complicated set of expressions, the only condition given is

$$\frac{3}{n_{42}} + \frac{1}{n_{41}} + \frac{1}{n_i} + \frac{1}{n_j} \leq 1$$

which does not depend on time parameters.

COMMENTS. Without the theorem given above, it is difficult to define optimal coefficients n_i in the set of relations (5), because all modules only have information about their own delays. By applying this theorem, it is sufficient to transform the relations (5) and solve the system of inequalities (7) for all possible sets Z^* once only. In this case, better results than in [3] are achieved.

The improvement of the arbitration performance is obtained by computing the coefficients in the proposed way without modification to the adopted hardware and algorithm of the arbitration network. According to the theorem, it is possible to replace one or more processors by others with different values of delays t_{ik} and τ_i without changing the coefficients n_i.

The actual value of the improvement depends on several parameters, e.g., whether odd or even parity is used, whether all possible arbitration numbers are allowed or only numbers chosen according to binomial codes $Bi(n,m)$ with depth n $(n<m)$ [7], whether the won/lost decision time τ_i is included in the arbitration process, etc..

However, it is simple to determine optimal coefficients for all these cases.

IV. APPLICATION

Let us discuss a situation in which each module takes its own decision after all modules participating in the arbitration process remove their busy signals from line E. Here, we have to consider only t_{ik} instead of $\text{Max}\{t_{ik}, \tau_i\}$ in formula (2). For a speedup of the arbitration process, it is our task to find new coefficients n_i. The first step is to determine the coefficients $C_i(Z)$ in formula (7) for all possible sets Z^* of processors that can compete for the next bus mastership.

The processor number i which corresponds to the arbitration number $A_i = <a_{i1}, ..., a_{iN}>$ is given in binary representation as

$$i = \sum_{k=1}^{N} 2^{N-k} a_{ik}. \qquad (8)$$

Below, we will use the following three definitions.
We define $A[l,k]$ as the sequence $a_l, ..., a_k$ of the arbitration number $A = <a_1, ..., a_N>$ where l and k satisfy $1 \leq l \leq k \leq N$.
We define $A[l,k]^1$ $(A[l,k]^0)$ of the arbitration number A as a 1-interval (0-interval) if all bits of the sequence

$A[l,k]$ are equal to "one" ("zero"), and not equal to a_{l-1} if $l>1$, and not equal to a_{k+1} if $k<N$.

We assign an ordering index to all intervals of A, starting from the most significant bit.

We define that the N-bit arbitration number B has a (c1,c2)-conflict with arbitration number A if A>B and there exist c1 and c2 so that

- c1 is the ordering index of the first interval $A[l,k]^1$ of arbitration number A that corresponds to a sequence $B[l,k]$ which contains at least one "zero" bit,

- c2 is the ordering index of the first interval $A[l,k]^0$ of arbitration number A that corresponds to a sequence $B[l,k]$ which contains at least one "one" bit.

Now we construct all conditions (7): Let $p_i= r_i$ if i in formula (8) is even, and $p_i=r_i-1$ else, where r_i is the number of 1-intervals of arbitration number A_i $(i=0,1,...,2^N-1)$. Let us form all possibilities for the following sets

$$U(m_i,g) = W(i,1+2\sum_{v=1}^{g-1} s_v, \ 2\sum_{v=1}^{g} s_v)$$

for every m_i $(m_i=1,2,...,p_i)$ and every sequence $<s_1,...,s_{mi}>$, where $s_g \in \{1,2,...,m_i\}$, $g=1,...,m_i$,

$$\sum_{g=1}^{m_i} s_g = p_i.$$

Here $W(i,c1,c2)$ is the set of all arbitration numbers which have a (c1,c2)-conflict with A_i. If $p_i=0$, then $m_0=0$ and $U(0,0)=\emptyset$.

The next relation gives us all coefficients for the system (7),

$$\sum_{i \in Z} \frac{C_i(Z)}{n_i} = \frac{r_i+m_i-p_i}{n_i} + \sum_{g=1}^{m_i} \frac{1}{n_j(ig)} \leq 1, \qquad (9)$$

where coefficient n_i corresponds to the arbitration number A_i $(i \neq 0)$, and coefficient $n_j(ig)$ corresponds to the arbitration number $B_j \in U(m_i,g)$ $(j \neq 0)$, which consists of sequences $B[l,k]$ with "zero" and "one" bits on the positions defined according to the definition 3). Now we have a construction rule for all coefficients $C_i(Z)$ in (7) and proceed to calculate n_i for a special example.

In the IEEE 896 standard (Futurebus) the arbitration numbers have a width of six bits and one more bit for parity. Therefore, here we have to find the coefficients n_i for $i=0,1,...,63$. Below, three different cases are compared for choosing integer coefficients n_i which fulfill the construction rule (9). For this consideration we do not take into account the analysis of a parity bit in the arbitration procedure.

1) We first take all coefficients n_i equal to 6. This choice corresponds to the setting of Taub [3].

2) Another choice for the coefficients n_i could be

$$n_i = \text{Max}\{r_i+2m_i-p_i, \ \underset{A_i \in U(m_i,g)}{\text{Max}} (r_j+2m_j-p_j)\}$$

if n_i is used in different inequalities (9). In this case we have

$$\begin{aligned}
n_i \ &= 6 \text{ for } i = 16, 17,..., 31, 36, 37, 38, 39, 41, 42, \\
&= 5 \text{ for } i = 8, 9,..., 15, 40, 43,..., 47, 50, 51, 53, \\
&= 4 \text{ for } i = 4,..., 7, 33, 34, 49, 52, 54, 55, 57, 58, \\
&= 3 \text{ for } i = 2, 3, 59, 61,
\end{aligned}$$

$$= 2 \text{ for } i = 1, 32, 48, 56, 60, 62,$$
$$= 1 \text{ for } i = 63,$$
$$= 0 \text{ for } i = 0,$$

and all coefficients n_i less or equal to 6.

3) Eventually, let us take

$$n_i = 9 \text{ for } i = 42,$$
$$= 6 \text{ for } i = 21, 37, 41, 43, 45, 53,$$
$$= 3 \text{ for } i = 2, 3, 59, 61,$$
$$= 2 \text{ for } i = 1, 32, 48, 56, 60, 62,$$
$$= 1 \text{ for } i = 63,$$
$$= 0 \text{ for } i = 0,$$
$$= 4 \text{ else.}$$

The average number of delay steps for the whole system, which has to be minimized under the constraints (9), is

$$F = \frac{\sum_{i=0}^{63} n_i}{64} .$$

Then we get **F1=6, F2=4.6, F3=3.9**, corresponding to the three cases given above. This shows that in cases 2) and 3) the average arbitration time is reduced considerably compared to the case 1) which applies to Futurebus [6].

V. CONCLUSIONS

In most multiprocessor applications, a global task is decomposed into a set of parallel cooperating processes. They can partly be executed independently and synchronized at certain instants. Usually, all these processes have to be finished before a common result can be achieved. To improve the performance of a multiprocessor system, it is required to reduce the total execution time of this global parallel task. The method that was proposed in this paper directly relates to this general case. It reveals two new and essential parameters that influence the execution speed of such types of parallel processing problems. The first one is the way of identifying the execution time of the global task. The second one is the usage of an appropriate coding for arbitration numbers that can be assigned to the individual processes.

The efficiency of the method was verified by application to one of the most important problems in multiprocessor design, bus arbitration. Here we considered the arbitration process as a global task which is distributed among all processors of a multiprocessor system. It was shown that it is possible to achieve the maximal performance of the arbitration network by choosing the time parameters according to the method presented. This can be done without any modifications to the hardware and arbitration algorithm used and can therefore also be applied immediately to many modern standard buses.

It is planned to use this method to speed up a global comparison network in the NERV multiprocessor system [10]. This network determines the maximum in a set of numbers which is asserted by all processors onto special bus arbitration lines. It is expected that the overall improvement achieved will be more than 10%, corresponding to a processing power of ≈ 40 CPUs.

REFERENCES

[1]   Borrill P.: "A Comparison of 32-bit Buses", IEEE Micro, Dec. 1985, pp. 71-79.

[2]   White G. P. "Bus Structure Eases Multiprocessors Integration", Computer Design, Vol. 23, 1984, No. 7, pp. 123-135.

[3]   Taub D.M.: "Arbitration and Control Acquisition in the Proposed IEEE 896 Futurebus", IEEE Micro, Vol. 4, 1984, pp. 28-41.

[4]   Khu A. "Integrated PLDs support Multibus II bus arbitration", EDN, January 7, 1988, pp. 165-172.

[5]   I896 Working Group of EWICS-TC10: i896 - A proposed Standard Backplane Bus Specification for Advanced Microcomputer Systems, Draft 5.2, 1983.

[6]   Taub D.M.: "Improved Control Acquisition Scheme for the IEEE 896 Futurebus", IEEE Micro, Vol. 7, No. 3, 1987, pp. 52-62.

[7]   Makhaniok M., Cherniavsky V., Männer R., Stucky O.: "Binomial Coding in Distributed Arbitration Systems"; to be publ. 1989.

[8]   Makhaniok M., Cherniavsky V., Männer R., Stucky O.: "Reduced Bus Acquisition Time in Distributed Arbitration Systems", to be publ. 1989.

[9]   Voronoi G.F.: "Sur quelques proprietés des formes quadratiques positives parfaites", J. reine und angew. Math., Vol. 133, 1908, pp. 79-178.

[10] Hauser R., Horner H., Männer R., Makhaniok M.: "Architectural Considerations for NERV - a General Purpose Neural Network Simulation System", Proc. WOPPLOT 89, Wildbad Kreuth, Germany, 1989

Flowshop and TSP

M.P. Pensini - G. Mauri - F. Gardin

Dipartimento di Scienze dell'Informazione
Università di Milano
v. Moretto da Brescia, 9 - 20133 Milano (Italy)

Abstract

In this paper we consider an NP-complete problem that it is particularly important in production processes scheduling: the *flowshop* problem. We propose a solution based on a parallel architecture, using Simulated Annealing (a combinatorial randomized technique), and a connectionist model: the Boltzmann machine. Then, we consider the methods used to implement an algorithm for the flowshop problem on a Boltzmann machine in order to improve the time computing performances. Our strategy is based on capability to reduce the original problem to another one (TSP), that is equivalent to the first; in fact, we will discuss the flowshop polynomial reduction to the asymmetric TSP (i.e. to Directed Hamiltonian Circuit, or DHC); then, we will investigate the equivalence between the DHC and the open symmetric TSP (or UnDirected Hamiltonian Path, UDHP). Finally, we will describe limits of the proposed approach and possible developments.

Introduction

Let define the flowshop scheduling problem as follows: given n jobs (with processing times for each job known) to process on m machines, every one from the first to the mth in the order, the flowshop problem consists of find out a jobs permutation so that a performance criterion is verified ([FRE82]). Depending on the performance criterion chosen and on the constraints imposed, we can have different cases of the same problem.

Particularly, we evaluate the performances using the completion time criterion, imposing that there is not any waiting time on a machine between the end of a job and the start of another one.

As Lenstra have demonstrated ([LEN77]), this is an NP-complete problem for $m > 3$ and it can be expressed in terms of an optimization problem.

Many heuristic methods have been studied to resolve the flowshop scheduling problem in operation research. One of the most important algorithms, the Johnson algorithm ([JOH54]), finds the optimal solution to a special flowshop case in $O(n \ log \ n)$ time (where n is the jobs number). This algorithm is often used as the first step to the resolution of more complex problems.

Another method, proposed by Campbell, Dudek and Smith ([CAM70]), achieves a good solution to the general flowshop problem; the solution is the optimal one in the 55 percent of cases.

It is only since the last few years that researchers in the neural network field have been carrying out studies on this argument (see [FOO88] and [FOO88a]).

We propose a neural approach, as opposite to the operational one, in order to provide methods for which computing time and quality solutions are improved.

Boltzmann machine is the network model considered; it is a stochastic model firstly proposed by Ackley, Hinton and Sejnowki ([HIN83]), and particularly adapted to solve combinatorial optimization problems. It allows a parallel implementation of the Simulated Annealing.

Simulated Annealing is a randomized optimization technique either similar to some physical annealing techniques or to local optimization ones.

We outline it, in order to show either the capacities, inherited by the parallel implementation, or the bounds, we try to overtake.

But the major issue is to transform the flowshop into a problem solvable on a Boltzmann machine. We can use the complexity theory to demonstrate the reducibility of the flowshop to the Travelling Salesman Problem (TSP).

Particularly, TSP is reducible to the problem of searching a hamiltonian circuit in a directed graph. Nevertheless, we cannot immediately implement an algorithm on Boltzmann machine: among the existence conditions of a Boltzmann machine there is also the request that the connection matrix of the nodes is symmetric.

Mathematically, it means that the graph has to be undirected.

The proof is in two phases:

1. the directed hamiltonian circuit (DHC) is a reducible to the direct hamiltonian path problem (DHP), see [LEN77].

2. the directed hamiltonian path can be reduced to an undirected hamiltonian path (UDHP).

Now, we can implement a resolution algorithm on Boltzmann machine. Proceeding from flowshop to UDHP, a series of problems on found solutions arises.

In fact, the solutions computed by Boltzmann machine also include values not being original problem solutions, and in order to improve the performances heuristic knowledge should be added to the network.

In par. 1 we will describe the Simulated Annealing algorithm, with reference to the local techniques, giving some mathematical results showing his effectiveness.

In par. 2 the Boltzmann machine architecture is described, and the common aspects to the Simulated Annealing underlined.

In par. 3, we will outline the flowshop problem and we report the complexity theory results related to TSP, at which (par. 4) the flowshop is reducible.

The par. 5 is central in our discussion: we will investigate the problem of transforming the TSP in open TSP (from DHC to DHP), then, the open asymmetric TSP in open symmetric TSP (from DHP to UDHP). In par. 6 we give an explanatory example.

Finally, in par. 7, we present some theoretical considerations.

1 Simulated Annealing and Combinatorial Optimization

The subject of Combinatorial Optimization (CO) consists in the searching of minimum or maximum values of a function. The problem can be characterized by a triple (R, F, K) where:

$R =$ finite solutions set

$F \subseteq R =$ finite admissible solutions set, i.e. the solutions set that meets the problem's constraints;

$K : F \to \mathbb{R} =$ cost function that assigns a real number to each solution.

The objective of the optimization is to find an admissible solution for which the cost function K is optimal (minimum or maximum), i.e. to find a solution $s \in F$ so that for every $sl \in F$:

$$K(s) \leq K(s\prime)$$

or

$$K(s) \geq K(s\prime)$$

Simulated Annealing (SA) is an approximated method based on a randomization technique, for large combinatorial problems.

The method originates from two different research fields, the theory of statistical mechanics and the CO: Kirkpatrick ([KIR83]) was the first to apply physical simulations techniques to CO problems.

Furthermore, there are analogies with some techniques which have already been used, particularly with local optimization techniques (see [JOH84]).

1.1 Local Optimization.

When an arbitrary solution to a CO problem is given, local optimization attempts to improve that solution by a series of incremental local changes. In order to define a local optimization method, first of all we need to perturb the solutions to obtain other ones. Now we have to define the concept of "neighborhood".

Generally, the set of solutions can be obtained in one perturbing step from a given solution A is called *neighborhood* of A; one of its elements is called *neighbor*.

This set is related to the method for perturbing the initial solution.

Choosing the new solution among the neighbors involves a heuristic criterion, usually the gradient descent criterion (the new solution corresponds to a lower cost function value).

Unfortunately, this technique gets stuck in a local optima, i.e. minimal cost function values with respect to the neighborhood considered alone, and not to the global set of solutions.

Since the local search depends on the initial solution, it is customary to carry out the process several times, starting from different randomly generated configurations, and to save the best result.

1.2   Simulated Annealing

Like the local optimization techniques, SA can be viewed as an optimization algorithm that continuously tries to transform the current solution into one of its neighbors by repeatedly applying the generation mechanism and an acceptance criterion.

The acceptance criterion is governed by a control parameter, which acts like the temperature in the physical annealing process. Apart from allowing improvements, the acceptance criterion also allows, in a limited way, deteriorations in the cost function, thus preventing the algorithm from getting stuck at local optima.

The criterion is based on the well-known Metropolis algorithm (see [MET53]), developed in the late 1940's for use in simulations of such situations as the behavior of gases in the presence of an external bath at a fixed temperature. In such algorithm, the probability of jumping from one state to another is fixed by the energy gap between the states and by the temperature value.

In our case, the states are the solutions and the gap energy is the cost difference between the solutions.

1.3   A mathematical description

SA can be mathematically described by the theory of the Markov Chains ([AAR85], [LAA86]). A Markov chain is a sequence of trials, where the outcome of a trial only depends on the previous one.

In the SA algorithm, trials correspond to the transitions from one solution into another.

If $X(t)$ is the configuration (the solution) obtained after k transitions, then the *transition probability* $T_{ij}(c)$ denotes the probability that the c-th transition ($c \in \mathbb{R}^+$) is a transition from configuration i to the configuration j at a given value of the control parameter:

$$
T_{ij}(c) = Pr\{X(c) = j \mid X(c-1) = i\}
$$

$$
= \begin{cases} G_{ij}(c)A_{ij}(c) & if\ i \neq j \\ 1 - \sum_{k \in S, k \neq i} G_{ik}(c)A_{ik}(c) & if\ i = j \end{cases} \tag{1.1}
$$

where $G_{ij}(c)$ is the *generation probability*, i.e. the probability of generating configuration j from i, $A_{ij}(c)$ is the *acceptance probability*,

The Markov chain is represented by the sequence of solutions generated by the algorithm. We can demonstrate ([FEL50]) that, if the Markov chain is irreducible and aperiodic, and if the matrices G and A satisfy a set of conditions, the Markov chain asymptotically converges in distribution to the optimal solutions set S_{ott} with probability 1. In other words:

$$\lim_{c \to 0} \lim_{k \to \infty} P_c\{X(k) \in S_{ott}\} = 1$$

For choosing the probabilities we give the following criteria:
- generation probability matrix $G_{ij}(c)$ is independent from c and is chosen uniformly with respect to the number of neighbors;
- the acceptance matrix $A_{ij}(c)$ has the form:

$$A_{ij}(c) = \begin{cases} \exp(-\Delta F_{ij}/c) & \text{if } \Delta F_{ij} > 0 \\ 1 & \text{if } \Delta F_{ij} \leq 0 \end{cases}$$

where $\Delta F_{ij} = F(j) - F(i)$.

The exponential function chosen is the Metropolis function (on which SA is based) for the Markov chains generation to simulate physical systems placed at finite temperature with many degrees of freedom ([MET53]).

It is possible to demonstrate ([AAR87]) that we also have the same result with a function like:

$$\frac{1}{1 + \exp(-\Delta F_{ij}/c)}$$

a sigmoid function for increasing values of Δ.

Usually, some conditions required for the convergence cannot be verified. The algorithm implementation on a computer needs a Markov chain of finite length. This approximation determines a final results, which may not be the optimal solution, but only an approximation.

Then, we define a parameters set of the algorithm, the so-called *cooling schedule*, composed by:

a. the initial value for the control parameter c_0.

b. the decrement factor of c; it defines the decay rate for c after each generated Markov chain.

c. the solutions number to generate for each Markov chain (i.e. the Markov chain length).

d. the termination criterion.

Setting up $c = 0$, the annealing algorithm is the analogous of the local search.

2 Simulated Annealing and neural networks: the Boltzmann Machines

2.1 Architecture

The Boltzmann machine is a stochastic neural network model as its activation function depends on a stochastic parameter which was firstly described by Hinton, Sejnowski and Ackley in 1983 ([HIN83]).
The network topology is similar to the Hopfield networks ([HOP85]); in fact, we can apply the Hopfield theorem on the convergence toward stable states. The topology is the following:

a) The units are symmetrically connected;
b) the units are updated one at time (sequential updating);
c) delay times on the connections are uneffective.

The units have binary states $(0, 1)$. So, if n is the unit number, it is always possible to represent the state of a Boltzmann machine by a binary vector belonging to $\mathbb{R}^n$.

As in the Hopfield networks, a number can be assigned to each global network state, that is the energy for that state: under some hypothesis the units can operate so that the energy function is minimized.
The expression for the energy is the following:

$$E = - \sum_{i<j} w_{ij} s_i s_j + \sum_i \theta_i s_i$$

where

w_{ij} = connection strength between i unit and j unit.
θ_i = threshold
s_i = state
A negative weight $-\theta_i$ can be considered between i unit and an added unit with state 1 always, in order to remove from the equation the threshold θ_i.

Then, we can write:

$$E = -\frac{1}{2} \sum_{ij} w_{ij} s_i s_j$$

The main difference between the Boltzmann machine and the Hopfield networks is the activation function. In fact, the activation function for the units is a stochastic function of the control parameter:

$$p_i = \frac{1}{1 + e^{-S_i/c}}$$

where S_i = state of the unit i, c is the parameter and

$$S_i = \sum_j w_{ij} s_j$$

This sigmoidal function is a good model of the cellular membrane response when it is stimulated by an electrical impulse (the curve is the electric potential change). The state S_i for a unit is the difference or energy "gap" when a state is modified.

$$\Delta E_i = E(s_i = 0) - E(s_i = 1) =$$
$$= \sum_j w_{ij} s_j = S_i$$

Like the SA, the probability function is governed by the control parameter c: for high values of c, the update is not depending on the state S_i (i.e., from the energy gap ΔE_i); while if c is near to 0 the unit will change his state only if $\Delta E_i < 0$. The function tends to be an heavy side function and the system to act like a linear threshold system (with deterministic activation function).

We noted that the dynamics of a Boltzmann machine requires that for high parameter values the updates are mainly randomized, while in correspondance to low parameter values the system tends to update the units only for negative ΔE values.

This is the same behavior found in SA dynamics: each activation is equivalent to a change in the solutions space in the annealing procedure. Furthermore, the acceptance of a new state value in Boltzmann machine depends on energy variation and on the control parameter, like the cost function variation in SA.

2.2 A mathematical model

Here we focus on the mathematical model of the concepts described above, to obtain the exact references that we will use while applicating the TSP concepts to neural networks.

A Boltzmann machine consists of N units, that can be represented by a undirected graph $G = (V, A)$, with $V = \{(x_0, x_1, \ldots, x_{N-1})\}$ nodes set that correspond to the units and $A \subseteq V \times V$ edges set that correspond to the connections among the units.
A binary value that is assigned to each node v_i corresponds to the i unit state.
A k configuration (state) of the Boltzmann machine is univocally defined by all nodes states. The v_i node state in the k configuration is noted as $r_k(v_i)$.
The configurations space is noted as S, with $|S| = 2^N$.
(v_i, v_j) will be active in the k configuration if

$$r_k(v_i) r_k(v_j) = 1$$

The connection strength between i unit and j unit is noted as $s(v_i, v_j)$ and the energy function is the following:

$$E_k = - \sum_{(v_i, v_j) \in A} s(v_i, v_j) r_k(v_i) r_k(v_j)$$

The k neighborhood is $S_k \subset S$ and is defined like the configurations set we can obtain from k changing one node state. The $k^{(i)} \in S_k$ configuration derived from changing v_i state has the following:

$$r_{k^{(i)}}(v_j) = \begin{cases} r_k(v_j) & j \neq i \\ 1 - r_k(v_j) & j = i \end{cases}$$

The energy gap between two configurations can be expressed by (see [AAR87]):

$$\Delta E_{kk^{(i)}} = E_{k^{(i)}} - E_k =$$

$$= (2r_k(v_i) - 1) \left\{ \sum_{(v_i, v_j) \in A_{v_i}} s(v_i, v_j) r_k(v_j) + s(v_i, v_i) \right\}$$

where A_{v_i} is the edges set affecting v_i.

The later expression is particularly important because it states that the energy variation depends only on the states of the nodes connected with v_i; then, it can be locally calculated, thus allowing a parallel execution.

2.3 The transition probability

The mathematical theory of the Markov chains is useful, as with SA, with Boltzmann machine to describe the system transitions.

In the Boltzmann machine, the trials are the configurations, whose values are calculated in two steps: first, a neighbor configuration $k^{(i)}$ is generated, then, if the new configuration is accepted, the new value is the $k^{(i)}$ value, otherwise the value is the k value.

As for SA, a transition probability T_{kl} is associated with the process, defining the probability of obtaining the l configuration having the k current configuration.

The transition probability has the form:

$$T_{kl}(c) = \begin{cases} T_{kk^{(i)}}(c) & l = k^{(i)} \\ 1 - \sum_{j=0}^{N-1} T_{kk^{(j)}}(c) & l = k \end{cases} \tag{2.1}$$

where

$$T_{kk^{(i)}}(c) = G_{kk^{(i)}}(c)A_{kk^{(i)}}(c)$$

with c real number given.

G is the generation probability and A is the acceptance probability: thus, the (2.1) is analogous of the (1.1).

Usually, the generation probability is chosen not to be dependent on c and on k and to be uniform to respect to nodes set, that is:

$$G_{kk^{(i)}}(c) = 1/N$$

The acceptance probability has the form:

$$A_{kk^{(i)}}(c) = \frac{1}{1 + \exp\{-\Delta E_{kk^{(i)}}/c\}}$$

Again with the given conditions (infinite Markov chain length, parameter value trending to 0) we can demonstrate that the system converges to stable states, meaning that it converges to a binary vector $\pi(c)$, i.e. a stable configuration. The output is this vector.

It is possible to demonstrate that this configuration is the optimal one ([LAA87], [AAR85]) and it represents a minimum energy function.

Since the conditions above cannot be met in practice, the final solution will only be a good approximation, depending on the cooling schedule defined for that problem.

3 Scheduling problem

Among the problems related to the automated manufacturing phases, the *scheduling problem* means ordering in time the operations on the jobs.
We can have different types of scheduling, depending on which relation between jobs and machines we define.
We give some definitions and concepts for the scheduling problem, as treated by Lenstra (in [LEN77]).

A *job* J_i $(i = 1, .., n)$ consists of a sequence of *operations*; the processing of a job on a machine is called *operation*; the jobs have to be processed through m machine M_k $(k = 1, .., m)$ without any break, in a given time.
Every machine can processes a job at a time only.

For every J_i we define:

n_i = operations number
v_i = machines order, i.e. an ordered machine $n_i$tuple
p_{ik} = processing time for the kth operation $(k = 1, \ldots, n_i)$
r_i = release time, when J_i can be processed
d_i = due date, i.e. the promised delivery date of J_i
s_i = starting time
C_i = completion time, the time at which processing of J_i finishes

Finally, we need to specify a function

$$f_i : \mathbb{N} \rightarrow \mathbb{R}$$

or *cost function*, in which the cost of J_i is usually based upon completion times.

Performance criteria are dependent on the objectives they want to meet, and they could be incompatible. We are interested in the following criterion:

$$min \ C_{max} = \text{minimize maximum completion time } (makespan)$$

3.1 Classification

Scheduling problems are defined by four parameters:

$$\{ \ n, \ m, \ l, \ k \ \}$$

where

n = jobs number;
m = machines number;

If $m > 1$, l will be one of the following:

$l = F$ flow-shop; $n_i = m$ $v_i = (M_1, \ldots , M_m)$ fixed $\forall \ J_i$, i.e. the machine order for all jobs is the same

$l = P$ permutation flow-shop; not only is the machine order the same for all jobs, but the job order is the same for each machine

$l = G$ job-shop; the general problem in which both n_i and v_i can change

$l = I$ parallel-shop; each job has to be processed on one and only one machine

k is the *optimality criterion*.

The class problem we will consider is a particular flowshop problem, that is:

$$n \ / \ m \ / \ F, \ no \ wait \ / \ C_{max}$$

The problem can be expressed as following:

each job among the n jobs $J_1 \ldots J_n$ have to be processed by m machines $M_1 \ldots M_m$ always in a given order; the job J_i $(i = 1..n)$ consists of m operations $O_{i1} O_{i2} .. O_{im}$;

O_{ik} denotes the operation on the ith job by the kth machine; p_{ik} is the processing time of ith job on kth machine; when the job processing starts, it has to finish without waiting time, that is:

$$C_i = S_i + \sum_k p_{ik}$$

We have to find out a processing order on each M_k so that the completion time is minimised.

3.2 TSP formulation

TSP is the search for the path through a given set of cities visiting each one once and only once and coming back to the starting point.
For a good resolution technique of this problem, see Carpaneto and Toth in [CAR80].
From a formal point of view, an analogous definition exists expressing the problem in terms of graph theory, and we have:

DEFINITION 3.1 — *Given an oriented graph $G = (V, A)$, with $V = \{1, \ldots, n\}$ set of cities, $A = \{(i, j) \mid i, j = 1, \ldots, n\}$ edges set and distance c_{ij} (weights) for every edge $(i, j) \in A$, TSP consists of finding out the hamiltonian circuit of minimum cost.*

If $c_{ij} = c_{ji}$ $\forall (i, j) \in A$, the TSP is called *symmetric*, otherwise it is *asymmetric*.

3.3 NP-completeness for the TSP

Before considering the relationship between flowshop and TSP, we recall some results about the NP-completeness for our problems, to point out the common properties.

THEOREM 3.1 — *Finding out a hamiltonian circuit in a directed graph (DHC) is a NP-complete problem (Karp 1972) ([KAR72]).*

THEOREM 3.2 — *Hamiltonian circuit problem is polynomially reducible to the TSP (Garey 1979) ([GAR79]).*

On the other side, we can demonstrate that $n/m/F, no\ wait/C_{max}$ is an NP-complete problem.

In fact, DHP is polynomially reducible to flowshop (see [KAN76]), then $n/m/F, no\ wait/C_{max}$ is as difficult as the directed hamiltonian path problem.

But $n/m/F, no\ wait/C_{max}$ is polynomially reducible to DHC, i.e., it is at least as difficult as the directed hamiltonian circuit.

Because both DHC (see the above theorem) and DHP (see [GAR79]) are NP-complete problems, $n/m/F, no\ wait/C_{max}$ is NP-complete too.

4 From $n/m/F, no\ wait/C_{max}$ to TSP

Let's go on to describe the results on the flowshop and TSP proposed by Lenstra ([LEN77]). With reference to the notation adopted, let n be the jobs number, and m be the machines number; we define c_{ij} the minimum break time length between s_i and s_j if J_j is scheduled immediately after J_i ($i = 1, \ldots, n \quad j = 1, \ldots, m$); let be

$$P_{hk} = \sum_{l=1}^{k} p_{hl} \qquad k = 1, \ldots, m \quad h = 1, \ldots, n \qquad (4.1)$$

Then:

$$c_{ij} = max_k(P_{ik} - P_{jk-1}) \qquad (4.2)$$

Finding a schedule for minimizing C_{max} is the same as resolving the TSP on $G = (V, A)$ with

$$V = \{0, \ldots, n\}$$

and cost matrix $C = [c_{ij}]$; job J_0 is an unreal job to represent the start and the stop point of the schedule, and where

$$n_0 = m$$

$$v_0(k) = M_k$$

$$p_{0k} = 0 \qquad (4.3)$$

We obtain an asymmetric TSP, with

$$c_{0i} = 0 \quad \forall i$$

$$c_{ii} = 0 \quad \forall i \tag{4.4}$$

The symmetric condition corresponds to an oriented (or directed) graph; the symmetric condition to an unoriented (or undirected) graph.

5 From asymmetric TSP to symmetric TSP

In the previous paragraph, we concluded that a $n/m/F, no\ wait/C_{max}$ is equivalent to an asymmetric TSP, the graph of which, $G = (V, A)$, has the following characteristics:

$$V = \{(v_0, v_1, \ldots, v_n)\}$$

$$|V| = n + 1$$

in which the node v_0 is the job J_0 the conditions for which are (4.3) and (4.4). In other words, J_0 represents the fictious job which is introduced in order to indicate the beginning and the end of the processing of the other jobs.

The solution for the scheduling problem is an ordered list of jobs

$$\{J_{\pi(0)=0}, J_{\pi(1)}, \ldots, J_{\pi(n)}\}$$

where $\pi(i)$ is the job in the i_th position.

The solution for the same problem, as expressed in terms of TSP with cost matrix $D = [d_{ij}]$, will be a permutation π^* of the nodes set like

$$\{v_{\pi^*(0)}, v_{\pi^*(1)}, \ldots, v_{\pi^*(n)}\}$$

for which is minimum the value of

$$\sum_{\substack{0 \leq i \leq n-1 \\ 1 \leq j \leq n}} d_{\pi^*(i)\ \pi^*(j)} + d_{\pi^*(n)\pi^*(0)=0}$$

if d_{rs} represents the cost of the edge $(r, s) \in A$, i.e. the distance between r and s.

The cost matrix is derived from (4.2), (4.3) and (4.4) and is asymmetric. We can observe that, since the problem solution is an ordered list of n jobs, and not of $n+1$ jobs, the TSP solution interesting for us can override that the first job J_0 is also the last one; in other words, we are interested in the value of $\pi(n)$ and not in the value of $\pi(n+1)$. Thus, for us it suffices to resolve an open TSP, i.e. to find a hamiltonian path (DHP) and not a hamiltonian circuit.

More precisely, and considering the previous note, we want to demonstrate that if a hamiltonian circuit in a directed graph exists, then a hamiltonian path in a undirected graph (UDHP) exists too, and the cost of the later is in this case the same of the first one.

We have demonstrated this result in two phases: in the first one we have demonstrated that

$$DHC \Leftrightarrow DHP$$

and in the second one that for every solution of DHP exists a solution of the undirected hamiltonian path (UDHP) with the same cost.

5.1 From DHC to DHP

Before considering the demonstration, we go back to the previous note.

We have formulated the terms of the problem, consisting of finding a permutation of the nodes set

$$\{(v_0, v_1, \ldots, v_n)\}$$

with the constraints

$$v_{\pi(0)} = v_0$$
$$v_{\pi(n+1)} = v_0 \tag{5.1}$$

Because this corresponds to a scheduling solution such as

$$\{J_0, J_{\pi(1)}, \ldots, J_{\pi(n)}, J_0\}$$

and because it is unimportant to know that the first and the last job are J_0 (J_0 is an added job for mathematical use, but without correspondent in reality), we can translate job J_0 in the last position into another job J_{n+1} for which will hold

$$v_{\pi(n+1)} = v_{n+1} \tag{5.2}$$

where v_{n+1} is the node corresponding to J_{n+1}.

Thus we increase the V cardinality, but we can consider the circuit problem as a path problem from v_0 to v_{n+1}.
It is necessary to resolve it because from here we will transform the asymmetric problem in a symmetric one.

Our task is to find the permutation of

$$\{(v_0, v_1, \ldots, v_{n+1})\}$$

such that

$$\sum_{\substack{0 \leq i \leq n \\ 1 \leq j \leq n+1}} d_{\pi(i)\,\pi(j)}$$

let be minimum, with

$$\pi(0) = v_0$$
$$\pi(n+1) = v_{n+1}$$

We are going now to demonstrate the first part of the theorem.

THEOREM 5.1 — *Given graph G, graph G' exists such that G has an oriented hamiltonian circuit if only if G' has an oriented hamiltonian path ([LEN77]).*

PROOF.
Given graph $G = (V, A)$, with V nodes set, A edges set and d_{ij} the cost for every edge $(i, j) \in A$. We choose a node $v' \in V$ and let $G' = (V', A')$ the graph for which

$V' = V \bigcup \{v''\} \qquad v'' \notin V$
$A' = \{(v, w) \mid \ (v, w) \in A \quad w \neq v'\} \bigcup \{(v, v'') \mid \ (v, v') \in A\}$

Moreover, if $(v, w) \in A'$ $w \neq v'$ and $t(x, y)$ is the cost for the edge $(x, y) \in A'$, let be

$$t(v, w) = d(v, w)$$

and for $(v, v'') \in A'$

$$t(v, v'') = d(v, v') \tag{a}$$

Because of its construction, G has a directed hamiltonian circuit if and only if G' has a directed hamiltonian path, and the cost of the path is the same of the cost of the circuit. C.V.D.

5.2 About the theorem.

Before, we concluded that $n/m/F, no\ wait/C_{max}$ is equivalent to a TSP represented by a graph with $n+1$ nodes and v_0 is the first.

Given what has just been demonstrated, we can consider a graph $G' = (V', A')$ with $|V'| = n + 2$ and with

$$v' = v_0$$
$$v'' = v_{n+1}$$

and A' as defined in the theorem.

Let $T = [t_{ij}]$ be the cost matrix of G'. To simplify, let

$$V' = \{0, 1, \ldots, n, n + 1\}$$

be the edges set for V' we have to minimize

$$\sum_{\substack{0 \le i \le n \quad 1 \le j \le n+1}} t_{\pi(i)\pi(j)}$$

under the constraints

$$\pi(0) = 0 \tag{5.3a}$$
$$\pi(n + 1) = n + 1 \tag{5.3b}$$
$$t_{ij} = 0 \qquad se\ (i = j) \vee (i = 0) \tag{5.4}$$
$$t_{i\ n+1} = t_{i0} \tag{5.5}$$

(5.3) derive from (5.1) and (5.2); (5.4) derive from (4.4) and the constraints (5.5) from (a) of the theorem (5.1).

$$t_{0\ n+1} = \infty \tag{5.6}$$

is to indicate (see [LAW85]) that the edge between first and last node doesn't exist.

From (5.3) we obtain that only the permutations among n vertices are important, since the first and the last position are fixed; this correspond to the first problem scheduling formulation.

5.3 From DHP to UDHP

We resume the input for the problem.

We have a graph $G = (V, A)$ with

$$|V| = n + 2$$

$$V = \{0, 1, \ldots, n, n + 1\}$$

and asymmetric cost matrix $D = [d_{ij}] = T \quad \forall (i, j) \in A$ subject to the constraints (5.4), (5.5), (5.6).
We want to demonstrate that there is a graph $G' = (V', A')$ the cost matrix of which is symmetric $B = [b_{ij}]$, such as

$$\forall \pi(V) \qquad \exists \pi^*(V') : \quad cost(\pi) = cost(\pi^*)$$

remembering that π is subject to (5.3); this is the same as demonstrating that if a hamiltonian path in an oriented graph exists, then there is a hamiltonian path in a not oriented graph with the same cost.

5.3.1 Building the graph for the NHDP

Firstly, we build the graph $G' = (V', A')$ and

$$V' = \{0, 1, \ldots, n, n + 1, \ldots, 2n - 1\} \qquad |V'| = 2n$$

We give the following definition:

DEFINITION 5.1 – *Given a node $i \in V$ with $i = 0, \ldots, n\text{-}1$, we call CORRESPONDING of i a node $i' \in V'$ such that:*

$$i' = (n + 1) + i \qquad (i.e.\ i' \equiv i \mod (n + 1))$$

Then $V' = \{0, 1, \ldots, n, n + 1 = 0', \ldots, (n - 1)'\}$ and if B is the cost matrix of V' :

$$B = 2n \times 2n = [b_{rc}] \qquad r, c = 0, \ldots, 2n - 1$$

We build B symmetric in the following way (we consider only the elements over the left diagonal, that is the elements b_{rc} with $r = 0 \ldots 2n - 1 \quad c = r \ldots 2n - 1$)

a) for $r = 0, \ldots, n$ and for $c = r, \ldots 2n - 1$

$$b_{rc} = d_{ij}$$

$$\text{with } (r = i) \wedge (c \equiv j \quad mod(n+1))$$

$$i, j = 0, \ldots, n \quad (i \neq 0) \wedge (j \neq n + 1)$$

$$b_{00'} = b_{0\ n+1} = d_{0\ n+1} = \infty$$

So $b_{rc} = d_{ij}$ if they are elements belonging to the same row, and if $c = j$ or if $c = j'$.

b) for $r = n + 1, \ldots, 2n \quad c = r, \ldots, 2n$

$$b_{rc} = d_{ij}$$

$$(r \equiv j \quad mod(n+1)) \wedge (c \equiv i \quad mod(n+1))$$

$$i, j = 0, \ldots, n$$

that is, if $r = j'$ e $c = i'$.

Notes

1) $b_{rc} = 0$ if $(r = c) \vee (r \equiv c \quad mod(n+1))$, that is, if they are elements on the same diagonal or if $r = c'$; these conditions can be derived from (5.4).

2) $b_{0c} = 0 \quad c \neq n + 1$, still derived from (5.4).

5.3.2   Proof

We can give now the proof.

THEOREM 5.2 — *If a hamiltonian path in an oriented graph (DHP) exists, then there is also a hamiltonian not oriented path (UDHP) with the same cost.*

PROOF.

We are interested in the nodes permutations π^* of V' such that

$$\pi^*(0) = 0$$
$$\pi^*(2n - 1) = 0'$$

We note that

$$d_{ij} = b_{ij'} + b_{j'j} \qquad i = 0, \ldots, n \qquad j = 1, \ldots, n - 1$$

$$d_{ij} = b_{ij}$$

$$i = 0, \ldots, n \qquad j = n, n+1$$

From 1), $b_{j'j} = 0$.

Then, setting

$$\pi'(i) = [\pi(i)]'$$

with

$$\pi(i) \neq n \qquad \forall i \in V$$

we obtain

$$d_{\pi(i)\pi(j)} = b_{\pi(i)\pi'(j)} + b_{\pi'(j)\pi(j)} \qquad se \ \pi(j) \neq n$$

$$= b_{\pi(i)\pi(j)} \qquad se \ \pi(j) = n \qquad (5.7)$$

and thus

$$\begin{array}{ccccccc}
d_{0\pi(1)} & + & d_{\pi(1)\pi(2)} & + & \cdots & + & d_{\pi(n)0} & = \\
\| & & & & & & \| & \\
(b_{0\pi'(1)} + b_{\pi'(1)\pi(1)}) & & \cdots & & \cdots & + & (b_{\pi(n)0'}) &
\end{array}$$

We have obtained a nodes permutation of V' the cost of which is equal to the nodes permutation cost of V. Q.E.D.

6 Example

We give an example with 5 jobs (here, we consider the flowshop already transformed in Asymmetric TSP, with the fictious job J_0): $0, 1, 2, 3, 4$. The cost matrix (the pairs are the arc cost of (x, y)) is:

		0	1	2	3	4
	0	0	0	0	0	0
$D =$	1	(10)	0	(12)	(13)	(14)
	2	(20)	(21)	0	(23)	(24)
	3	(30)	(31)	(32)	0	(34)
	4	(40)	(41)	(42)	(43)	0

In the STSP corresponding we have $2 \cdot 5 - 1 = 9$ edges:

$$\{0, 1, 2, 3, 4, 5 = 0', 6 = 1', 7 = 2', 8 = 3'\}$$

and connection matrix:

	0	1	2	3	4	0'	1'	2'	3'
0	0	0	0	0	0	∞	0	0	0
1		0	(12)	(13)	(14)	(10)	0	(12)	(13)
2			0	(23)	(24)	(20)	(21)	0	(23)
$C = $ 3				0	(34)	(30)	(31)	(32)	0
4					0	(40)	(41)	(42)	(43)
0'						0	(10)	(20)	(30)
1'							0	(21)	(31)
2'								0	(32)
3'									0

Let $\pi = (0, 2, 4, 1, 3)$ be a solution for the first problem, i.e.:

$$\pi(0) = 0$$

$$\pi(1) = 2$$

$$\pi(2) = 4$$

$$\pi(3) = 1$$

$$\pi(4) = 3$$

Then, the solution cost is:

$$d_{02} + d_{24} + d_{41} + d_{13} + d_{30}$$

In the STSP corresponding the solution is:

$$\pi^P = (0, 2', 2, 4, 1', 1, 3', 3, 0')$$

and its cost:

$$c_{02'} + c_{2'2} + c_{24} + c_{41'} + c_{1'1} + c_{13'} + c_{3'3} + c_{30'} \tag{5.8}$$

From C matrix we obtain:

$$c_{02'} = c_{2'2} = c_{1'1} = c_{3'3} = 0$$

then

$$(5.8) \quad \begin{aligned} &= c_{24} + c_{41'} + c_{13'} + c_{30'} \\ &= d_{24} + d_{41} + d_{13} + d_{30} \end{aligned}$$

7   Conclusion

The results obtained can be expressed as:

1. connection weights matrix is symmetric;
2. the solution space S_{TSP} for the ATSP is included in the solutions space S calculated by Boltzmann machine;

However, the solutions computed by the Boltzmann machine differ from S_{TSP}. Not only because the nodes' updating rule allows to accept configurations which do not correspond to problem solutions (by constraints relaxation), but also because we really solve a problem which doesn't correspond any longer to the initial one.

In other words, we can only guarantee that the S solutions network space includes the solutions of S_{TSP}, but a solution calculated by the Boltzmann machine doesn't necessarily belong to S_{TSP} (as stated by the second theorem).

Generally, approximated solutions can exist (like those computed by the Boltzmann machine) that cannot be transformed into ATSP solutions and then in scheduling solutions.

In fact, we note that S (solutions network space) includes two subsets: S_F, the solutions set for the Boltzmann machine and S_{TSP}, the final set for the scheduling problem.
Moreover, S_F is confined by the typical constraint set (the connection weights), and by the added constraints (in order to force the network to converge towards final configurations coherent with the problem to be solved), whereas S_{TSP} heuristic conditions on S_F are necessary in order to compute a solution belonging to S_{TSP}.

In conclusion, we need to increase the network *knowledge* about the problem, by increasing the number of variables in the conditions set expression so the S_{TSP} subset (for which the (5.7) is verified) can be extracted from S_F.

In our work the theoretic results have been more important than the simulation results and we haven't examined the aspects related to Boltzmann machine implementation closely.

Moreover, our tests show that the solutions computed by Boltzmann machine can be compared with the solutions estimated "good enough" when obtained by traditional operation research techniques (greedy algorithm, exaustive search).

We think this approach should be investigated in order to take advantage of Boltzmann machine parallel architecture.

References

[AAR85] Aarts, van Laarhoven, *Statistical cooling: a general approach to combinatorial optimisation problems*, Philips Journal Res., 40, p.193 (1985).

[AAR87] Aarts, Korst, *Solving Travelling Salesman Problem with massively parallel architectures based on the Boltzmann machines*, submitted for publication in Europ. Journ. Op. Res. (1987).

[CAM70] Campbell, Dudek, Smith, *A heuristic algorithm for the n-job, m-machine sequencing problem*, Mgmt. Sci., 16, B630 (1970).

[CAR80] Carpaneto, Toth, *Some new branching and bounding criteria for the asymmetric TSP*, Mgmt. Sci., 26, p.736 (1980).

[DAN77] Dannenbring, *An evaluation of flowshop sequencing heuristics*, Mgmt. Sci., 23, p.1174 (1977).

[FEL50] Feller, *An introduction to probability theory and applications*, vol. 1, Wiley, NY (1950).

[FOO88] Foo, Takefuji, *Stochastic Neural Networks for solving Job-Shop scheduling* Parti 1,2, in IEEE International Conference on Neural Networks, San Diego, CA, vol.II, p.275 (1988)

[FOO88a] Foo, Takefuji, *Integer linear programming neural networks for job-shop scheduling* in IEEE International Conference on Neural Networks, San Diego, CA, vol.II, p.341 (1988)

[FRE82] French, *Sequencing and scheduling: an introduction to the mathematics of the job-shop*, J.Wiley and Sons, Chichester (1982).

[GAR79] Garey, Johnson, *Computers and intractability: a guide to the theory of NP-completeness*, W.H.Freeman and Co., S. Francisco (1979).

[HIN83] Hinton, Sejnowski, Ackley, *Boltzmann machines: constraint satisfaction networks that learn*, Tech. Rep. No. CMU-CS-84-119, Pittsburgh, PA: Carnegie Mellon University, Dep. C.S. (1983).

[HOP85] Hopfield, Tank, *Neural computation of decisions in optimization problems*, Biological Cybernetics, 52 (1985).

[JOH54] Johnson, *Optimal two- and three-stage production schedules with setup times included*, Nav. Res. Logist. Q., 1, p. 61 (1954).

[JOH84] Johnson, Aragon, McGeoch, Schevon, *Optimization by simulated annealing: an experimental evaluation (part 1)* (1984).

[KAR72] Karp, *Reducibility among combinatorial problems*, Miller and Thatcher eds., Complexity of computer computations, Plenum Press, NY, p.85 (1972).

[KIR83] Kirkpatrick, Gelatt, Vecchi, *Optimization by Simulated Annealing*, Science, 220, p.671 (1983).

[KOR87] Korst, Aarts, *Combinatorial optimization on a Boltzmann machine* submitted for publication in *Journal of Parallel and Distributed Computing*, (1987)

[LAA87] Laarhoven, Aarts, *Simulated Annealing: theory and applications*, Kluwer Academic Publishers, Dordrecht, The Netherlands (1987).

[LEN77] Lenstra, *Sequencing by enumerative methods*, Matematisch Centrum, Amsterdam (1977).

[LAW85] Lawler, Lenstra, Kan, Shmoys, *The Travelling Salesman Problem: a guided tour of combinatorial optimization*, Wiley, Chichester (1985).

[MET53] Metropolis, Rosenbluth N., Rosenbluth M.N., Teller A.H., Teller E., *Equation of state calculations for fast computing machines*, Journ. of Chemical Phys., 6, p.1087 (1953).

Architectural Considerations for NERV - a General Purpose Neural Network Simulation System

R. Hauser[1], H. Horner[2], R. Männer[1], M. Makhaniok[3]

1) Physics Institute, University of Heidelberg, Philosophenweg 12, D-6900 Heidelberg, West Germany
2) Institute of Theoretical Physics, University of Heidelberg, Philosophenweg 19, D-6900 Heidelberg, West Germany
3) Institute of Technical Cybernetics, BSSR Academy of Sciences, Surganov Str. 6, 220605 Minsk, USSR
maen@dhdmpi50.bitnet

1. Introduction

Research in artificial neural networks was triggered over 40 years ago, mainly by investigations of the structure of the brain, of the visual system of animals and men [1], and of the features of neurons and their interconnections [2]. It was recognized that a computer with an architecture comparable to that of the brain can be extremely powerful for special applications where traditional computers are very weak. The most important example is pattern recognition in acoustic signals or images. Obviously a huge new field could be opened since neural networks do not have to be programmed but are trained for their task by teaching or by self-organization. Moreover, the ability of neural networks to adapt itself to a changing environment looked very promising.

Many of these principal features of artificial neural networks have been demonstrated by simple applications of small networks [3]. However, the interest in this field dropped rapidly after it was shown by an theoretical analysis [4] that the abilities of networks with a simple architecture, as were generally used at that time, are strongly limited to solving a small class of problems exclusively. Only a few researchers continued the work [e.g., 5, 6, 7] and were eventually able to overcome these limitations by exploiting more complicated neural network architectures and appropriate learning algorithms [8]. The renewed interest in neural networks was reinforced due to the general availability of powerful computers which today allow to simulate small neural networks.

This paper describes NERV, a very powerful and flexible simulation system for neural networks. It is shown which problems arise in the simulation of neural networks, how a "natural" computer architecture would look like that maps easily to this application, how this architecture has to be modified for feasibility, and which special hardware features are required for setting up a well balanced system. This is followed by a detailed description of the hardware architecture of NERV and an application example.

2. Models of neural networks

Research in artificial neural networks is motivated by the hope that some of the powerful features of biological neural systems can be imitated technically. All artificial neural networks are therefore designed to resemble in some basic respects their biological counterparts [9].

Extremely simplified, biological neural networks consist of a high number of neurons with dense interconnections. In the human brain there are about 10^{11} neurons where each one is connected to roughly 10^3 until 10^4 other neurons, i.e., there are more than 10^{14} interconnections called synapses. With this architecture, a neural network shows a dynamical behavior on a slow and a fast time scale, which can be connected to learning and to evaluation, respectively.

The fast dynamic behavior is dominated by the features of the neurons. Although a single neuron has a very complicated behavior which is not yet understood in detail, it can basically be said to switch between an active and a passive state only. Which one is taken depends on the state of the connected neurons. Their relative importance for the neuron considered is given by the individual synaptic strength. Essentially, each neuron sums up the weighted incoming excitations and inhibitions and determines by means of some nonlinear function its new state. This is done by all neurons locally and asynchronously which gives rise to a massively parallel processing of neuron states. If an initial network state is set up, e.g., by asserting some input pattern, the system evolves according to its inherent fast dynamics. It can be easily seen that biological neural networks are feed-back systems. Because the typical neuron switching time is in the order of 1 ms and a typical pattern recognition process, e.g., the recognition of a face, takes $\approx 10^{-1}$ s, about 100 processing steps are required. If, however, each neuron forwards its current state to only 10 disjunct others, 10^{100} neurons would be required for a feed-forward system. Although there are some simple applications of feed-forward interconnections in biological and artificial neural networks [10], most of the powerful information processing features are thought to be connected with the structure of neural networks as dynamical systems with nonlinear feed-back. Basically, these systems can evolve in three ways after setting up an initial state. They may converge to a stable state, to a sequence of states that are taken on periodically, or they may not converge at all and show chaotic behavior. Pattern recognition is thought to be a consequence of the first process. In this model, known patterns represent attractors. Presenting an unknown pattern corresponds to starting the network dynamics at the corresponding point in the state space. This point belongs to one of the attracting regions and the network state evolves towards the corresponding attractor, the most similar known pattern. The convergence therefore makes the recognition process generally very insensitive against pattern distortions, e.g., against fuzzy, incomplete or noisy input data.

The slow dynamic behavior is dominated by the features of the synapses. Synaptic strengths are not fixed but are modified according to a "learning" rule. One of the simplest rules for non-supervised learning states that the synaptic strength between two neurons is increased by a small amount if both neurons are in the same state [11]. A learning rule for supervised learning states that the synaptic strength between neurons are adjusted so that the difference between the actual and the required network output state is minimized [8]. In many applications the learning phase can be separated from the evaluation phase, e.g., if a fixed set of sample patterns is given. Some applications, however, require continuous learning which is overlaid to the evaluation.

Obviously, there are many parameters that determine the actual behavior of an artificial neural network. Some refer to the network architecture, e.g.,

- whether the network is homogeneously connected so that there are no topological differences between neurons or whether it is structured into layers or another way,
- whether neurons in homogeneous regions are fully connected so that there is one synapse between each two neurons or whether the synapses are diluted,
- whether synapses are symmetric or may have different strengths in both directions, and
- the possible range of synaptic strengths from, e.g., 32 bits down to clipped synapses with 1 bit resolution only.

Some parameters refer to the network dynamics, e.g., the way the neuron update process is determined. Examples are

- the Hopfield dynamic [12] where only one neuron per step is allowed to change its state,
- the dynamic of the strongest local field [13] where the very neuron with the currently highest excitation value is selected for updating, or
- the Little dynamic [14] where all neurons are allowed to change their state independently.

Other parameters influence the learning behavior of the network, e.g., the type of learning rule used, which usually has to correspond to statistical features of the sampling patterns, and the speed of learning. In addition, there are global parameters like the network size or a "temperature" which determines the statistical behavior of the network.

Unfortunately, today a theoretical analysis of neural networks is only possible in the most simple cases. The study of the capabilities of artificial neural networks or their suitability for certain applications is generally only possible via simulations. However, even with todays supercomputers it is not feasible to investigate the huge parameter space of larger neural networks to some extent. This is demonstrated by an example, the determination of the storage capacity of an associative memory.

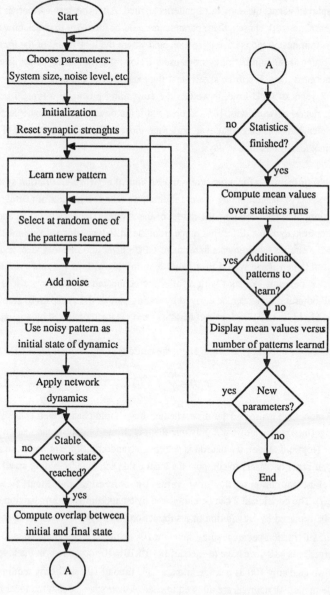

Fig. 1: Flow chart of a typical simulation task

Here, a fully connected neural network is to be investigated for the maximum number of patterns that can be stored and retrieved without errors. A flow chart of this task is given in Fig. 1. During the network setup phase, all system parameters have to be predefined. The learning of sample patterns is prepared by initializing the synaptic weights. Next the first pattern is learned. The same pattern is distorted, e.g., by adding noise, taken as the initial network state, and the network dynamics are applied until convergence. The similarity between the original and the recovered pattern is measured, e.g., by computing the Hamming distance between both. If more patterns would have been learned already, one could now select another one to do some statistics. After computing mean values over the statistics runs, new patterns could be learned until the retrieval process becomes unreliable. The whole procedure results, e.g., in a diagram where the computed mean values are displayed versus the number of patterns learned. After this result is output for a certain set of parameters of the neural network chosen, these parameters could be modified to check how the results depend on them. Obviously, four nested loops have to be executed where the loop count of the innermost three loops is each time in the order of the number of neurons used. Below, the computing power is estimated which is required for the innermost loop. It will be shown that the execution of the example program would require about 10^9 s, i.e., 32 years of CPU time, if we use the computing power of a typical microprocessor (10 MIPS), if we have a neural network with 10^4 neurons, and if the outermost loop is neglected, i.e., if no system parameters are changed. It is clear that even under the most optimistic assumptions (1 GFLOPS) todays supercomputers cannot lessen this time to days.

3. Derivation of the requirements for a general purpose neural network simulation system

For an estimation of the computing power required, we take two models of neural network dynamics with very different computational structure, the model of the strongest local field (SLF) and the Little model (the better known Hopfield dynamic differs not much from the SLF dynamics as mentioned below). It will be shown that they lead to similar requirements in terms of CPU power and memory size, and that the same requirements are also obtained if we would take usual learning algorithms for this estimation. We assume $N=10^4$ fully connected neurons. We start with a fully interconnected network since it is the most general setup from which all other networks can be derived by setting part of the synapses to zero. Each neuron has only two states $s_i(t) \in \{1,-1\}$, $i=1...N$. Each pair of neurons i and j is interconnected by a synapse with strength J_{ij}.

We consider the evaluation phase first. Here, the neurons compute their local field $h_i(t)$ according to

$$h_i(t) = \sum_{j=1}^{N} J_{ij} s_j(t).$$

In the SLF model, all neurons determine whether they should change their state. This is the case for all neurons i with $\mathrm{sign}(h_i(t)) \neq s_i(t)$. However, only one them is allowed to do so, the neuron with the highest value of $h_i(t)$ (in the Hopfield model, the neuron is selected at random). Since at most one neuron k changes its state at each step, it is not necessary to compute the local field each time by adding up all terms $J_{ij} s_j(t)$. Instead, it is sufficient to compute the new value by correcting the local fields according to $h_i(t+1) = h_i(t) + 2J_{ik} s_k(t)$. The coefficient 2 can be eliminated by an appropriate normalization of h_i, the multiplication by s_k can be replaced by an addition or a subtraction since $s_k = \pm 1$. Since we therefore have essentially one addition or subtraction per step, since there are 10^4 neurons to be simulated, and since the number of steps until convergence is $\approx 10^4$, we have to execute $1 \times 10^4 \times 10^4 = 10^8$ operations in the innermost simulation loop (see Fig. 1). If we take only 100 ns per operation, a CPU time of 10 s is already required.

In the Little model, all neurons are allowed to change their state, so that the local field has indeed to be added up of the terms $J_{ij}s_j(t)$. Thus, the number of operations that are required per step is increased by the factor $N=10^4$. On the other hand, the systems converges much faster, e.g., in 10 steps. The total computing time until convergence would therefore be $10^4 \times 10^4 \times 10 = 10^9$, i.e., increased only by a factor of 10.

The learning phase usually requires roughly the same computing time. This can be seen easily for one of the simplest learning rules, Hebb learning. Here, the synaptic strengths are increased by a small amount if the two interconnected neurons are in the same state {+1 or -1}. For the execution of this learning algorithm, all elements J_{ij} of the synaptic matrix have to be scanned for possible updates for each new pattern. The time required for this operation is $O(N^2)$ which is comparable to the convergence time in Hopfield, SLF, and even Little models, according to the estimations given above. There are, however, more complicated learning algorithms like back-propagation [8] which require much more computational effort.

Taking into account that the statistics loop over learned patterns as well as the learning loop scales with the number of neurons (10^4), it is obvious that an extreme computing power is required which can only be provided by a powerful parallel processor. Besides, the requirements for main memory are also considerable. Its size is essentially given by the volume of data that have to be stored for the synaptic matrix J_{ij}, since J_{ij} consists of N^2 elements. Apart from the number of neurons used, this volume depends on the resolution of the synaptic values. In our example synapses are stored as 16 bit values for 10^4 fully interconnected neurons. Thus a memory size of $10^4 \times 10^4 \times 2$ Bytes = 200 MBytes is necessary. This is much larger than the size of local data or program code. Unfortunately, all synaptic values are accessed in the innermost simulation loop so that caching would be no solution and the synaptic matrix has to be stored completely in high speed memory.

Apart from the basic abilities to store all required data and to execute all computations at a reasonable speed, a general purpose simulation system must be easy to handle. It is a very powerful tool not only for research in computer science but also in physics, biology, and other fields. Users are therefore generally neither trained nor interested in special problems of parallel processors so that the system should look like a usual computer as far as possible. This concerns first its programmability. Due to the huge number of possible architectures, learning , and dynamic algorithms, users have to be able to implement new software in a common high level language. Most output data which are computed during simulations have to be presented graphically so that a graphical user interface could best be used also for system control. It is clear that two operation modes have to be provided. One is interactive simulation control in which the user can set up a neural network, get first results very fast, and can easily adjust network parameters as required. The other one is some kind of batch processing for production runs which are used to compute statistically significant results and which may require hours, days, or weeks of processing time. User interface and program development tools of the NERV system are described elsewhere [15, 16].

4. The architecture of the NERV system

As pointed out above, the operation of biological neural networks is based on a massively parallel and asynchronous repeated update of neurons (evaluation) and synapses (learning). Only information is used which is locally available. In some neural network models, global information is used in addition. Examples are the Hopfield model, where exactly one neuron is globally selected at random, the SLF model, where the neuron with the globally strongest local field is selected, or models using a global temperature parameter to control the convergence speed of the network. In fully connected networks, the state of all neurons has additionally to be available to each neuron. However, each neuron only requires its respective part of the synaptic matrix.

A "natural" architecture for a neural network simulation processor would therefore be like shown in Fig. 2. To each neuron, one processor element is assigned which executes the time consuming learning and dynamics algorithms on a local CPU. The program, which is the same for each processor element, is stored in a local memory as well as all data that are required only locally. One example are the strengths of the synapses which connect the respective neuron with others. They are given by one column of the synaptic matrix J_{ij}. Another example is the actual neuron state s_j which is required by each processor element and therefore has to be copied into all local memories. On one hand, the decentral storage of multiple copies of the neuron state allows to operate the processor elements independently as long as the neuron state does not change. On the other hand, each neuron update has to be done synchronously in all local memories.

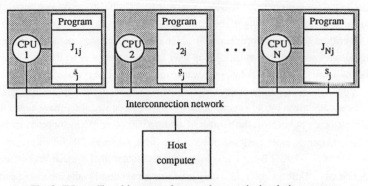

Fig. 2: "Natural" architecture of a neural network simulation system

This requires an interconnection network which also allows the implementation of the other global operations required, e.g., the selection of one processor element at random or according to some global criterion. In addition, it has to support various synchronization operations. In most neural network models it is required to synchronize all processor elements after each step in the innermost simulation loop for the next neuron update. After convergence, the sequential part of the simulation program has to be continued. Whereas this could be done by one of the processor elements, it is much simpler to assign this task to an host computer which controls the processor elements. This host can be a commercially available system that provides usual peripherals, a graphic user interface, and software development tools. The synchronization between a processor element and the host has also to be provided by the interconnection network.

This "natural" architecture is, however, not feasible for the required network size of 10^4 neurons if an usual microprocessor is used as the CPU of the processor elements. Today, a realistic multiprocessor system may consist of up to several hundred microprocessors, so that one processor element has to simulate multiple neurons. The only system changes required by this modification is that now multiple columns of the synaptic matrix have to be stored in the local memory of the processor elements (Fig. 3). For the NERV system it is planned to use up to 320 processor elements in a single crate. The estimations below are therefore done for a number of processor elements $P=100$.

Since all processor elements are topologically identical and since the host computer has to access them in the same way, a "natural" choice for the interconnection network is a homogeneous structure that provides the same transfer capabilities for each connected processor element and the host computer. One examples is a full connection where one link is provided for each pair of processor elements, another one is a bus system. However, in our example a full connection would require $P(P-1) \approx 10^4$ links which is not feasible. A common bus, on the other hand, seems to be the bottleneck of a multiprocessor system with hundreds of CPUs. This is indeed true, but it is shown below that it is possible to remove this bottleneck by minor bus extensions.

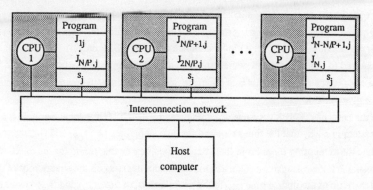

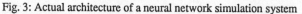

Fig. 3: Actual architecture of a neural network simulation system

Tab. 1: Execution times for time critical operations

1) Computation of the local field:

model	strongest field	Little
update operation	$h(i) := h(i) + 2J_{ij}s_j$	$h(i) := \sum_{j=1}^{N} J_{ij}s_j$
for N/P neurons	$\dfrac{N}{P}t_{upd}$	$\dfrac{N}{P}N\,t_{upd}$

2) Selection of the neuron with the strongest local field (SLF only):

model	strongest field	Little
select local strongest	$\dfrac{N}{P}t_{lsel}$	
select global strongest	$P\,t_{gsel}$	

3) Transfer of new neuron state(s)

model	strongest field	Little
number of neurons	1	N
transfer to all PEs	$P\,t_{xfer}$	$P N\,t_{xfer}$

We consider again the two neural network models used above, i.e., the SLF model and the Little model. Here it can easily be estimated

- the time required to calculate the actual local field for all simulated neurons from t_{upd}, the time required for a single neuron;
- the time required to select locally, i.e., within a processor element, the neuron with the strongest local

field from t_{lsel}, the time required for a local compare operation (SLF only);
- the time required to select globally the processor element containing the neuron with the strongest local field from t_{gsel}, the time required for a global compare operation (SLF only);
- the time required to transfer updated neuron states to all processor elements from t_{xfer}, the time required for a single bus transfer (for SLF, this is at most one neuron, for Little, there are O(N) updated neurons).
These times are given in Tab. 1.

Since the costs of the NERV system are dominated by the CPUs and local memories, it is reasonable to design the system in a way that the times required for bus operations, i.e., t_{gsel} and t_{xfer}, are both considerably smaller than the computing time. From theses conditions, reasonable limits for the maximum number of processor elements P follow as given in Tab. 2. This maximum depends on the square root of the number of neurons N and the ratio of the update time t_{upd} to the global selection time t_{gsel} and the transfer time t_{xfer}, respectively. Essentially, the update operation consists of adding or subtracting a synaptic strength to the current local field, and the global selection is a combination of one bus transfer and a compare operation. t_{upd} and t_{gsel} are therefore comparable to the transfer time so that both ratios are in the order of 1.

Tab. 2: Conditions for a balanced system and resulting max number of processor elements

model	strongest field	Little
GSel time < CPU time	$Pt_{gsel} < \dfrac{N}{P}t_{upd} \Rightarrow P < \sqrt{N\dfrac{t_{upd}}{t_{gsel}}}$	
Xfer time < CPU time	$Pt_{xfer} < \dfrac{N}{P}t_{upd} \Rightarrow P < \sqrt{N\dfrac{t_{upd}}{t_{xfer}}}$	$PNt_{xfer} < \dfrac{N}{P}Nt_{upd} \Rightarrow P < \sqrt{N\dfrac{t_{upd}}{t_{xfer}}}$

With $N=10^4$, the number of processor elements would therefore be limited to ≈ 100, a number considerably smaller than planned. Because of the square root, t_{gsel} as well as t_{xfer} would have to be decreased by a factor of 100 to get a sufficiently high maximum number of processor elements. This is not feasible. Alternatively, the time required for global selection and the time required for the neuron state update could be decreased by a factor of $P=100$, if both could be done in a single bus cycle. Thus, the bus system has to be extended by a global compare logic as well as by the ability to execute broadcast transfers. The capabilities required for the interconnection network are given in Tab. 3.

Tab. 3: Functions required for the interconnection network

operation	initiated by
normal read/write transfers	host computer
broadcast write transfers	host computer and processor elements
global compare	processor elements
processor synchronization	host computer and processor elements
interrupt	processor elements

5. NERV setup

The actual setup for the NERV system was chosen according to the considerations above. It consists of a VME based multiprocessor with up to 320 processor elements, an Apple MacintoshII PC as the host computer, and an intelligent interface between both (Fig. 4). The interface function is executed by a commercially available VME processor board, an ELTEC Eurocom-5 [17]. It is called the master processor since it is responsible not only for transfer and buffering of programs and data, but also for controlling and synchronizing the processor elements.

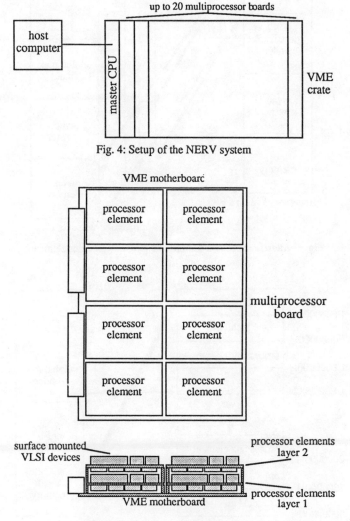

Fig. 4: Setup of the NERV system

Fig. 5: Setup of a multiprocessor board

A multiprocessor board of the NERV system consists of a VME motherboard with the bus interface logic. In addition, the motherboard carries subboards with autonomous processor elements. They have been designed to provide only the absolute minimum of functions in order to get a high package density. Up to 16 processor elements fit onto a single motherboard (Fig. 5). Each one consists of a 68020 CPU operating at

25 MHz, 1 MByte of static RAM, and interface and control logic. A floating point coprocessor is not required for the computations mentioned above, neither a memory management unit.

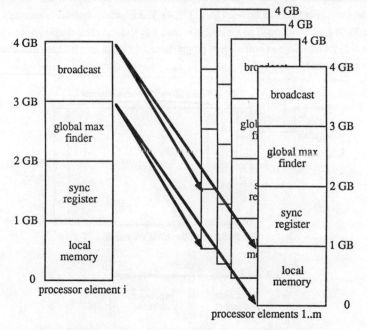

Fig. 6: Address space and address mapping of processor elements

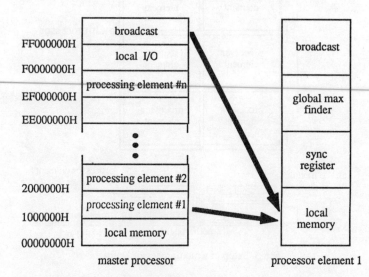

Fig. 7: Address space and address mapping of the master processor

The broadcast feature which has been added to the VME bus protocols is initiated by executing a write transfer into a certain part of the address space of the processor element (Fig. 6) or of the master processor (Fig. 7). This causes a simultaneous write operation to all local memories. The data transfer acknowledge signal of the VME bus cannot be used to signal a successfully terminated broadcast transfer, because the transfer has to be extended until all processor elements have acknowledged. This requires a logical AND of the individual acknowledge signals. For that purpose, a free bus line on the P2 connector was used as an broadcast acknowledge signal. It is a open collector line, whose HIGH level signals a successfully terminated broadcast write transfer. An extended version of this protocol is currently under consideration as a standard for the VME bus [18]. The typical broadcast time is about 1 μs. With this, the communication time both for Hopfield and Little type network computations is negligible against the computing time of a processor element. The processor elements use only broadcast transfer to communicate with each other and with the master processor.

For most theoretical models it is required that the processor elements are synchronized after each iteration step and that the master processor is notified of the termination of a number of iteration steps. For this, another free open collector line on the VME bus is used. This line can be set by each processor element via a register. On the VME bus, this line becomes HIGH only when all processor elements did set their register to 1. In this case, an interrupt is generated on the master processor. It can then do the required action, e.g., start the next computation step or read out simulation results. The usual VME bus interrupt cannot be used for this purpose, because it does not allow the required synchronization of processor elements.

The operation of the global compare logic is similar to that of the priority arbitration logic of Futurebus+ [19] or the OR-network of the Connection Machine [20]. Basically, a binary coded value, e.g., a random number or the value of the local field, is asserted onto a set of open collector bus lines. Because all processor elements do so in parallel, the global OR of all values is built on the bus lines. A combinatorial logic on each processor element compares this global value to the local value. This is done bit sequentially, starting with the most significant bit. If a difference is found, the processor element does not have the highest priority and will negate its value. With unique values, this procedure ensures that only one processor element remains connect to the comparison bus and is selected. In the NERV system, we use four bus lines for this purpose.

The current compare logic uses full binary codes with a width of four bits. Wider words are compared by software-chaining of 4 bit nibbles each. Currently, 13 μs are required to compare 32 bit values, where 8 μs are used by the compare hardware and 5 μs by software overhead. For realistic values, e.g., $N=10^4$, $P=320$, a global compare has to be executed every 30 μs. The overall system efficiency therefore amounts to 30 μs/(30 μs+13 μs), i.e., it is below 70%. By using binomial coding for the compare values instead of full binary coding, however, it is very simple to speed up the compare process with only minor hardware modifications [21, 22] (binomial codes with a width of four bits can easily be generated from full binary codes via software table look-up). If an additional fifth bus line and adjustable delays are used, it is possible to reduce the compare time to 2 μs + 5 μs. This increases the overall efficiency to over 80%. In a 320 processor system, this effect corresponds to 38 processors that would not be used without binomial coding.

6. Conclusions and further work

Operations have been discussed which are typical for a general purpose simulation system for neural networks. From that, a set of requirements regarding the computational speed, the memory size, and the interconnection structure has been derived. It was shown that a multiprocessor architecture can fulfil these requirements and that even 320 processor elements can be used on a single VME bus if the bus protocols are extended by a few features. These extensions are a broadcast transfer capability that allows to write neuron updates simultaneously into the local memories of all processor elements, a global compare logic that allows to select one processor element according to a global value, and a synchronization line for the synchronization of the processor elements and the master processor.

We presented in some detail the hardware architecture of the NERV system which was developed according to these considerations. Its characteristics are given in Tab. 4 and Tab. 5.

Tab. 4: Architectural characteristics of the NERV system

architecture	multiprocessor and host computer
multiprocessor system	single VMEbus crate
host computer	Macintosh II
peak perfomance	1300 MIPS
max number of processors	320
total system memory	320 MBytes static RAM (no wait state)
processors/VME board	16
CPU	68020, 25 MHz
local memory	1 MByte static RAM (no wait state)
neuron update hardware support	VMEbus broadcast
neuron selection hardware support	global compare logic
programming language	Modula-2
user interface	simulation package with graphical surface

Tab. 5: Architectural characteristics of the NERV system

max number of updates/s	10 Mio. (programmed in Modula-2)
	100 Mio. (programmed in assembler)
types of networks	all (freely programmable)
max number of neurons	13,000 $\left.\begin{array}{l}\text{fully connected network}\\\text{with 16 bit synapses}\end{array}\right.$
max number of synapses	150 Mio.
max number of neurons	300 Mio. $\left.\begin{array}{l}\text{sparsly connected network}\\\text{with clipped synapses}\end{array}\right.$
max number of synapses	6 Bio.

Up to now, most commercially available simulation systems for neural networks run on workstations with specialized coprocessors. However, their computing power is much too small for an comprehensive evaluation of various network architectures for a specific application. The multiprocessor structure of the NERV system can overcome this bottleneck. Its hardware and software architecture is powerful enough for the simulation of neural networks with over 10^4 neurons and an arbitrary architecture, e.g., a full connection. A full size NERV system with $\approx$1300 MIPS and 320 MBytes of fast RAM in a single VME crate provides an extremely powerful personal simulation system for neural networks. Such a system would allow to execute simulation runs that require more computing power than is usually offered by supercomputing centers to individual users. Moreover, as a general purpose system, NERV will offer an user friendly way to specify the network architecture and all network parameters. Since the NERV system is potentially applied by researchers and developers from many fields as physics, biology, engineering etc., we envisage a graphical user interface that supports the setup of a specific network, allows to control a simulation run interactively, and provides a graphically display of results. An appropriate extension of the available NERV software is under development.

The NERV system is currently running as a small prototype and has been tested completely. The kernel of an operating system, utility software, standard simulation programs, and a graphical user interface [16] have been developed and are in the testing phase. This software package is currently being extended to support a wider range of applications. We hope that we can extend the system in the next year to its full size.

7. References

[1]   Hubel D.H., Wiesel T.N.: Brain Mechanism of Vision; *Sci. Am.*, Vol. **241** (1979) 150-162

[2]   McCulloch W.S., Pitts W: A Logical Calculus of the Ideas Imminent in Nervous Activity; *Bull. Math. Biophysics*, Vol. **5** (1943) 115-133

[3]   Widrow B., Hoff M.E.: Adaptive Switching Circuits; Proc. 1960 IRE WESCON (1960) 96-104

[4]   Minsky M.L., Papert S.A.: Perceptrons; MIT Press, Cambridge, MA (1969)

[5]   Kohonen T.: Self-Organisation and Associative Memory; Springer, Berlin (1984)

[6]   Grossberg S. (Ed.): Neural Networks and Natural Intelligence; MIT Press, Cambridge, MA (1988)

[7]   Rumelhart D.E., McClelland J.L.: Parallel Distributed Processing, Vol. **1 + 2**; MIT Press, Cambridge, MA (1986)

[8]   Le Cun Y.: Une procedure d'apprentissage pour reseau a seuil assymetrique; *Proc. Cognitiva*, Vol. **85** (1985) 599-604

[9]   Lippmann R.P.: An Introduction to Computing with Neural Nets; *Comp. Arch. News*, Vol. **16**, No. 1 (1988) 7-25

[10] Linsker R.: From Basic Network Principles to Neural Architecture; *Proc. Nat'l Acad. Sci. USA*, Vol. **83** (1986) 7508-7512, 8390-8394, 8779-8783

[11] Hebb D.O.: The Organization of Behavior; Wiley, New York, NY (1949)

[12] Hopfield J.J.: Neural Networks and Physical Systems with Emergent Collective Computational Abilities; *Proc. Nat'l Acad.Sci.*, Vol. **79** (1982) 2554-2558

[13] Horner H.; in Haken H. (Ed.); Proc. Int'l Symp. on Synergetics, Schloß Elmau, West Germany (1987)

[14] Little W.A., Shaw G.L.: *Math. Biosci.*, Vol. **39** (1978) 281

[15] Männer R., Horner H., Hauser R., Genthner A.: Multiprocessor Simulation of Neural Networks with NERV; Proc. Supercomputing ΄89, Reno, NV (1989)

[16] Lange R., Genthner A., Männer R., Horner H.: Application Development Tools and User Interface for the Neural Network Simulation Multiprocessor NERV; subm. to CompEuro 90, Tel-Aviv, Israel (1990)

[17] Eltec: EUROCOM-5 Hardware Manual; Eltec Elektronik, Mainz, West Germany (1987)

[18] ESONE VMEbus working group; priv. comm. (1988)

[19] Taub D.M.: Improved Control Acquisition Scheme for the IEEE 869 Futurebus; IEEE Micro, Vol. **7**, No. 6 (1987) 52-62

[20] Hillis D.W.: The Connection Machine; MIT Press, Cambridge, MA (1985)

[21] Makhaniok M., Cherniavsky V., Männer R., Stucky O.: Binomial Coding in Distributed Arbitration Systems; subm. to *Electr. Lett.* (1989)

[22] Makhaniok M., Cherniavsky V., Männer R., Stucky O.: Effective Implementation of Distributed Arbitration in Multiprocessor Systems; Proc. WOPPLOT 89, Wildbad Kreuth, Germany (1989)

Associative Neural Networks in Analog VLSI: Advantages of Decrementing Algorithms.

V. G. Dobson, Department of Experimental Psychology
Oxford University, Oxford, OX1 3UD. U.K.

1. Introduction.

This report is concerned with evaluating associative network algorithms as candidates for analog implementation in VLSI. The primary advantage of analog neural networks is in greater miniaturisation, enabling greater component densities and the construction of larger and more powerful networks on a single VLSI chip. This, in turn, provides much greater speed and lower power consumption. However, one cost of miniaturization is progressive loss of accuracy, as link signals become smaller relative to the noise, or random fluctuation due to inherent manufacturing imprecision [1]. Clearly, the neural network algorithms which are least susceptible to these fluctuations are preferred, because they enable most efficient exploitation of the potential of analog implementations.

There are three basic associative network algorithms, referred to below as Decrementing [2-6], Incrementing [7-11], and Correlative [11-21]; Incrementing and Decrementing algorithms will be referred to collectively as 'Conjunctive' algorithms (Appendix I). The operation and performances of these algorithms will be compared over a comprehensive range of network architectures, associative tasks, and vector structures and implementational precision.

In associative network architectures each network element, or unit, is addressed by each input line via a matrix of modifiable links. They can be used to associate vectors of activity on the input lines with vectors of activity over the units (hetero-association). Alternatively, input vectors can be associated with a single active unit (pattern-classification), sometimes using winner-take-all circuits to facilitate recall. In auto-associative architectures units address one another via the link matrix, and activity vectors are associated with themselves (auto-association). In each architecture, algorithms may be required to activate only those units which have been associated with the current input vector (exact-match selection), or to activate units associated with the vectors most similar to a noisy input (best-fit selection). The vectors associated may be sparse or dense, with low or high proportions of active components, respectively; and they may contain positive or negative noise (additional or missing active components). Crucial differences in performance also emerge between networks with 'ideal' circuits, specified exactly in accordance with algorithmic requirements, and networks with imprecisely specified circuits, in which link signal values vary randomly about the mean.

Preview of findings.

The general conclusions of this report can be summarised as follows. In ideal circuits, the performances of Incrementing and Decrementing algorithms are identical for all associative tasks. However, in imprecisely specified circuits, Decrementing algorithms are significantly more robust, particularly for exact-match selection from input vectors which contain no positive noise. In ideal circuits, Conjunctive and Correlative algorithms give identical performance at pattern-classification tasks. In imprecisely specified circuits of practical size, Decrementing algorithms are preferred, because they are significantly simpler to implement, and less error prone. In ideal circuits performing auto- and hetero-associative tasks Correlative algorithms provide best performance with dense patterns, but Decrementing algorithms are superior over a range of sparse patterns. This range increases in imprecise circuits, because of the greater cost-effectiveness of Decrementing algorithms.

The practical advantages of Decrementing algorithms, in pattern-classification and sparse vector association tasks, are strategic. This is because pattern-classification tasks offer the most immediate beneficial applications for analog associative networks, and because many constraint-satisfaction and relaxation-labelling tasks are most appropriately represented in terms of sparse vectors.

Subsequent sections of this report will first review a simple model of the effects of manufacturing imprecision on discriminations between accumulated signals in analog associative networks. This model will then be used to explain why Decrementing networks perform more reliably than Incrementing networks at pattern-classification with imprecise circuits, and to show that these advantages generalise to hetero-associative and auto-associative tasks also. Subsequent sections show that Decrementing networks provide more reliable performance than Correlative networks at pattern-classification, with imprecise circuits, and at associative tasks involving sparse pattern vectors with both ideal and imprecise circuits.

2. Discriminations between sums of imprecisely specified signals.

The effects of manufacturing imprecision on signal processing in associative networks can be understood by making the convenient assumption that individual link signals have a mean of $'S'$, and are normally distributed, with an s.d.(standard deviation) of σ. This means that the distribution of signal totals over units receiving from L active links will have a mean of $L*S$ and an s.d. of $\sigma\sqrt{L}$. Error rates in thresholding and winner-take-all operations can be predicted from the value of $\sigma\sqrt{L}$.

For example, let it be assumed that the criteria for distinguishing between L and L + 1 signals are set at the midpoints between them, i.e. at L + 0.5, and that the s.d. is also set to $\sigma\sqrt{L}$ = 0.5. According to the standard normal distribution, if $\sigma\sqrt{L}$ = 0.5, then some 16% of units receiving L link signals will have signal totals exceeding L*S + 0.5, and the same proportion will have signal totals below L^S - 0.5, these will thus be classified spuriously as having received L + 1 or L - 1 signals respectively. Error rates can be reduced to below 2% by reducing the s.d. to $\sigma\sqrt{L}$ < 0.25.

Figure 1 shows the values of σ required to restrict error rates to 16% and 2% in discriminations between L and L + 1 link signals for L values ranging from 0 to 100. As L increases, link signal imprecision, σ must be reduced, to keep error rates constant, and reductions in σ require increased cost, or increased silicon area, or both [1]. Clearly, algorithms which minimise L, the number of signals which each unit must sum in order to determine which activity state it should adopt, will provide more compact and cost-effective implementations.

Figure 1.

Distinguishing between L and L + 1 link signals:
Minimum S.D. of individual signals required to limit error rates to 16% and 2%

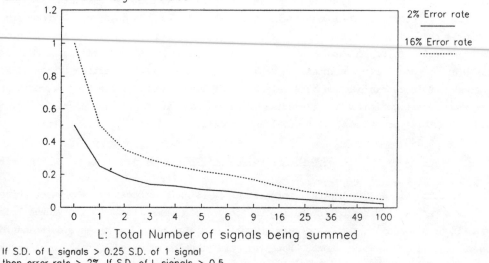

If S.D. of L signals > 0.25 S.D. of 1 signal
then error rate > 2%. If S.D. of L signals > 0.5
S.D. of 1 signal then error rate > 16%

In fact, as subsequent sections describe, there are wide variations between algorithms in the L-values that they require. Correlative algorithms require units to sum signals from each of the other N - 1 units in the network. Incrementing algorithms require summation of signals from all A active elements in the input vector. Decrementing algorithms merely require the detection of $\geq$ 1 signals, for exact-match tasks. For best-fit tasks, Decrementing algorithms only require units to sum the number of mismatches, M', between active input elements and stored vectors. As the units giving best-fits receive the smallest numbers of mismatch signals, the L-values of the different algorithms associating the same vectors can be ranked as follows (N - 1) > A > M' > 1, with the Decrementing algorithms requiring summation and comparison of the smallest numbers of signals.

3. Performance of Pattern-classification tasks by networks using Incrementing and Decrementing algorithms.

This section will first outline the operation of Incrementing and Decrementing algorithms performing pattern-classification with binary vectors in ideal circuits, and then go on to show that differences in performance emerge with imprecisely specified signals, and that these differences generalise to the other associative tasks.

In Pattern-classification, each unit is trained to respond to only one vector (its 'stored' vector), by modifying its links from the input lines according to Incrementing or Decrementing link modification algorithms.

In Incrementing algorithms, all links are initially disabled and inoperative. During training, the links from active input lines to the unit(s) that they are required to activate, are enabled, or incremented from zero, to give a standardised signal of 'S'. Assuming that there are 'A' active lines in each input vector, each trained unit receives a total signal of A*S from its stored vector, and exact-match selection is provided by setting unit thresholds to Θ = A*S. Best-fit selection on input vectors with negative noise requires thresholds to be varied with the number of active input lines. Selection of best-fit vectors with positive noise, requires the threshold to be ramped until only one unit remains active. A winner-take-all circuit may be used to assist selection [22,33].

In Decrementing networks, all links are initially enabled, producing standardised signals of 'S'. During training, links from active input lines are disabled, or decremented to zero gain, so that each unit receives no signals from its stored vector, and exact-match selection is provided by setting unit thresholds to a negative threshold of Θ = 0. This means that they are spontaneously active unless receiving one or more signals, like NOR gates. Best–fit selection with negative noise is performed automatically by the exact–match algorithm, with no need for the variable threshold algorithm required by the Incrementing network. Best–fit selection with positive noise can be performed by ramping the global threshold. However, it can also be provided by allowing accumulated unit signals to decay locally at equal rates. The unit with the fewest mismatch signals falls below threshold first, and inhibits the other units via a winner–take–all circuit [3–5].

Differing error rates at best–fit selection by Conjunctive algorithms.

While the Incrementing algorithm selects units with the largest number of 'match' signals, the Decrementing algorithm selects units with the smallest number of 'mismatch' signals. As Figure 1 indicates, the accurate summation and comparison of large numbers of signals is more sensitive to individual signal fluctuation than for small numbers of signals. In best–fit selection tasks, both Decrementing and Incrementing networks can make omission and commission errors. However, Decrementing networks require significantly lower levels of link signal precision to deliver equal reliability in performance. In networks with A = 100 active input lines, and up to 7 positive noise signals, the minimum s.d. in link signals required by Decrementing networks to discriminate L from L + 1 signals with 2% error rates varies from $\sigma \leq$ 0.25 to $\sigma \leq$ 0.1 (Figure 1). This is from 4 to 10 times more than that demanded by the corresponding Incrementing networks. Generally, a discrimination between $\Theta + 1$ and $\Theta + 2$ signals which requires a link signal s.d. of σ in an Incrementing network, can be performed by a Decrementing network with a link signal s.d. of $\sigma\sqrt{\Theta}$.

Differences between Conjunctive algorithms in omission error rates at exact–match selection tasks.

In exact–match tasks, fluctuations in individual signals can cause Incrementing networks to make both omission, or commission errors. Variations in threshold which reduce the rate of one type of error invariably increase the rate of the other error type, and the combined error rate is minimised when they are equal. However, assuming that all link modifications have been effective, and that input vectors contain no positive noise, individual signal fluctuations cannot cause errors of omission in exact–match tasks in Decrementing networks. Moreover, errors of commission can only occur where single signals are too small to be detected. Consequently, they can be effectively eliminated from Decrementing networks performing exact–match tasks, without risk of causing errors of omission, either by increasing the mean signal magnitude S, or by reducing signal detection thresholds [2–5].

For example, to distinguish between an exact–match signal from A = 100 active links and a sub-threshold signal from A − 1 = 99 links, with error of commission and omission rates both at 2%, the Incrementing network requires a threshold of Θ = 99.5 and individual link signals with a mean of S = 1, and an s.d. of σ = 0.025 (Table 1 : L = 100). The corresponding task for a Decrementing network is simply to distinguish between L = 1 and L = 0 signals with 2% errors of commission, and this requires an individual link signal s.d. of σ' = 0.5 and a negative threshold of Θ = 0. Errors of commission can be reduced to the 4 s.d. level of 0.003%, either by increasing the mean signal, S to 2, or by reducing σ' to 0.25, without increasing errors of omission. Note that control over link signal accuracy, σ', is only required below the mean, variations above the mean do not affect error rates. These arguments apply primarily to exact-match selection from input vectors which are complete or incomplete (no positive noise).

Small amplitude signals in Incrementing algorithms.

Two further problems stem from the progressive decrease in individual link signal magnitudes as Incrementing networks are scaled upwards in size. Firstly, if total unit signals are restricted to the 0-5v range, then, to represent threshold signals of Θ = 100, for example, on a single unit, individual link signals must be reduced to a mean of S' ≤ 0.05v. Secondly, the difference between Θ and Θ − 1 of such signals falls substantially below this mean S', because of the exponential compression of signals which is typical of VLSI implementations. Both of these processes operate in Incrementing networks to make the difference between Θ and Θ − 1 decline to ever smaller levels, as Θ increases, making discrimination more vulnerable to background noise.

There are no corresponding constraints on mean link signal magnitude, S, as Decrementing networks are scaled up in size. For exact-match selection, S could be so large that units are saturated by a single signal. It is important that individual mismatch signals are at least large enough to be detected, and so the exponential relative amplification of small signals works to the advantage of Decrementing networks in this instance. For best-fit selection, performance is also assisted by large, readily distinguishable signal levels, and would usually be unimpaired if units were saturated by as few as 10 mismatch signals. However, large signals are not essential to Decrementing networks, they can operate with S values as small as those required in Incrementing networks, in which case they can use units which saturate at signal totals significantly lower than that required by the Incrementing algorithm (Θ = A). This provides potential benefits in terms of reduced power requirements and heat output.

Thresholding Problems in Incrementing algorithms.

There is a third set of problems which arise only in Incrementing networks and which also increase with scale. These problems stem from the need to vary local unit thresholds with global activity levels. For example, in order to cope optimally with incomplete patterns, thresholds should be calculated afresh for each input vector, so as to vary with A, the number of active lines; for exact-match selection, $\Theta = A*S$. However, if the s.d. in the individual link signals received by the threshold modulation system is the same as the s.d. in the signals to the network units, then the threshold setting process itself becomes a major source of error, effectively increasing the s.d. of the distribution of total signals to units by $\sqrt{2}$.

A further asymmetry between Incrementing and Decrementing algorithms for pattern-classification emerges in networks using input vectors with different numbers of 'don't-care' lines [6,29]. Such variations mean that different units require different numbers of signals to provide an exact-match to their stored vector. This prevents Incrementing networks from reliably setting local unit thresholds simply according to global activity levels. Instead, each unit must be provided with its own individual threshold setting circuits, doubling the numbers of signals integrated by each unit, and correspondingly increasing error rates or precision requirements. Decrementing algorithms, on the other hand, cope with variations in numbers of don't-care lines automatically.

Differences between Conjunctive algorithms performing hetero-associative and auto-associative tasks.

More formal descriptions of Incrementing and Decrementing algorithms are provided in Appendix 1. In hetero-associative architectures, each input vector is paired with a 'target vector' containing a number of active units. Units may be active in several target vectors, and so receive modified links from the active lines in each associated input vector. If vectors are generated randomly, it is possible for units to receive modified links from all active lines in input vectors with which they have not been associated as a whole. However, such spurious activations can be prevented by systematic coding strategies, or they can be eliminated by selective and automatic re-coding as errors occur [27,28,33].

The higher error-rates of Incrementing algorithms at pattern-classification in imprecise circuits generalise to hetero- and auto-associative tasks. However, as these tasks require the activation of more than one unit, they provide more opportunities for errors of omission. In exact-match selection this can only increase error rates in Incrementing networks, as Decrementing networks cannot make errors of omission.

In Incrementing algorithms performing auto-associative tasks on vectors without positive noise, special problems still arise from the need to monitor the numbers of active units in the network, and to feedback these global measures as thresholds to the same units. This requires accurate matching of local and global signals, in magnitude, phase, and time constants. With input vectors containing positive noise there is an additional requirement to ramp or oscillate thresholds according to the optimal algorithms [10,11]. These ramping procedures are also error sensitive.

In Decrementing auto-associative networks, problems of synchronising and matching threshold signals do not arise for input vectors in which positive noise is absent. However, input vectors with positive noise may initially inhibit some, or all, units. There are two procedures for recall from vectors with positive noise. Firstly, global 'thresholds' may be ramped with the number of positive noise signals. Note that these Decrementing 'thresholds', representing numbers of noise signals, are much smaller than the corresponding Incrementing algorithm thresholds representing the total numbers of active units, and are correspondingly less sensitive to noise. Alternatively, global thresholds may be fixed close to $\Theta = 0$ and the accumulated signals on each unit may be allowed to decay locally, at equal rates, so that the units receiving fewest noise signals are the first to fall below the fixed global threshold, and recover to increase the inhibition of the non-pattern units. This local algorithm is as efficient as threshold ramping in Incrementing algorithms [4,5], and avoids any additional errors caused by temporal mismatches between global threshold and local link signals.

The efficient operation of the threshold modulation system in Incrementing networks relies heavily upon the assumption of equal link gains. Consequently, if the numbers of links between elements vary randomly, or links are randomly ablated, the performance of Incrementing networks is impaired much more than that of the corresponding Decrementing networks, particularly at exact-match selection tasks. Comparisons between the performances of associative network algorithms in randomly generated circuits have been presented elsewhere [3-5].

4. Performance of associative tasks by Decrementing and Correlative algorithms.

Previous sections have been concerned with Decrementing and Incrementing associative network algorithms in which unit activities and link gains can be represented by binary values, and link gains are irreversibly modified by one or more conjunctions of activity in the link elements. In this section, the operation of Correlative [12-21,36] algorithms will be described, and their performance compared with that of Conjunctive, specifically Decrementing, networks.

In their simplest forms [12–14], Correlative networks   associate bipolar (+1) vectors by adding the product of the activities of the two units forming each link to the link gain. This means that every link gain is modified by +1 by every association, and the storage of V random vector associations requires link gains to range between limits of V and –V in integer steps, with an s.d. of about √V. During pattern recall, each link transmits a signal equal to the product of its input line activity and its gains. Each unit totals these signals from all N – 1 of the other units in the network, setting its activity state to match the sign of the total.

Pattern–classification tasks.

In pattern–classification tasks, each unit is associated with only one input vector (V = 1), so that, in Correlative systems, link signals are +1, the performances of Conjunctive and Correlative networks in ideal circuits are the same. In practical implementations, Correlative networks, suffer from two problems which do not arise in Decrementing networks. The first problem is that   input vectors with unequal proportions of active and passive components require special bias circuits to be dedicated to each unit. The second problem arises from the need   to generate precisely matched signals of 1 or –1. This causes significant complications in analog implementations, where the magnitudes of oppositely signed signals are difficult to match precisely,   because of differing charge mobilities in p– and n– channel transistors [29,30,35]. These two additional sources of error mean that Correlative algorithms are more susceptible to errors in pattern–classification tasks than Decrementing algorithms. For these reasons, analog implementations of pattern– classification networks using Correlative algorithms, have been followed by implementations using Conjunctive algorithms [31–34].

Hetero– and auto–associative tasks.

Comparisons between the performances of Conjunctive and Correlative algorithms in ideal circuits are complicated by the variety of alternative Correlative algorithms [12–21,36], and differences in sensitivity to A/N, the proportion of active components in   the N–component vectors being associated, and to differences between performances in infinite networks, and in networks of practical size (e.g. where N < 10,000).

In tasks involving 'dense' randomly generated vectors where A = 0.5 N, networks using the standard Hopfield–type Correlative algorithm [14–16] can store 5–7 times more associations than the corresponding Conjunctive   network, and can recall these vectors robustly despite signal imprecision. However, with 'sparse' vectors of A = $\log_2 N$, the standard Hopfield algorithm stores only a small fraction of the patterns which can be stored by a Conjunctive algorithm [7,9,15,16]. Moreover, the optimal information storage capacity of networks using Decrementing algorithms to store sparse vectors is several times greater than that of networks using standard Hopfield algorithms to store dense vectors [5]. The performance curves of the two algorithms crossover at about A = 0.3 N.

Modified Correlative algorithms using iterative training routines [20,21,36] have been shown to improve significantly on the performance of the standard Hopfield network with dense patterns. Theoretical studies [19] suggest that it may also be possible for modified Correlative algorithms to equal the performance of Conjunctive networks with sparse patterns [7,9].

Figure 2.

Numbers of sparse vectors stored with 99%
accuracy by Decrementing and Correlative
algorithms in 1000-unit network

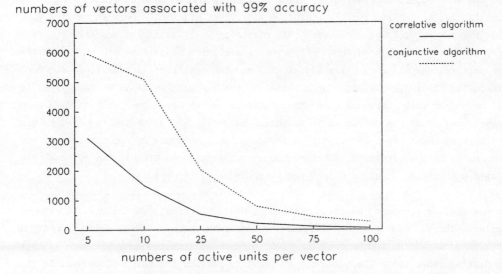

numbers of vectors associated with 99% accuracy

However, while modified Correlative algorithms using local rules [17,18] have improved on the performance of the standard Hopfield algorithm in sparse vectors, particularly in infinite networks, they remain significantly less powerful than the corresponding Conjunctive algorithms in networks of practical size for VLSI implementation. For example, Figure 2 compares the performance of Decrementing and Correlative algorithms in networks with N = 1000 units, with sparse vectors with from

5 to 100 active units per vector, and error rates at 1% of the number of active units. The performance of the Correlative algorithm is taken from Fig 1 in Buhmann et al. [17]. The performance of the Decrementing algorithm is derived in Appendix II below [5]. Figure 2 indicates that, in networks of practical size, the Decrementing algorithm can store 2-3 times as many vectors with 0.5% – 10% of units active.

Direct comparisons are complicated by variations in relative performance with network size, and by differences in error-rate computations. In Decrementing algorithms, all errors are errors of commission. However commission errors do not appear to have been included in the accuracy calculations of the Correlative network [17], and so these may be underestimates. While Conjunctive networks give best performance over a range of sparse vector activity values with $A = O[\log_2 N]$, the range of activity values over which they provide better performance than Correlative networks shrinks as network size increases, eventually becoming insignificant in infinitely large networks [17,18]. However, in networks of practical dimensions, as Figure 2 indicates, Conjunctive networks are clearly preferred for sparse vectors.

It may be possible to improve on the performance of modified Correlative algorithms with sparse vectors, by using the link matrices produced by algorithms using local rules [17,18] as a basis for the iterative algorithms which have been successful with dense vectors [20,21,36]. However, this possibility is mainly of academic interest here, because it has been shown [19] that Correlative algorithms cannot exceed the best performance of Conjunctive algorithms, and the resultant link matrices are much more costly to implement in analog hardware. For associative tasks where $V \gg 1$, it will be clear that signals in Correlative networks must be specified with even greater precision than those in Incrementing networks, and units are required to sum many more of these precisely specified signals, taking account of opposite signs, and to compare them accurately with threshold. Furthermore, as Correlative networks are 'scaled-up' in size, their accuracy requirements increase rapidly.

For example, a standard Hopfield network of N = 1000 units, storing V = 1000 random bipolar vectors, requires link signals to represent values with an s.d. of about 33 and a range of ±1000, and each of the 1000 units must linearly sum 999 such signals and distinguish between the outcomes, in order to provide optimal performance. If the informational cost of storing the link matrix values is taken into account in evaluating performance of alternative algorithms [22], then this Correlative algorithm 'costs' about 5-10 times as many bits per link as the equivalent Decrementing algorithm using binary 1-bit links. In analog implementations these precision costs translate into greater silicon area [1], and further limitations on the size of Correlative networks which can be implemented on a single analog chip.

In ideal circuits, Decrementing algorithms provide better performance than modified and standard Correlative algorithms over a strategic range of sparse vectors. However, they are also computationally much simpler than the Correlative algorithms, and so provide more cost-effective and miniaturised implementations with imprecise analog circuits. This means that, at vector densities at which the performance of Correlative algorithms is theoretically superior, the Decrementing algorithm may still offer more cost-effective practical implementations. It may also be profitable to re-code dense vectors into sparse formats more suitable for Decrementing algorithms.

As vector densities are progressively increased, the greater cost-effectiveness of Decrementing algorithms can be traded off against the relatively improving capacity of Correlative networks until performance curves crossover. The exact location of this crossover point depends on the network size, and a range of technical factors, including coding strategies, and limits or costs of manufacturing precision, and is a topic for further research.

6. Computational Strategies and Algorithms for Associative networks.

The relative simplicity of Decrementing algorithms, compared with their Incrementing equivalents, and the differing performances of Conjunctive and Correlative algorithms in sparse and dense pattern sets, deserve some further words of explanation.

Both Incrementing and Decrementing Conjunctive algorithms operate by utilising the existence of 'hard' NAND constraints in their stored pattern sets. Unmodified links indicate the presence of NAND constraints in the training set, i.e. activity conjunctions which do not occur. The pattern recall process is simplified by the fact that signals from unmodified links reliably indicate the presence of NAND constraints, or mismatches between current and stored vectors, and the number of such unmodified link signals varies with the Hamming distance between the two vectors. There is a qualitative difference between the information conveyed by unmodified link signals in Decrementing networks, and modified link signals in Incrementing networks. These differences are related to the asymmetries between the processes of falsifying and confirming universal statements in logic and science [3-5].

A unit receiving one or more unmodified link signals in a Decrementing network is reliably informed that it has never been associated with the current input vector. However, by themselves, one or more signals from modified links in an Incrementing network do not indicate whether or not there has been an association between the unit and the whole input vector, and the unit requires further information on total global and local activity levels, thresholds etc. to resolve this ambiguity. Thus

Decrementing algorithms represent hard NAND constraints directly in mismatch signals from unmodified links, while Incrementing algorithms represent them indirectly, in match signals from modified links, and must then use threshold logics on the summed match signals to extract mismatch information [3-5].

The explanation for the differences in performance between Conjunctive and Correlative algorithms is that they adopt different computational strategies, exploiting quite different kinds of information in their pattern sets. Conjunctive algorithms dedicate their links to represent the presence of hard and reliable NAND constraints in the pattern set by their unmodified state, and they dedicate their units to change state on detection of one or more unmodified link signals. Correlative algorithms utilise 'soft' probabilistic constraints, using individual links to represent correlations in activity between their elements, and requiring units to integrate information from all links in parallel to arrive at their appropriate activity state.

Conjunctive algorithms work best with sparse pattern sets because these are rich in 'hard' NAND constraints, and each association modifies such small numbers of links that many random vectors can be stored without saturating the link matrix. The remaining unmodified links convey individually reliable and 'hard', NAND mismatch signals. Pattern-classification tasks involve a form of sparse encoding, as only one unit is active in each recalled vector, and so these are suitable for Conjunctive algorithms. The disadvantages of Conjunctive algorithms emerge when storing sets of dense random patterns, which may contain very few NAND constraints, and where each association modifies a large proportion of the links, rapidly saturating the link matrix.

Correlative algorithms are at a disadvantage with random sparse patterns, where most units are in the passive state, because they are unable to fully exploit NAND constraints to identify these passive units. An individual signal representing a 'hard' NAND constraint does not receive special priority, and can easily be submerged by the sum of signals from the other N - 1 links representing 'soft' correlations. In dense patterns, hard NAND constraints may be few or absent, and the only constraints available are probabilistic, so that the only feasible approach is to use Correlative algorithms. The relative unreliability of individual probabilistic signals is compensated for by integrating them all in parallel for each unit.

These arguments suggest that Conjunctive and Correlative associative algorithms are not simply different points on a continuum of algorithms adapted for sparse and dense vectors respectively [14,15,19]. They represent quite different computational strategies, and exploit quite different information sources in their pattern sets. In sparse pattern sets, where NAND constraints are plentiful, the Conjunctive algorithms are significantly simpler and more efficient. In dense pattern sets, where NAND constraints are absent, only the sophisticated Correlative algorithms can be effective.

Coding strategies.

While the performance of alternative associative algorithms with randomly generated pattern sets provides useful indications of overall capacities, evaluations must allow for the fact that, in practical applications, pattern sets are unlikely to be completely random, and performance may be significantly improved by systematically re-coding raw data according to the associative algorithm being used. For example, naturally sparse data may be re-coded into dense format to suit Correlative algorithms [20,21].

Similarly, it may often be prudent to re-code vectors from dense to sparse formats, in order to take advantage of the simplicity and economy of Decrementing algorithms. The goal of the sparsification process is to maximise the numbers of NAND constraints in the pattern set. Sparsely encoded patterns are common in both biological and artificial information processing systems. Relaxation labelling tasks always involve sparse vectors in which label units are divided into sets, with only one unit in each set active in each vector [24-26], and the feature-detecting modules of mammalian sensory systems appear to generate sparse representations [23]. A wealth of sparsification procedures have been proposed [27,28,33], including adaptive algorithms for selective higher order encoding which increase the proportions of NAND constraints where required. These codes can markedly improve the performance of Conjunctive algorithms, enabling more associations to be stored than with random patterns, and eliminating risk of retrieval errors with perfect input vectors.

Error rates in all networks can be reduced by encoding strategies which increase Hamming distances between input vectors. However, these strategies can also be counter-productive, incurring costs in the form of increased numbers of input lines, and larger link matrices. For example, in Conjunctive networks, antagonistic, or complementary encoding doubles the number of units and/or input lines, and increases the proportion of active lines to half. This increases the number of mismatch signals produced by negative noise, in both Incrementing and Decrementing networks. However, it also increases average threshold, together with the link signal accuracy required for signal-noise discrimination in Incrementing networks. Generally, encoding strategies which improve performance in Incrementing or Correlative networks in sparse vectors, also benefit Decrementing networks correspondingly, and so do not reduce the differentials between them.

5. Manufacturing precision and applications in analog implementations

The advent of digital RAM chips with 16 megabytes of memory raises the possibility that analogue associative networks with 10**4 fully interconnected units will be implemented in the near future. PRONNs (Programmable Read Only Neural Networks) [1] with realistic device densities of $2*10^9$ links/cm and 0.1u lines have already been implemented. However, there is a size/accuracy/ cost trade-off in analog VLSI component manufacture. The dimensions of smaller and more densely packed components are more difficult to specify precisely, and their signals are more vulnerable to background noise. In the simplest PRONNs, with tightly-packed one-bit weights, technological noise depends on gate output voltage spread, and resistor conductance spread, and has been estimated to be 15% of the maximum signal [34]. Hopfield [1] has described accuracies of 10% in resistive connections as typical, and defined 'great accuracy' as 'large compared with 3 - 4 bits'. Difficulties in manufacturing transistors with standardised gains have also been described by Carver Mead [35].

As Figure 1 indicates, individual link fluctuation rates of $\sigma = 0.15$ produce 2% error rates in discriminations between L = 2 and 3 signals, and 16% error rates where L = 11 and 12 signals. As L, the total number of signals received by a unit, is increased, the absolute value of the S.D. of the noise also increases (S.D. = $\sigma\sqrt{L}$). This places a premium on recall algorithms which minimise L, the numbers of signals which units must sum and compare in order to determine their activity states. As described above, associative network algorithms differ significantly in the L levels which they require. In Correlative algorithms L = [O]N, the number of units in the network. In Incrementing algorithms L = A, the number of active elements in the input vector. In Decrementing algorithms L $\geq$ 1, for exact-match tasks, and L = M', the smallest number of mismatch signals, for best-fit tasks. These L-values can be ranked in order of magnitude N $>$ A $>$ M' $>$ 1, with the Decrementing algorithms requiring summation and comparison of the smallest numbers of signals, and correspondingly higher tolerance of link signal imprecision.

An impression of the speed with which best-fit selection can be performed by analog implementations of Decrementing algorithms can be obtained from the performance of a chip designed to implement an auto-associative Correlative algorithm [29]. In this case, asymmetries between p- and n-transistors caused inhibitory signals to be 4-6 times stronger than excitatory signals, units receiving the fewest inhibitory (mismatch) signals recovered first and prevented recovery in other units by the auto-associative winner-take-all circuits. This process took from 50-600 nanoseconds.

In Correlative networks performing hetero- and auto-associative tasks, the problems stemming from the need to match signals of opposite sign can be solved by using separate pairs of output and input lines to emit and receive positive and negative signals [12,30]. However, this quadruples the number of links required in analog implementations, compared with the equivalent software models, and this must be taken into account when comparing the silicon areas required to implement Correlative and Conjunctive models. Moreover, as Correlative links are required to generate a wide range of values, they are likely to be significantly larger than the equivalent binary link in a Conjunctive network. For pattern-classification tasks, the p- and n-channel matching problem has been avoided by the use of Conjunctive algorithms [31-34]. Some of the practical advantages of Decrementing Conjunctive algorithms using signals to represent mismatches rather than matches have recently been confirmed for pattern-classification tasks [33].

Computationally, the Decrementing algorithm is simpler and more direct, and most tolerant of link signal fluctuations due to inherent manufacturing imprecision. It exploits the exponential properties of analog VLSI device physics most effectively, both in the relative amplification of small numbers of mismatch signals, and in the natural tendency of all signals to decay exponentially. Negative, decrementing processes, such as disabling links, and inhibiting unit activity, are simpler to implement because they are self-terminating, and can be mediated by 'avalanche' processes which do not require guidance to a particular end-point value, or subsequent maintenance at that value. Negative thresholds are also a safety feature in auto-associative circuits, preventing any opportunity for growth of noise activity by positive feedback.

Applications of Analogue Neural Network Associative Memories.

Analog implementations of neural networks provide significantly greater densities of units and links, and, in integrated circuits, they are suited to very fast pattern-classification tasks where input vectors are compared to a set of stored vectors in parallel, and the best match, or all close matches are computed. This involves parallel processing of large quantities of low precision data such as that encountered in image processing. Analogue associative memories have numerous applications in areas involving pattern recognition, vector quantization, associative search and image compression, e.g. for FAX transmission. Analog neural network chips [31] have already been shown to perform computationally-intensive tasks such as line-thinning and feature extraction with recognition accuracies comparable to the best obtained by conventional techniques, with a potential 20-fold speed increase over software equivalents. [31,36]

The Decrementing network algorithm provides the most cost-effective analog implementation for pattern-classification tasks, and for auto- and hetero-associative tasks with sparse pattern sets. This is strategically important because, while most practical applications so far have been for pattern–classification, a significant number of problems are most readily encoded in a sparse format. For example, in many constraint satisfaction problems in image and language analysis, only small proportions of label units can be active simultaneously. [24–26].

7. Summary and Conclusion.

Analog VLSI implementations of neural networks offer significantly greater component densities enabling the construction of powerful networks on a single chip, with significant benefits in speed and cost. As the number of links in a network increases with the square of the number of units, larger, more powerful networks must operate with smaller, more compact links. However, the smaller the area that individual links occupy, the greater the random fluctuations in signal values between links due to manufacturing imprecision. Consequently, the preferred neural network algorithms for analog VLSI implementation are those which are disrupted least by circuit imprecision, and in which error–rates increase least rapidly as networks are 'scaled–up' in size.

This report has reviewed the operation of associative networks using Conjunctive and Correlative algorithms at associative tasks with positive and negative noise, and in sparse and dense vectors. The basic conclusion is that the Decrementing algorithm is least complex, computationally, requiring units to perform fewer, simpler arithmetical operations, and so is least susceptible to the errors due to manufacturing imprecision. In Correlative and Incrementing algorithms, the numbers and precision of arithmetical operations required scale up with increasing network size, as does vulnerability to imprecision.

In exact–match selection tasks, Decrementing networks only need to distinguish between some signals and no signals, like a Boolean NOR–gate. Imprecise signals cannot make errors of omission, and so the algorithm can be perfected to prevent errors of commission. In best–fit selection, the units with stored vectors closest to the input vector receive fewest inhibitory mismatch signals. These signals can be allowed to decay locally and independently, so that units with the least inhibition recover soonest to inhibit units receiving more mismatch signals. The advantage of distinguishing and selecting units receiving fewest signals is that this leaves room for the signals themselves to be both larger and less precisely specified, and therefore more robust against background noise and manufacturing imprecision. Alternatively, signals may remain small, and the use of a smaller range of signals can be converted into lower capacitances or power consumption.

The direct and local nature of Decrementing algorithms leads to optimal simplicity and explicit exploitation of VLSI device physics. Incrementing networks require global threshold calculations which are not only an additional error source but, in auto-associative architectures, must be precisely matched in phase and time-constants. Similarly, units in Correlative networks must compare the sums of their positive and negative inputs, and equal signals of opposite sign should match and cancel out precisely.

In ideal circuits, the Incrementing and Decrementing 'Conjunctive' associative network algorithms provide equivalent performances for all tasks and network architectures. In ideal circuits performing pattern-classification tasks, Conjunctive and Correlative algorithms also provide exactly equivalent performances. In ideal circuits performing hetero- and auto-associative tasks, Conjunctive algorithms can store more random sparse vectors while Correlative algorithms can store more random dense vectors.

In imprecise circuits, the Decrementing algorithm provides significantly better performance for a given accuracy and manufacturing cost than the corresponding Incrementing algorithm. The Decrementing algorithm is also so much simpler to implement than the corresponding Correlative algorithm that it remains more cost-effective in vector densities outside the range in which it is theoretically more powerful, and it becomes profitable to re-code dense vectors into sparse formats suitable for processing in Decrementing networks. The greater efficiency of Decrementing networks in sparse vectors is of strategic value, because of the number of constraint satisfaction and relaxation labelling tasks which can be expressed most directly in sparse vectors. The high efficiency of decrementing networks at pattern-classification tasks is also important, because these are the tasks with the most immediate beneficial applications.

ACKNOWLEDGEMENTS

I would like to thank A. Treves, R. Payne and G. Littlewort for helpful discussions. This work was supported by MoD RARDE.

REFERENCES

[1] Hopfield J. (1989) Neural networks: algorithms and microhardware. IJCNN International Joint Conference on Neural Networks, Washington D.C. June 18-22. Tutorial 7.

[2 Dobson V.G. (1975) Pattern learning and the control of behaviour in all-inhibitory neural network hierarchies. Perception 4, 35-50.

[3] Dobson V.G. (1989) Hetero-associative networks using link-enabling vs. link-disabling local modification rules. In I. Aleksander (Ed.) Neural Computing Architectures (MIT:Cambridge) 279-304.

[4] Dobson V.G. (1987) Superior accuracy of Decrementing over Incrementing associative networks in initially random connectivities. Journal of Intelligent Systems 1, 43-78.

[5] Dobson, V.G. (1989) Decrementing associative networks. In L. Personnaz & G. Dreyfus (Eds.) Neural Networks from Models to Applications. (IDSET: Paris) 316-325.

[6] Dobson V.G. (In press) Visual search with Decrementing Network Competitive Windows. In D. Brogan (Ed.) First International Conference on Visual Search. (Taylor & Francis: London).

[7] Willshaw D.J., Buneman O.P. & Longuet-Higgins H.C. (1969) Non-holographic associative memory. Nature (Lond.) 222, 960-962.

[8] Marr D. (1969) A theory of cerebellar function. Journal of Physiology, 202, 437-470.

[9] Palm G. (1982) Neural Assemblies. (Springer-Verlag: Berlin).

[10] Gardner-Medwin A.R. (1976) The recall of events through learning associations through their parts. Proc. Roy. Soc. (Lond.) B, 194, 375-402.

[11] Lansner A. & Ekeberg O. (1985) Reliability and speed of recall in an associative network. IEEE PAMI-7, 4, 490-499.

[12] Kohonen T. (1988) Self-organization and Associative Memory (Springer-Verlag: Berlin).

[13] Amari S. (1988) Statistical neurodynamics of various versions of correlative

associative memory. IEEE 2nd. International Conference on Neural Networks; San Diego, July 24-27 I, 607-613.

[14] Hopfield J.J. (1982) Neural networks and physical systems with emergent computational abilities. Proceedings National Academy Science U.S.A., 79, 2554-2558.

[15] Forshaw M.R.B. (1987) Pattern storage and associative memory in quasi-neural networks. WOPPLOT '86: Parallel Processing - Lecture Notes in Computer Science 253 (Springer-Verlag: Berlin) 184-197.

[16] Amit D.J. (1987) Properties of models of simple neural networks, Heidelberg Colloquium on Glassy Dynamics. (Springer-Verlag: Berlin).

[17] Buhmann J., Ritter H., and Schulten   K. (1989) On sparsely-coded associative memories. In L. Personnaz & G. Dreyfus (Eds.) Neural Networks from Models to Applications. (IDSET: Paris) 360-371.

[18] Tsodyks M. V. & Feigel'man M.V. (1988) The enhanced storage capacity in neural networks. Europhysics Letters, 6 (2), 101-105.

[19] Gardner E. (1987) Maximum storage capacity in neural networks. Europhysics Letters, 4 (4) 481-485.

[20] Gardner E., Stroud N. & Wallace D.J. (1989) Training with noise and the storage of correlated patterns in a neural network model. J.Phys. A:Math.Gen. 22, 2019.

[21] Gardner E., Stroud N. & Wallace D.J. (1987) Training with noise: Applications to word and text storage. In R. Eckmiller & C. v.d.Malsberg,   Neural Computers (Springer-Verlag: Berlin).

[22] Lazzaro J., Ryckebusch M.A. & Mead C.A. (1989) Winner-take-all networks of O(N) complexity. In D. S. Touretsky, Neural Information Processing systems (Morgan Kaufman: Palo Alto Ca.) 703-711.

[22] Minsky M.L. & Papert S.A. (1988) Perceptrons. (MIT: Cambridge).

[23] Barlow H.B. (1972) Single units and perception: a neuron doctrine for perceptual psychology. Perception 1, 371-394.

[24] Winston P.H. (1984) Artificial Intelligence (Addison-Wesley: Reading Ma.).

[25] Hopfield J.J. & Tank D.W. (1985) Neural computation of decisions in optimisation problems. Biological Cybernetics 52, 141–152.

[26] Kitler J. & Illingworth J. (1985) Relaxation labelling algorithms – a review. Image and Vision Computing 3:4 206–216.

[27] Ekeberg O. & Lansner A. (1989) Automatic generation of representations in a probabilistic artificial neural network. In L. Personnaz & G. Dreyfus Neural Networks from Models to Applications. (IDSET: Paris) 178–186.

[28] Venta O. & Kohonen T.A. (1988) Content–addressing software method for emulation of neural networks. IEEE 2nd. International Conference on Neural Networks, San Diego, July 24–27, I 191–198.

[29] Graf H.P., Jackel L.D. & Hubbard W.E. (1988) VLSI implementation of a neural network model. Computer, 21, 3, 41–49.

[30] Verleynsen M., Sirletti B. & Jespers P. (1989) A new CMOS architecture for neural networks. In J.G. Delgado–Frias & W.R. Moore (Eds.) VLSI for Artificial Intelligence (KAP: Norwell Ma.) 209–218.

[31] Jackel L.D., Graf H.P., Hubbard W., Denker J.S., Henderson D. & Guyon I. (1988) An application of neural chips: handwritten digit recognition. IEEE 2nd. International Conference on Neural Networks, San Diego, July 24–27, II 107–115.

[32] Ruckert U. & Goser K. (1989) VLSI design of associative networks. In J.G. Delgado–Frias & W.R. Moore (Eds.), VLSI for Artificial Intelligence (KAP: Norwell.Ma.) 227–235.

[33] Baum E.B., Moody J. & Wilczek F. (1988) Internal representations for associative memory. Biol. Cybern. 59, 217–228.

[34] Dechambost E. & Sonrier M. (1989) Implementing classifications on one bit connection analog programmable neural network. International Joint Conference on Neural Networks, Washington D.C. June 18–22 IJCNN II 579.

[35] Mead C.A. (1989) Analog VLSI and Neural systems (Addison Wesley: Reading Ma.).

[36] Tarrasenko L., Seifert B.G., Tombs J.N., Reynolds J.H. & Murray A.F. (1989) Neural network architectures for associative memory. First IEE International Conference on Artificial Neural Networks. London U.K. 16–18.

Appendix I: Conjunctive Associative Network Algorithms.

A. Decrementing Algorithms.

1. Storage Rules:

$$W_{ij} = \{ \begin{array}{l} 1 \text{ if } \Sigma_V X_i^V * X_j^V = 0 \quad (X_j^V \text{ is jth unit of Vth vector}) \\ \quad\quad\quad\quad\quad\quad\quad\quad (X_x = 1/0) \\ 0 \text{ otherwise} \quad\quad\quad (W_{ij} \text{ is link from } X_i \text{ to } X_j) \end{array}$$

2. Recall Rules:

(i) exact-match selection (with no positive noise)

$$X_j(t+1) = \{ \begin{array}{l} 1 \text{ if } \Sigma_i W_{ij} * X_i(t) = 0 \\ 0 \text{ otherwise} \end{array}$$

(ii) best-fit selection (using local signal decay)

$$S_j(t) = D*S_j(t-1) + \Sigma_i W_{ij} * X_i(t)$$

where ...

$$X_j(t+1) = \{ \begin{array}{l} 1 \text{ if } S_j(t) < \Theta \\ 0 \text{ if } S_j(t) >= \Theta \end{array}$$

S_j ... Accumulated Signal on X_j
Θ ... Threshold
D ... Decay factor (1>D>0

In analogue circuits, the key pattern is presented for one time interval, while units are disconnected from ground, then L links connected in parallel to supply voltage Vs would produce the potential S_L at unit Xj:

$$S_L = Vs (1 - e^{-\gamma L})$$

where γ is the time constant for Xj

The threshold Θ, is set between ground and the potential produced by one link, so the time t, required by a unit with potential S_L to fall below Θ can be derived from:

$$\Theta = S_L e^{-ct}$$
$$t = 1/c \ln(S_L/\Theta)$$

where $e^{-c} = D$
and $1 \gg \Theta > 0$

B. Incrementing Algorithms.

1. Storage Rules:

$$W_{ij} = \{ \begin{array}{l} 0 \text{ if } \Sigma_V X_i^V * X_j^V = 0 \\ 1 \text{ otherwise} \end{array}$$

2. Recall Rules: exact-match selection (with no positive noise)

$$X_j(t+1) = \{ \begin{array}{l} 1 \text{ if } \Sigma_i W_{ij} * X_i(t) = \Theta \\ 0 \text{ otherwise} \end{array}$$

(if Θ = A, the number of active input units)

Appendix II: Associative storage of random vectors.

In Decrementing networks associating random vectors, the error rate per vector retrieved 'E' is given by:

$$E = p^a*(N - A) \qquad\qquad ... (1)$$

where N is the number of units per array, A is the number of active units per stored target vector, a is the number of active units in the key vector, and p is the proportion of modified links.

p is given by:
$$p = 1 - ((N^2 - A^2)/N^2)^V \qquad\qquad ... (2)$$

where V is the number of stored vector associations.

Information storage capacity for sparse and dense vectors

The information storage capacity of Decrementing networks is optimal for sparse random pattern sets with Log(2)N active units per vector. For N = 1000 this is approximately 10 active units per vector. I, the information required to specify one binary vector is given by:

$$I = Log_2 \frac{N!}{(N-A)! * A!} \text{ bits} \qquad\qquad ... (3)$$

Note that sparse random vectors are disproportionately rich in information per active cell relative to dense vectors. While the information required to specify a dense vector with A = N/2 active units cannot exceed N bits (2 bits per active unit), sparse vectors where A = 1 contain Log(2)N bits per active unit, and vectors where A = 0.11N convey about N/2 bits. In a 1000 unit network, a sparse 10 unit vector represents about 80 bits of information, while a dense 500 unit vector represents about a 1000 bits. The sparse vector conveys about 4 times more information per active unit than the dense vector.

Decrementing algorithms are also efficient in proportions of links modified during associations of sparse vectors. In a 1000-unit network, associations of random sparse vectors with A = 10 active units involve only 0.01% of all links, and 7000 such vectors can be associated before the error rate exceeds 1 per pattern. However, dense vectors, with A = 500 active units, involve 25% of all links, and only about 20 vectors can be stored before the same error rate is reached. Consequently, information storage capacity with dense vectors is only about 4% of that with sparse vectors in this network.

Connectionist models of utterance production

Hans — Jürgen Eikmeyer

1. Linguistic modelling

From a methodological point of view linguistics is not a uniform discipline. At present, at least two methodological proposals are rivalling, the classical paradigm of *structural linguistics* and the field of *procedural linguistics*.

Structural linguistics in its version called generative grammar tries to specify a finite mechanism which enumerates the set of grammatical sentences of a natural language. This set is supposed to be infinite. In order to enumerate the set of sentences structural linguistics has established several levels of linguistic structure shown in (1).

(1)

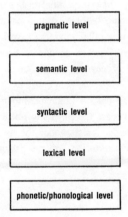

Each of these levels specifies its own structural units and states well-formedness conditions for these units. These units are sounds, allophones or phonemes on the phonetic/phonological level, syllables and words or morphemes and lexemes on the lexical level[1], clause, phrases or sentences on the syntactic level, propositions on the semantic level and contexts or situations on the pragmatic level.

Structural linguistics sets up relations among the units of the levels, relating complexes of lower level units to simple units on higher levels. Thus, the levels are arranged in the hierarchical structure shown in (1).

Structural linguistics explicitly excludes processes of language processing and language production as legitimate objects of linguistic research. This dogma is attacked by procedural linguistics (cf. Derwing 1980). Since linguistic units are the objects manipulated by linguistic processes one cannot be success-fully studied without the other. Thus, procedural linguistics does not call into question structural levels and units, it only claims to integrate them into a description of linguistic processes.

Structural linguistics nevertheless frequently uses metaphors suggesting dynamics such as as e.g. a grammar generating a sentence or an automaton accepting a sentence. This dynamic parlance, how-ever, does not refer to production and interpretation processes but to abstract processes needed for enumerating sets of linguistic expressions. If several such processes are studied together, they are exclusively supplied with a sequential schedule induced by the hierarchy given in (1).

I adhere to the procedural linguistic paradigm. In the following I will completely ignore processes of language interpretation and focus on a specific model for one aspect of utterance production, namely some types of errors occurring in spoken language. Spoken language is also constantly ignored by structural linguistics. It mainly focusses upon written language, although spoken language is prior to written language onto- and phylogenetically. Moreover, spoken language differs considerably from the written variant, especially on the lexical and syntactic levels.

Taking processes of language production and interpretation into account leads to regarding linguistics as a special branch of cognitive psychology. As an immediate consequence, the restriction to sequential process schedules has to be given up. K. Bock (1982) has collected convincing psychological evidence, that the processes of utterance production have a parallel organization. Consequently, the models used for the description of these processes should be of a parallel nature in order to be psychologically real or adequate. This does not mean, that a parallel model is psychologically adequate *per se*. It simply means, that a sequential model never will have a chance to become psychologically correct, while a parallel one might correctly describe the processual organization.

This may serve as a justification for a paper on linguistic modelling given on a workshop on paral-lelism. In the following sections I will discuss the linguistic phenomena to be described and give some examples for them. Then a parallel, connectionist model for these phenomena will be presented and evaluated. The paper concludes with the exposition and discussion of a tool for computer simulations of the model.

2. Repairs, errors and slips of the tongue

Spoken language has quite a few syntactic patterns not to be found in written language. One of these patterns – *repairs* – have been of special interest for both ethnomethodological and psycholinguistic research and we[2] chose them as our topic of research as well. We try to model them from the per-spective of their production, with regard to their automatic analysis (parsing, cf. Lisken & Rieser 1987), and within a dynamic theory of grammar (cf. Kindt & Laubenstein 1989). Levelt (1983, p.45) gives the following English example for a repair, which he constructed according to the laws empirically observed:

221

(2)   Go from left again to uh ... , from pink again to blue

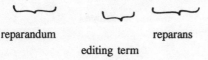

        reparandum   reparans
                        editing term

The main constructional parts of repairs are the *reparandum*, i.e. that part of the utterance which has to be repaired or corrected, the *reparans*, i.e. that part which introduces the correction, and the *editing terms* in front of the reparans. A wide range of variation is allowed at the position of editing terms. Editing terms are optional, i.e. they may be completely missing. They can as well be realized by a hesitation ("uh"), a pause ("...") or a sequence of single admissible editing terms as above. Editing terms may be ordinary lexical units ("no") or complete syntactic units ("excuse me", "no, this is wrong").

For my present argument only a more or less trivial observation on repairs is important, namely the point that errors are integral parts of repairs. Without an error it would simply not be necessary to repair anything at all. In the example above the speaker had erroneously used the word "left" instead of the word "pink" and, consequently, he had to correct himself. In case one wants to model the production of repairs, it is thus necessary to provide a model for the production of errors. This will constrain the class of feasible models as will be argued below.

Sometimes speakers notice an error while they are still planning an utterance and before they have uttered something erroneously. In such cases − so-called *covert repairs* −, the reparandum is missing. However, these cases are no contradictions to the observation stated above. Even in covert repairs there is still an error involved, although it cannot be traced in the utterance.

There is a second reason for focussing on the description of errors. This reason has to do with the methodology of procedural linguistics. One of the main problems of procedural linguistics is that processes of language production and interpretation are not directly observable. These processes go on within the heads of speakers and as long as they go on smoothly, one has no great chances to gather any evidence about them. However, in case language processes run into problems one gets a chance for their empirical investigations. Errors so to speak create windows concerning processes of language production, even if these windows are open for just very short time spans. There is a slogan frequently used for linguistic research under this perspective, the slogan "speech errors as linguistic evidence". It is the title of a volume containing several case studies applying the methodological trick advocated here (Fromkin 1973). Errors can be observed on all linguistic levels. If they become manifest on low level units they are called *slips of the tongue*.

It is possible to classify slips of the tongue according to the effects they have on two linguistic units involved. In an *exchange* error the two units occur in the opposite order as usual, i.e. they interchange their respective positions. An *anticipation* error is characterized by a unit which has to occur later but which appears earlier as well at its normal position. *Perseveration* errors are opposites to anticipations since the first structural unit occurs once more later on. In *substitution* errors, finally, some unit overwrites the correct one. The following list (cf. Schade 1987) gives examples for these types of slips of the tongue with respect to the levels of syllable parts (3a), syllables (3b) and words (3c).

(3)

type of slip	example	source

(3a)

exchange	let's have a look	P.Molzberger
	→ het's lave a look	at WOPPLOT 89
anticipation	nach Innsbruck von München	Meringer (1895)
	→ nach Minnsbruck ...	
perseveration	blaues Bauklötzchen	Forschergruppe
	→ blaues Blauklötzchen	Kohärenz (1987)
substitution	department	Dell (1986)
	→ jepartment	

(3b)

exchange	self-destruct induction	Dell (1986)
	→ self-instruct deduction	
anticipation	my car towed	Dell (1986)
	→ my tow towed	
perseveration	explain .. rule insertion	Dell (1986)
	→ explain ... rule exsertion	
substitution	my teeth have improved	Fromkin (1973)
	→ my teeth have reproved	

(3c)

exchange	die Venus von Milo	Meringer (1895)
	→ die Milo von Venus	
substitution	but Nixon in China	Fromkin (1973)
	is an historical event	
	→ ... an hysterical event	

Starting with repairs we have now narrowed down the range of phenomena to the production of slips of the tongue. They serve as prototypical cases for the study of the production of errors.

3. Modelling slips of the tongue

In order to simulate the production of errors one has to select a suitable model. Neural network or connectionist models are known to be able to describe the production of errors. A correct model of errors should not make the assumption, that in case of errors processes of a special kind are invoked. Errors will emerge from a somewhat "crazy" parameter setting. The machinery for bringing them about should be the same as the one producing ordinary results. According to Dell (1986) connectionist models have exactly this property. With respect to symbol processing models, however, it is largely unknown how they can describe errors, although these models are the mainstream models in computational linguistics and AI at present. Deciding upon a connectionist model for the production of errors and slips of the tongue has the additional advantage, that these models are parallel *per se*. No extra measures have to be taken to parallelize them. The decision on connectionist models is not a dogmatic one. I am ready to change it as soon as symbol processing models offer mechanisms for modelling errors.

A connectionist model consists of a set of nodes or processors which are characterized by their activation, usually some numeric value. The single nodes are linked to other nodes by excitatory or inhibitory connections having different strengths or weights. These connections are used to spread any node's activation to the nodes linked to it. Inhibition or activation received from other nodes determines the new activation a node will receive in the next state of the model. Calculations of the new activation are done according to some suitably chosen activation function. Simulation in a connectionist model usually proceeds in cycles. In each cycle all nodes are allowed to transfer their activation to connected nodes and thus all nodes mutually change their activation value.

In order to design a connectionist model for the simulation of slips of the tongue one has to set up the nodes and their topology, i.e. the connections. With respect to slips of the tongue on the level of syllables and syllable parts one should have a look at structuralist linguistics' proposals for the structure of words. According to one of these proposal words consist of one or more syllables and each syllable consists of an optional *onset* (usually a consonant cluster), a *nucleus* (a vowel) and an optional *coda* (again a consonant cluster)[3]. The structure of the word "Wildbad" e.g. can be described by the following tree diagram:

(4)

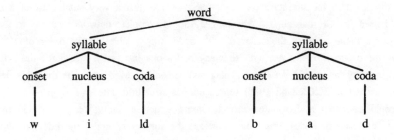

This diagram depicts what linguistis call *syntagmatic* relations. Such relations describe the admissible sequences of structural units for making up higher order units. A second kind of linguistically relevant relations, the *paradigmatic* relations, describe the possible choices at certain structural positions. With respect to syllable structure the paradigmatic relations allow all vowels (and diphthongs) to occur at the nucleus position, and all of a suitable set of consonant clusters to occur at the onset and coda position, respectively. Connectionistically the syntagmatic and paradigmatic relations of word and syllable structure are accounted for in a network topology as proposed by Schade (1987), shown in (5).

(5)

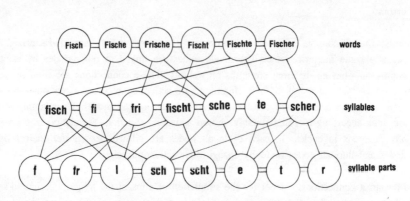

The paradigmatic relations holding are reflected by *lateral inhibition* among all nodes representing units of the same class. This is indicated by the double lines connecting the respective nodes in (5). The syntagmatic relations are accounted for by mutual activation of the relevant nodes. The word "Frische" e.g. activates its two syllables "fri" and "sche", the syllable "sche" activates its onset "sch" and its coda "e", and vice versa. Mutual activation is depicted by a single line in (5). It has the effect of *positive feedback* discussed in the literature (cf. Dell 1985). (5) is just a very small part of a huge network one would need for German slips of the tongue. For a complete network one has to face a number of nodes for syllable parts in the order of 10^2, nodes for syllables in the order of 10^3 and nodes for words in the order of 10^4. Even with so many nodes one has just a model for the syllabic structure of words, nothing being said yet about syntax and semantics. Moreover, the model is only a local model (cf. Rumelhart & McClelland 1986) representing a structural unit by a single node in the network. If we could construct a distributed model, where structural units are represented by an activation pattern on a cluster of nodes, this would increase the number of nodes by orders of magnitude.

With a network topology as above one can simulate errors from two different sources. A possible source for errors is the speaker's focus on a second, different topic while he produces a word. Focussing on a different topic can be simulated by externally activating or priming nodes related to the other topic. This is an indirect way of modelling the *causae* of errors which is applied primarily because the models proposed here are far from complete. Ultimately one would like to model the source for an error as well. However, errors becoming manifest on the syllable level e.g. may have their sources in higher levels as semantics or pragmatics, which are presently not integrated into the model.

The second source for errors is the production rate as one can observe in connection with tongue twisters. The faster one utters a tongue twister the more likely are slips of the tongue to occur. The production rate can be simulated by manipulating the number of cycles allowed for the activation to spread. The more cycles the simulation takes, the slower is the production rate. We made relevant experiments with networks like (5). According to Schade (1987) here are some of the results:

(6)
experiment 1:

    no. of cycles resulting utterance

        5 fi-schers---fritz---
        4 fi-schers---fitz---

experiment 2:

    no. of cycles resulting utterance

        6 fi-schers---fritz---fischt---fri-sche---fi-sche---
        5 fi-schers---fritz---fischt---fi-sche---fi-sche---

In (6) a single hyphen represents the end of a syllable, three consecutive hyphens represent the end of a word. The slips have been marked by underlining. The errors simulated by these experiments correspond to errors observable in reality.

In presenting these results I skipped one important point which I want to address now. Despite of the parallel nature of the connectionist model it has to produce a *sequence* of syllable parts which make up the resulting utterance. This is simply because the organizational pattern of natural language occurrences is inherently sequential: the sounds of a word are pronounced one after the other, words of an utterance appear in a sequence and texts are made out of consecutive sentences (or paragraphs or chapters). In principle, sequentialization can be brought about by an extra machinery controlling the network's processing externally as e.g. in (Dell 1986). The resulting model is hybrid, adding some symbol processing capabilities to the network model. A second, more conservative, solution will try to model sequentialization within the expressive limits of connectionist models, i.e. in terms of nodes, connections and activation functions. Such a solution has been proposed in Schade (1987). It uses special nodes, the *control nodes*, which encode the structure of syllables starting with an onset and followed by a nucleus and a coda. The control nodes are organized in a chain, the onset node activating the nucleus node, the nucleus node activating the coda node and the coda node activating the onset. The activation function chosen guarantees, that exactly one of the three control nodes has a high activation whereas the others are not activated at all. The control nodes are moreover linked by positive feedback to the syllable-part nodes belonging to the respective class, the coda node e.g. activating and being activated by all vowels and diphtongs. As the experiments shown in (6) demonstrate, this model works satisfactorily.

This conservative solution has another advantage. It correctly describes the so-called *initialness effect* (cf. Dell 1986): slips of the tongue mostly occur in the initial sounds of words and syllables. A connectionist model of slips of the tongue should be able to simulate this effect in case it aims at psychological adequacy. The model discussed here does so, since it uses only positive feedback and no extra machinery disturbes this effect. When activation is injected into a word node in networks like (5), it takes some time before this activation has spread into the corresponding nodes for syllables and for syllable parts. It takes still more time before the activation arriving there is fed back into the nodes for words and syllables. In other words, positive feedback relations take some time to come into action. Hence, the first syllable of a word and its onset have a relatively low activation at the beginning of a word's production. In fact, their activation may be lower than the activation of a rivalling syllable or syllable-part node. This is demonstrated in (7), where the activation of the mono—syllabic word node "fritz" and its corresponding syllable node "fritz_s" are shown at different points of time within the production of the word "fritz":

(7)
experiment 3:

node	activation	point of time
fritz	0.955793	when producing the onset 'fr'
fritz_s	0.589064	
fritz	0.993685	when producing the nucleus 'i'
fritz_s	0.938664	
fritz	0.999290	when producing the coda 'tz'
fritz_s	0.988458	

These results thus demonstrate, that connectionist models with positive feedback automatically account for the initialness effect. This concludes the discussion of structural and procedural aspects of a connectionist models for the production of slips of the tongue. The next topic will be the simulation of such models with computers.

4. Computer simulation of connectionist models

Simulations with connectionist models can hardly be carried out with paper and pencil. Computers have proven to be suitable tools for such an enterprise. It is possible to implement a connectionist simulation in which programming language so ever. However, in the last decade it has been demonstrated by researchers of software engineering, that it has some advantages to build special tools for special purposes. The technique applied is to establish a level of conceptual abstraction on top of some programming language. Some conceptually motivated abstract representation is specified together with a processor model manipulating this representation. The new conceptual level allows the user of such a tool to ideally think in terms of the conceptual field and not in terms of a universal purpose programming language.

Applying this technique to the field of simulation of network models leads to the idea to define a *network representation language* (NRL) and a *network compiler* (NC). The task of the NRL is to allow for the formulation of network descriptions. These descriptions will be translated by the NC into a program formulated in a target programming language (cf. fig. 8). This target language can also be used for the realization of the NC itself. The user of such a pair of a NRL and a NC does not have to worry about the target language, he need not even have any idea about the programming language used. He is just confronted with the NRL rich enough to offer the expressive means needed for the conceptual field in question, i.e. network simulation.

(8)

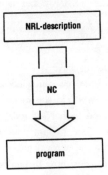

A NRL is more than a program library added to some programming language. With respect to network simulations there are some other proposals as e.g. the famous Rochester simulator (cf. Goddard 1987). This simulator consists of a library of C-programs which its user can arrange in order to make up his own simulation. Using e.g. the Rochester simulator requires the user to have some command over the programming language C. It demands for the user to know details about the implementation of the conceptual entities. Thus, a user of such a program library is constantly switching between thinking in terms of the concepts he is interested in and in terms of the programming language. Although there is nothing wrong with such a solution of a program library in principle, I suggest that a solution with a NRL and a NC is easier to handle.

Our system *BiKonnex* (cf. Eikmeyer & Polzin, to appear) of a NRL and a NC has been realized as a prototype in the programming language CheOPS (cf. Eikmeyer 1987), an object-oriented extension of PROLOG. The times needed for actual simulations with this prototype are desparately long. This was expected before we started with realizing BiKonnex. Nevertheless, there was a good reason to start with CheOPS/PROLOG, the reason being simply that the ultimate expressive power of the NRL needed is difficult to estimate. The reason lies in the rapid development of different kinds of networks for different purposes at present. Consequently, the conceptual field is not stable and uniform and one has to experiment with different proposals. Realizing and changing a network compiler in PROLOG, a language with good capabilities with respect to symbol processing, can be done much faster than in other languages. Moreover, we had lots of experience with the simulation of connectionist models in CheOPS directly which we could profit from. However, this is not where we want to stop. Currently, a first version of the NC in classical C is implemented, for the future we think of an implementation in parallel C for transputer systems. We can then use a parallel machine for the simulation of a parallel model and we can even try to construct a parallel parser for the NRL. Moreover, transputers are relatively cheap and they are easily expandable and adaptable to the computing power needed.

Some remarks should be made with respect to the NRL used in BiKonnex. It has two components, a declarative one for the representation of the network topology and a procedural one for the specification of the simulation process to be performed. Concerning the declarative aspect we added the concept of *clusters* to network models. A cluster is a set of nodes which share some property, be it a structural one or a procedural one. Structural properties shared by nodes may e.g. concern the topology. The more regular the topology of a network is, the easier it is to describe in BiKonnex (see the example below). Procedural properties shared by nodes may e.g. arise from the fact, that they are submitted to the same procedure during a simulation.

In order to demonstrate the BiKonnex NRL let us have a look at an example, the famous little red riding hood (cf.Jones & Hoskins 1987). When little red riding hood is walking through the forest, she may meet the woodcutter, her grandma or the wolf. If she meets her grandma, she has to approach her, kiss her on the cheeks and offer some food to her. If today's little red riding hood meets the woodcutter, she will approach him, offer some food to him and start flirting. As in the old days of fairy tales, little red riding hood will still run away, scream and look for the woodcutter if she meets the wolf. However, little red riding hood has to learn how to recognize the three other characters by a combination of features. The wolf has big ears, big eyes and big teeth. Grandmas usually have big eyes, they are kind and wrinkled. The woodcutter has big ears, he is kind as well and handsome. Little red riding hood's learning to behave apprpriately can be simulated by a learning network using backpropagation. It has six input nodes – the features mentioned above – and seven output nodes – the seven actions little red riding hood can perform. Adding a hidden layer of three nodes results in a network topology as shown in (9). The BiKonnex representation of this topology is stated in (10).

(9)

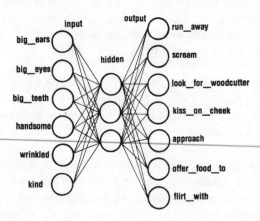

(10)
cluster input of big_ears, big_eyes, big_teeth,
          kind, wrinkled, handsome;
     activating hidden by [0, 0.2];
end

cluster hidden of 3 nodes;
     activating output by [0, 0.2];
end

```
cluster output of run__away, scream, look__for__woodcutter,
                    kiss__on__cheek, approach,
                    offer__food__to, flirt__with;
end
```

As shown in (10), clusters can either be defined by specifying the number of nodes the cluster consists of (cf. the hidden layer), or by giving a vector of names for the nodes belonging to the cluster (cf. the input and output clusters). Cluster names can be used to refer to all its nodes at once as e.g. in 'activating output by [0, 0.2]'. Including this line in the hidden layer definition will establish excitatory connections from all nodes of the hidden cluster to all nodes in the output cluster (21 connections). The interval specified in this line sets the weight of every such connection to a pseudo random number from the given interval. Intervals can be used at all places in specifications where numbers are expected. Reference can always be established by vectors of cluster and/or node names. Value assignments may also use vector notation, numeric operations on vectors are defined componentwise.

This leads us to the procedural aspects expressible in the NRL. The main control structure is a loop with a specified number of cycles. Functions, especially activation functions, and procedures can be defined and called. As a prerequisite, variables are supported. The programming environment of BiKonnex offers commands for the compilation of NRL expressions into CheOPS code. Compiled code can be included by other simulation programs. Thus, BiKonnex can be regarded as a special purpose programming language.

5. Conclusions

This paper has demonstrated, that linguistics phenomena as errors, especially slips of the tongue, can be successfully modelled with parallel connectionist models. Errors are of central interest for procedural linguistics, that branch of linguistics which aims at a description of language processes. Parallel models are well suited for this aim since their processual organization is exactly the organization found in real cognitive processes of utterance production. At present, only connectionist models can describe error production to a certain extent. Symbol processing models still have to invent suitable techniques. Since connectionist modelling requires computer simulation it is useful to provide special tools as the BiKonnex system.

6. Notes

1. This view of the lexical level is linguistically naive since it neglects the level of morphology. This level is excluded here because the model for slips of the tongue proposed adopts the view of a syllabic structure of words which is rivalling with the view of a morphemic structure.
2. This pronoun refers to the people working together in a research project named "Gesprochenes Deutsch" which is sponsored by the Deutsche Forschungsgemeinschaft as one of the projects of the Forschergruppe "Kohärenz" established at Bielefeld University, Faculty of Linguistics and Literary Sciences.
3. In fact, this proposal is one of the older ones. Nevertheless, it seems to be suited for a model of slips of the tongue.

7. References

Bock, K., 1982: Towards a cognitive Psychology of Syntax : Information Processing Contributions to Sentence Formulation. In: *PsychologicalReview* 89, 1-47.

Dell, G.S., 1985: Positive Feedback in Hierarchical Connectionist Models: Applications to Language Production. In: *Cognitive Science,*9, 3-23.

Dell, G.S., 1986: A Spreading-Activation Theory of Retrieval in Sentence Production. In: *Psychological Review*, 9, 3-23.

Derwing, B., 1980: Against autonomous linguistics. In: T.A. Perry (ed.), *Evidence and Argumentation in Linguistics*. Berlin:de Gruyter.

Eikmeyer, H.−J., 1987: *CheOPS: An Object-oriented Programming Environment in C-Prolog.* Kolibri 4, Universität Bielefeld.

Eikmeyer, H.−J. & T. Polzın, (to appear): *Simulating networks with BiKonnex.* Kolibri, Universität Bielefeld.

Fromkin, V.A., 1973: *Speech Errors as Linguistic Evidence.* The Hague:Mouton.

Forschergruppe Kohärenz (Hrsg.), 1987: 'n Gebilde oder was' Daten zum Diskurs über Modellwelten. Kolibri 2, Universität Bielefeld.

Goddard, N., 1987: The Rochester Connectionist Simulator. User Manual. University of Rochester.

Jones, W.P. & J. Hoskins, 1987: Back−Propagation. A generalized delta learning rule. In: *Byte*, October 1987, p. 155-162.

Kindt, W. & U. Laubenstein, 1989: *Reparaturen und Koordinationsstrukturen.* Kolibri 20, Universität Bielefeld.

Levelt, W.J.M., 1983: Monitoring and self−repair in speech. In: *Cognition*, 14, 41−104.

Lisken, S. & H. Rieser, 1987: Ein Parser in C für simulierte Reparaturen. MS, Universität Bielefeld.

Meringer, R & K. Mayer, 1895: *Versprechen und Verlesen: Eine psychologisch-linguistische Studie.* Stuttgart:Göschen.

Rumelhart, D. & J. McClelland, 1986: *Parallel Distributed Processing : Explorations in the Microstructure of Cognition.* Vol. 1 : Foundations. Cambridge : MIT.

Schade, U., 1987: 'Fischers Fritz fischt fische Fische' − Konnektionistische Modelle der Satzproduktion. Kolibri 6, Universität Bielefeld.

SELF-ORGANIZATION OF INFORMATIONAL SYSTEMS

G.J. Dalenoort

Institute for Experimental Psychology
University of Groningen, The Netherlands

1. Introduction, epistemological issues

Our intellectual tradition seems to provide clear criteria for the distinction between natural and artificial systems, but it is not so easy to give formal criteria, that do not depend on preconceived or antropomorphic notions. Moreover, there exist numerous examples of mixed nature, where a human constructor uses natural processes to reach his goal: the farmer sowing seeds, the brewer making beer, the physician healing a patient, the physical chemist growing a crystal. In each of these cases the constructor creates the favourable conditions under which natural processes will - hopefully - follow a desired course in time.

Our intuitive, antropomorphic notions depend strongly on the nature of the systems concerned, and on the choice of the boundaries between them and their environment. For example, if we lock up a programmer and his computer in a room, providing all the natural needs for both both systems, such as fresh air, food, liquid, and electric current, we may consider the contents of the room as a self-organizing system, hopefully producing a computer program. The computer reacts on the commands typed in by the programmer, and the programmer reacts on the information appearing on the screen of the monitor.

A more sophisticated observer would observe that the goal of this self-organizing system resides in the head of the human subsystem. But if we lock up two persons in a room to solve a problem, the situation is more symmetrical. If neither of the subsystems has a clear representation of a goal to be reached, and nevertheless some high-level order emerges, we have what we may rightly call self-organization. Other examples are biological evolution, morphogenesis (the prime example of self-organization), and crystal growth.

For such systems where no single subsystem has a representation of a goal, it is impossible to indicate a clear constructor, or he must be found in some abstract entity. An analogy may perhaps be found in Fermat's principle for the path a lightbeam will "take" from a point A to a point B. Assuming it will reach B at all, the path is the line that takes the least time. Many of our optical devices are constructed such that there are many equivalent pathes, with equal shortest times, like the optical lens. How can we in a formal sense tell the difference with a human ? Perhaps the goals we pursue are also nothing else than a special configuration of brain variables, and our experience of choosing a goal just is an illusion.

Another important factor for considering a system as self-organizing, is the level of aggregation, or level of description. At the level of atoms or biological cells we generally do not consider a biological system as goal-directed. If we do so for the level of biological cells, the context usually is that of biological evolution; most biologists would consider biological cells as causal systems. In any case biological language shows a strange mixture of goal-directed and causal terminology, and the former terms are usually used under the assumption that a causal terminology can be found in the framework of biological evolution.

We can conclude that there is quite some freedom in viewing systems as self-organizatory or as being artificial, depending on the choices of the boundaries and of the level of aggregation, but also that we have rather strong preferences determined by cultural conditions. Sometimes a new look may be obtained on a system by considering it together with part of its environment as a self-organizing system.

There exists a variety of types of self-organization. The growth of a crystal, or the establishment of a surface of a liquid in a container (in a gravitational field) may be considered as processes of self-organization with little possibility of macroscopic variation. Of course, the combinatorial number of possibilities for the final positions of the particles is enormous. But since the particles are of a very limited number of types, and since the type of interactions relevant for the process of self-organization is very simple, there is very little effect at the macroscopic level.

Processes of self-organization may be considered from the point of view of a balance of competition and cooperation. For various types of systems, there is a large variety in the respective degrees of either. In the case of biological self-organization, that variation depends on the choice of the level of description: the biochemical level, the cellular level, the system as a whole, the level of populations, corresponding to the types of building-stones and their types of interaction. In social groups, the members compete in trying to obtain advantages from their group membership, but they cooperate in securing the existence of the group.

For systems largely consisting of the same types of subsystems, it also is a natural step (cognitively) to view the control as collective. There is no single unit or place where a plan for the course of action is stored. The action at the global level is the integrated result of all the individual actions. Also, in most cases, many individual actions may be going on at the same time, in parallel. Therefore, we can characterize processes of self-organization as collective and parallel. This applies to all kinds of self-organization.

The nature of the current digital computer makes that any informational process running on it must be strictly foreseen, implicitly or explicitly. Programs on such computers run according to an algorithm, of which every step must be well-defined. This applies to the level of the program. Of course, unforeseen effects may be obtained by variation of input. If the program is to use the input, the format of the latter must comply with the program, but at a higher level things may occur which we cannot - could not - predict. This is especially clear for processes of so-called deterministic chaos. Such processes are described by mathematical equations that depend on some parameters in very special ways: a small change in those parameters causes large effects in the global behaviour of the system. In contrast, in neural networks we are interested in the opposite effect: how the system can maintain global behaviour and characteristics despite large variations in the input.

Thus, information-processing on - current - digital computers, interpreted at the level of the program by itself, is a constructed process, and by no means self-organizational. As argued before, the construction of a computer program by a programmer may be seen as self-organization of the system of programmer and computer.

In Section 3 we shall further elaborate on this aspect, we now must first discuss information-processing from a somewhat more formal point of view.

Processes of self-organization give rise to a number of phenomena that are of epistemological interest. They are part of the set of basic issues that any epistemology must handle: the relation between goal-directed behavior and our urge to interpret processes in terms of cause and effect, the related issue of scientific reduction, and the question what is 'explanation'. To some extent the third issue encompasses the first two in so far these are part of the explanation. I have dealt with these issues in more detail recently, (Dalenoort, 1989a) and have not to add to that discussion at present. Therefore, I restrict myself to a few recapitulating remarks.

We may characterize self-organization as types of processes in which global order emerges out of a relatively large number of interactions that are localized or short in duration. The global order often involves goal-directed behaviour, and the interactions mostly refer to a lower level of aggregation and are seen as causal. They may also be goal-directed at a more individual level, as for example in the case of self-organization in social groups. Explanation of this relation can be given as follows. Of all types of local interactions, or inter-related sets of local interactions, those will have survival value that are most successful in the competition for resources, and of which the products have a positive influence on the occurrence of the interactions. As a consequence this process can be seen as aimed at survival, which concept is inherently goal-directed. With respect to scientific reduction we can follow this line of thought further: the products of the local interactions have new properties, that cannot be observed in the elements taking part in the interactions. In attempting to explain the relation between both, we enter into what traditionally has been considered as a problem of scientific reduction. I have argued before (Dalenoort 1987a) that it is very misleading to see this cognitive process as one-sided: that it must be directed at explaining the high-level properties in terms of the low-level ones. From the cognitive point of view the explanation usually goes in both directions: from the system as a whole to the parts, and the other way round. The process of explanation is cyclic and understanding thereby increases in degree. That many sciences are able afterwards to formal-ize the relation between different levels in one direction, and thereby add to further insight, should not lead to the conclusion that understanding goes only from the lower level to the higher. One may compare the situation with the observation of the process of a house being built, and the experience of living in it when it is ready.

2. Information-processing as classification

Any system can be described in a variety of ways, depending on the point of view chosen. An observer, and describer, of a system must choose which aspects of it are to be central in his or her account. This to large extent also determines the representation language and the representations that are most suitable for the purpose (Dalenoort 1990). What is 'suitable' also depends on our cognitive apparatus, on the way we understand systems and processes. In the Introduction we already have give the example of the constructor of a computer program, where we may either consider the programmer as one system, interacting with another system, the computer with its software. We may also consider the program to be constructed as a separate system, with the computer and its software as the tools for construction. This perhaps is the most widespread view that programmers have of themselves at work.

We now turn to two other types of representations, physical descriptions of the world, and informational ones. Our aim is to characterize the latter one in more elementary terms, those of classification.

In physical representations, for every system a complete set of variables must be specified, or given. The completeness is determined by the question whether the mathematical equations, containing these variables and no other ones, determine the possible values in a unique and consistent manner. Any precise model must be expressible in terms of such an equation, and if the condition is not satisfied, the model is incomplete, or remains at a qualitative level.

For a mechanical system the variables in question are mass, position, velocity, and external forces, for all subsystems, or for the system as a whole. Of course, if the subsystems, or the system, may not be considered as point masses, also their shape and the points where external forces apply, must be given. For electrical systems also electric currents and magnetic fields are part of the complete set. In principle the values of all variables are given with infinite precision. The lack of such precision, due to measurement errors in experiments, are handled by means of statistics, which is not a part of the idealized physical description. (In quantummechanical representations the statistics is part of the description.)

In the real world lack of precision is an inherent part of life. If we say that someone is in Amsterdam, the position is in many cases sufficiently determined, for example if the speaker is in Paris. However, if I ask someone where another person is, and both of us already know that he is in Amsterdam, and we know of each other that we know it, the answer "He is in Amsterdam" would at least be surprising.

What is called 'lack of precision' in some types of descriptions of the world, corresponds to a statement in physics like: "the closed system is an idealization". For short times we may consider some systems as closed systems, closed for energy, closed for matter. The influence of small deviations from being closed, is expressed by saying that there are perturbations. Perturbations are also necessary, they can serve to initiate changes. The best example is that of mutations in the process of biological evolution. But also in society and culture many important ideas and developmemts were initially considered as perturbations. Of course not all *avant-garde* movements turn out to leave noteworthy traces. In general only a tiny proportion survives, or is remembered.

What mechanisms have organisms developed to cope with imprecision and perturbations ? There is a whole range of physical mechanisms available that may be employed: threshold effects, chemical selectivity, different degrees of adhesion between different sorts of material, membranes that are selective with respect to the types of molecules they let through, phase transitions, hysteresis, and others. All of such mechanisms and phenomena may be used by organisms to make *classifications*. Classification is the basic operation of information-processing. In describing a system, we may choose how we want to describe it. We may choose the mechanisms and phenomena just mentioned, but we may also attempt to describe it in terms of the *effects* of those mechanisms and phenomena. Then the more abstract concept of classification is essential. Making, changing, equating, elaborating, and using classsifications can then be seen *as* information-processing.

The simplest classification has two classes of equal weights, that is, there is a probability of a half for an element to be in either class. This classification corresponds to an amount of information of one bit, the unit in which amount of information is generally measured.

The advantage of the concept of classification over that of information, is that it is more basic, and that a well-developed mathematical formalism for classification is available. (For further thought on this point see earlier articles, Dalenoort 1979, 1985.)

An important research task with respect to a given system is to find out whether the physical phenomena at a given level of aggregation can serve as a basis for a description in terms of classifications. The transitions between states belonging to different classes thus constitute information-processing.

As a consequence of these arguments, we can state that it would be wrong to say of any system that it *is* an information-processing system, if this is meant in an exclusive sense. At the same time it is, for example, a physical system, and it may also be a biological system. It would be better to say that the behaviour of the system is such that we can construct *informational representations* of it in a natural manner. It is up to us, the observers, whether we want to use the processes at lower levels of aggregation as concepts for a description in terms of information, or classifications. Of course one can always choose "artificial" criteria to enforce a description in terms of classification, but then the explanatory value may be low or nil.

Another important research task is to explain in physical, chemical, biological terms, or whatever, how the processes on which we base the classifications emerge from the processes at a lower level of aggregation. At the lower level we may have atoms, or molecules, or cells, or individuals. They only interact in a mutual fashion. Under certain conditions their interactions have a collective effect which becomes manifest as some ordering at the higher level. We then have self-organization. We can thus state that information-processing, as described at one level, is a consequence of self-organization at a lower level.

If a system is designed in a strict manner, it is unlikely to come up with new behaviour. The system has been preprogrammed, and the types of behaviour are fixed. It is unlikely that we can catch in the detailed design the "interesting" cases, for example, a machine that would be creative in a manner comparable to that of man. The course of interactions at the lower level must not be so constrained that the "interesting" possibilities are ruled out. The task is rather to constrain the processes such that useful self-organization can take place. My surmise is that neural networks offer that possibility, and that production systems, to mention a dominant kind of systems, do not. Also, for many 'connectionist systems', as neural-network type systems have recently been named, it is doubtful whether they can be considered as self-organizational in spite of the fact that they are to serve as models for brain processes.

There do not exist sharp boundaries between the different sorts of systems. But the basic type of operation in constructed systems does not seem to leave sufficient room for unforeseen order to emerge. On the other hand, it is a big problem to discover the constraints that will leave sufficient "freedom" for interesting phenomena to emerge.

It is unlikely that sophisticated forms of learning are realizable by designing systems at only one level of aggregation. Current practice of programming in artificial intelligence tries to follow that road. Learning has for a long time been a subject to be avoided. The approach advocated here is to include the underlying levels of aggregation, and to discover the constraints that must be imposed for the desired type of learning to take place. We can then take nature as an example, e.g. the development of perception and cognition in children. It is unlikely that we shall be able to design systems or programs that can do the same as humans can, for example in perception and natural language. Given the current state of the art, it does not seem likely either that we can make programs that learn, if the processes of learning are designed only at our level of operation.

Let us now consider different levels of aggregation, the nature of the processes taking place, and see how well they can be described in terms of classification on one hand, or as physical, chemical, or biological process on the other hand. One aim of our survey also is to discover whether alternative natural ways of information-processing would be possible.

The levels of atoms and molecules

On their mutual encountering, atoms may become attached to each other, to form molecules. Although one might state that an atom makes classifications with respect to other atoms as far as the point of interaction is concerned, we usually do not consider this as classification. The main reason probably is that the physical and chemical framework provide sufficient possibilities for description. This state of affairs is recognizable in more general: if a deterministic framework for description is available, as physics and chemistry provide, there is no need or urge, to describe the processes in terms of information. This is of course in agreement with Shannon's information theory, wich qualifies information in terms of probabilities and averages, for an observer. Even though Shannon's measures have a formal resemblance to the physical expressions for entropy, the absence of an observer in the physical domains itself, makes the equation of entropy to lack of information questionable.

The same holds true for interactions between molecules. Although complex molecules can show interesting chains of reactions, these in themselves do

238

not seem to have been described in terms of classification or information. Only when they occur within the framework of biological cells, the molecular processes obtain a "deeper meaning". Descriptions in terms of reaction rates, stability of loops due to catalytic interactions, interaction strengths between loops, constitute the usual terminology for their processes.

The level of biological cells

The biological cell probably is the lowest element in the hierarchy of natural systems (up to human societies) for which informational representations have been important. This is certified by such terminology as "genetic code", "the amount of information in the DNA molecule", etc. The process of the synthesis according to the code of the DNA molecule can also well be described in terms of classification. Molecular structures are being attached to each other in very specific fashions, leading to structures with specific tasks for the survival of the cell.

The level of biological tissue

This is perhaps the level least well understood of the natural hierarchy. It involves all biological processes of morphogenesis and self-regulation of organisms. If biological tissue becomes damaged, it often has a great capacity to restore its structure, or to create a modified structure good enough for survival. Examples are superficial wounds of our skin, and the tail of the lizard. Embryology and descriptive anatomy are the biological approaches at one level of description. Understanding in the epistemological sense, as mentioned at the end of Section 1, involves relating these levels to the underlying levels of biological cells, and of the "genetic code". the genetic code manifests itself in the specific structure of tissue, the shape and function of organs, and of organisms as a whole. Even the social behaviour of individuals in a population is to a large extent determined by their genetic code, viz. ants and bees. It is this relation between the higher levels and the biochemical level that is least well understood, as stated before. Informational representations play an important role in the descriptions of the processes involved. A description purely in terms of physical, chemical, and biological interactions woud be very opaque, right impossible for our cognitive abilities.

The levels of organisms and populations

At these levels it is hardly necessary to give many examples to show the importance of informational representations. Especially for higher organisms, with nervous systems, they are indispensible. Nervous systems have developed in biological evolution to enable organisms to do information-processing themselves, to make classifications. It is not only used by us to describe the processes involved. The natural activity of nervous systems is making, handling, and comparing classifications. (See further Dalenoort, 1987b) The interactions between individuals, the processes in societies of animals and humans are to a large extent described in terms of exchange of information. It would be a hard job to describe your reading this text in purely physical terms indeed.

3. Cognitive Systems: Learning as Self-Organization of Informational Systems

In Section 1 we have compared the notion of self-organization to that of construction. In Section 2 we have characterized informational systems, or rather informational representations, in terms of the process of making and using classifications. In this section we attempt to integrate the two aspects, in answering the question how informational systems change their criteria for classification, or can make new classifications. If this proceeds through processes of self-organization it may be called learning, otherwise the system must be classified as being programmed. Of course mixtures between the two types of changes are possible.

In a set of objects, or systems, in which global order emerges through self-organization, there is no explicit plan or map of that order. Also, there is no agent directing the actions of the elements, at most there may occur facilitation or inhibition of specific interactions between specific elements, independent of evaluation of the process as a whole. As a consequence the process of self-organization is purely dependent on random encounters between elements. An encounter may have an effect, or not, and this may be influenced by the presence of a catalyst, or by other factors, such as temperature, an appropriate solvent, by actions like shaking, etc.

If in a system physical, chemical, or biological processes occur that we can interpret as classifications, we can describe the system in terms of an informational representation. We may also say that the system processes information. With the concept of classification we have available a simple and clear criterion for deciding if a system processes information. We want to have a similarly simple and clear criterion for deciding if a system may be called *cognitive*. As far as I know, no such criterion is in general use, although proposals have been made. For example, K. Lorenz (1973) equated life and cognition in stating 'Life itself is a cognitive process', and Maturana and Varela (1980) have created the general formalism of autopoiesis to also capture the concept of cognition (see also Fürlinger, 1986, p. 306).

It seems useful to reserve the term *cognitive* for the more complex types of information-processing. Within the present approach we can find a criterion in the combination of the concepts of self-organization and classification. If a system changes its criteria for classification, or starts to make new types of classification in a process of self-organization, we can say that it *learns*. Learning seems generally accepted to be a phenomenon typical for cognition. Therefore, we may well make it the defining property of cognition, with as a basis the interpretation of information-processing in terms of classification as given.

It is possible to refine the classifications under discussion, by posing additional criteria on the types of changes. For example, one may require that the changes add to the survival value of the system, or contribute to specific goals of the system. This latter case may apply to human behaviour, where not all goal-directed or purposeful behaviour is directly related to survival. It is also useful for a more pragmatic approach, for example if one takes as a goal the finding of food, which generally is accepted to be a characteristic goal of biological systems. If the system adapts its behaviour to the circumstances in its food-finding behaviour, we can say that it changes its classifications, so that it is cognitive.

In psychology, the term *learning* has also been used in the restricted form of 'conditioning'. It is a matter of taste whether one wants to consider this type of learning as change of classification. It may well be, for example in the case where an animal, or a child, learns to recognize burning nettles, or hot surfaces, i.e. learns to classify them as plants and objects to be avoided.

An important criterion is whether the classification learned is always the same, for all animals, in all environments, or whether it is dependent on the environment. For example, we certainly want to consider the process of a bird learning to find its way back to its nest as cognitive, but the comparable process in bees not. The reason is that in the bee the process is likely to be automatic, it orients with respect to the sun, and it does not learn the features of its environment, or only in a very restricted sense. Its capacity to orient seems rather to be the result of an evolutionary process, not of individual learning. Of course this only holds until the time we discover that the bee does learn.

From this example we can see that the process of changing, or creating new classifications, must occur in individuals, if we want to call it 'learning'. If it occurs in a population, we should not call it learning, although the population can said to make clasifications. But this distinction between individuals and populations with respect to classification, leads to new problems: why could we not consider an organism as a population of biological cells ? Why could we not say that the classification an individual acquires, is the acquisition of that population of cells ? This certainly is a way of describing neural processes in brains, which can be represented as to operate by competition and cooperation, concepts which very well apply to populations (cf. Edelman 1987).

If we want to call a process of change and acquisition of classifications 'learning', and if we want to use a representation in terms of the concept of 'population', we must insist on the population forming an organic whole. The elements of such a whole are not just more or less arbitrary collections of elements, they are functionally interrelated with respect to survival. This is much less the case with, say, a population of deer or ants. In the latter, an arbitrary number, provided it is not too large, may be eliminated, because in principle each member fulfills the same (type of) function in the whole. We may also say that the organism as collection of cells constitutes one level of stability in the hierarchy of aggregation, and the population another one (cf. Dalenoort 1989b).

Apart from such subtle problems for natural systems, we have obtained a clear criterion for distinguishing between systems that learn by self-organization and those that acquire knowledge by being programmed. This insight may also help us to find the way to create artificial systems that can learn in an autonomous fashion, i.e. by self-organization.

Let us now consider the method that during biological evolution has led to the emergence of learning, in cognitive systems. It is perhaps possible to imagine more or less homogeneous chemical systems to which the above considerations and criteria for classifications and changes in classifications apply, but nature apparently has followed another road in accomplishing cognition. The road nature has followed, is that of creation of cells, in each of which to large extent the chemical processes can take place independently of what happens in the environment of the cell. The membrane of the cell is selective, and can allow in specific chemicals, and let other ones out. This method has proven so successful, that it is difficult to think of other methods for reaching comparable results. In digital computers alternative methods for information-processing have shown their feasibility, but these have not come into existence by self-organization. Moreover, there are good reasons to doubt whether we will be able by traditional methods of computer-programming to obtain systems that can do information-processing of a kind and quality comparable to that performed by humans.

Brains, or neural networks, are excellent instruments for classification. They can also change and acquire classifications, or equivalently, they can *learn*. This is the property *par excellence* of brains. Within experimental cognitive psychology the topic of making and learning classifications has been the subject of elaborate experimentation, e.g. in concept learning. It also has been studied in young children, see e.g. K. Nelson, 1977. (For a discussion of this aspect in relation to development of concepts, see Dalenoort, 1987b.)

We come to the conclusion that until now cognition is only possible in organisms with nervous systems. As soon as we shall be able to make computers also self-organizing, we add a new dimension to the class of cognitive systems, albeit that it will only have been able to emerge from the platform of natural cognition.

Acknowledgement

This paper was presented at WOPPLOT-89, (Workshop on Parallel Processing, Logic, Organization, and Technology), held at Wildbad-Kreuth, Bavaria, W. Germany, 25-31 July 1989. The final version was completed during the participation of the author in the Project "Mind and Brain" at the Centre for Interdiscipolinary Research, of the University of Bielefeld, W. Germany, during the academic year 1989-1990. The services of the Centre are gratefully acknowledged.

References

Dalenoort, G.J. (1979), Deterministic versus teleological explanation in general systems theory, *6th Intern Congress of Logic, Methodology, and Philosophy of Science*, Hannover, August 1979, Abstracts in Section 10.

Dalenoort, G.J. (1985), Representations of cognitive systems, *Cognitive Systems* 1, 3-16.

Dalenoort, G.J. (1987a), Is physics reductionistic ?, in: Wm.J. Baker e.a. (Eds.), *Current Issues in Theoretical Psychology*, North-Holland Publ., Amsterdam.

Dalenoort, G.J. (1987b), Development of concepts, *Cognitive Systems* 2, 123-140.

Dalenoort, G.J. (1989a), The paradigm of self-organization: studies of autonomous systems, in G.J. Dalenoort (Ed.), The paradigm of self-organization, Gordon and Breach, London.

Edelman, G.M. (1987), *Neural Darwinism*, Basic Books, New York.

Fürlinger, A. (1986), On concepts of "cognition" in biology, *Cognitive Systems* 1, 305-322.

Lorenz, K. (1973), *Die Rückseite des Spiegels*, Piper, Munich.

Maturana, H. and Varela, F.J. (1980), *Autopoiesis and Cognition, the Realization of the Living*, Reidel, Dordrecht.

Nelson, K. (1973), Some criteria for the cognitive primacy of categorization and its functional basis, in: P.C. Wason & P.N. Johnson- Laird (Eds.), *Thinking*, Cambridge Univ. Press, p 223-238.

Address: G.J. Dalenoort, Inst. for Experimental Psychology University of Groningen, P.O. Box 72, 9700 AB Groningen, The Netherlands.

Distributed Semantic Representation of Word Meanings.

Burghard B. Rieger

Department of Computational Linguistics

FBII: LDV/CL – UNIVERSITÄT TRIER

Postfach 3825, D-5500 Trier

Germany

1 Why should it be aimed at? or the semiotic problem.

Although our understanding of the bunch of complex intellectual activities subsumed under the notion of *cognition* is still very limited, particularly in how knowledge is acquired from texts and what processes are responsible for it, recent achievements in wordsemantics, conceptual structuring, and knowledge representation within the intersection of cognitive psychology, artificial intelligence and computational linguistics appear to offer promising results. Their seminal combination is likely to gain momentum in the future in a wide range of disciplines and applications concerned with *natural language understanding* by machine, opening up new vistas to overcome the traditional duality of mind and matter in models of meaning.

1.1   The dualism of the rationalistic tradition of thought—as exemplified in its notions of some independent (objective) reality and the (subjective) conception of it—has been, and still appears to be, the common ground and widely accepted frame for modelling the semantics of natural language. According to this view, the meaning of a language term is conceived as something static which is somehow related to (and partly derivable from) certain other entities, called signs, the term is composed of. As a sign and its meaning is to be related by some function, called interpretation, language *terms*, composed of *signs*, and related *meanings* are understood to form structured sets of entities which at the same time belong to the (objective) reality and its (subjectively) interpretable representation of it.

According to this conception, these very sets of related entities (or parts thereof) are believed to picture reality and its recognized structuredness. Therefore, some knowledge of it has to be presupposed and accessible in order to let signs and their meanings be identified as part of language terms whose interpretations will have to be derived. Hence, *understanding* of language expressions can basically be identified with a process of matching some input strings with supposedly predefined configurations of word meaning and/or world structure whose representations have to be available to the understanding system (natural or artificial) as provided by its particular (though limited) *knowledge*. The so-called *cognitive paradigm*[1] of advanced structural and particularly procedural linguistics can easily be traced back to stem from this fundamental duality, according to which natural language understanding can be modelled as the knowledge-based processing of information.

1.2 Subscribing to this notion of understanding, however, tends to be tantamount to accepting certain presuppositions of theoretical linguistics (and particularly some of its model-theoretical semantics) which will have to be overcome. They may be exemplified by the representational means developed and used so far in cognive psychology (*CP*), artificial intelligence (*AI*), and computational linguistics (*CL*).

These approaches employ mostly graph-theoretical formats of trees and nets (*Fig.* 1). Their nodes/vertices and the arcs/edges between them are meant to depict entities of variant ontological status, like e.g. objects, properties, relations, processes, meanings, etc., or classes thereof, like e.g. concepts, types, variables, slots, etc., to form larger representational structures, like e.g. frames, scenes, scripts, etc. which are to be specified by the kind of labels attached (and/or functions related) to them.

Depending on the theoretical framework provided by the respective disciplines'

---

[1]The proper domain of study is the structure of the knowledge possessed by an individual who uses a language. In adopting this mentalist individual-oriented stance, the *cognitive paradigm* sets itself off both from approaches that deal with the text itself as the central focus and those that consider social interaction to be primary. This knowledge can be understood as formal rules concerning structures of symbols. In hypothesizing that the relevant aspects of language knowledge can be characterized in formal structures, the *cognitive paradigm* is in disagreement with views such as *phenomenology* which argue that there is an ultimate limitation in the power of formalization and that the most important aspects of language lie outside its limits. (Winograd 1983, pp.20–21)

Figure 1: Graph-theoretical formats for *categorial*-type representations of knowledge and meaning.

epistomological basis or performative goals, these formats converge in being essentially *symbolic* representations of a *categorial*-type[2] format. Roughly, this can
be characterized as consisting of sets of related entities of some well-defined, pre-
structured kind (the *model*-structure) which are (to be) associated with sets of
some other kind of entities (the *sign*-labels) or aggregates thereof, the latters having a well-established, pre-defined meaning which can therefore function as the
formers' interpretation (*Fig.* 2). Accordingly, *word meaning* and/or *world knowledge* is uniformly represented as a (more or less complex) labelled graph (*Fig.* 3)
with the (tacid) understanding that associating its vertices and edges with symbols
from some established system of sign-entity-relationship (like e.g. that of natural
language) will render these graph-theoretical configurations a model of structures
or properties which are believed to be those of either the sign-system that provided
the graphs' labels or the system of entities that was to be depicted.

Although there seems no other way to describe and discuss any of the semantic characteristics and properties of meaning outside of and independent from its
symbolic representation and *declarative predication*, the mere application of these
techniques in semantic representations will only repeat the process on another propositional level, it will not, however, provide the *semiotic* answer to how signs may
function as symbols the way the do for cognitive system (natural or artificial), and
why a predicate can be declared of anything and be interpreted and understood the

---

[2]Behind all the theories of linguistic structure that have been presented in the twentieth century, there is a common set of assumptions about the nature of the structural units. This set of
assumptions can be called 'categorial view'. It includes the implicid assertion that all linguistic
units are categories which are discrete, invariant, qualitatively distinct, conjunctively defined, [and]
composed of atomic primes. (Labov 1973, p.342)

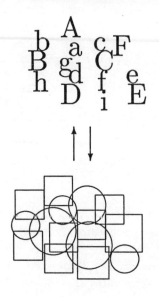

Figure 2:
Rationalistic *Entity-sign-relations* presuppose a sign-independent, pre-structured, and stabel reality of entities which are to be associated with labels composed from a set of language signs which denote independently established meanings.

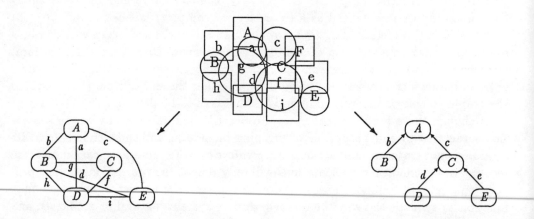

Figure 3: *Symbolic* representations and graph-theoretical modelling of meaning and/or knowledge in *CL* and *AI* presuppose an established *sign-entity-relationship* of well-structured language meanings and/or world objects.

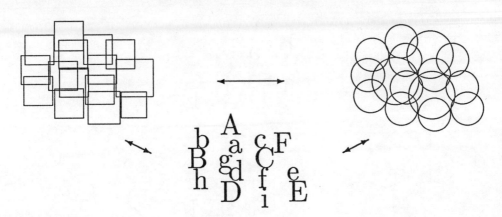

Figure 4: Unstructured, yet related realms of possible (*proto-semiotic*) environments (top) to an information processing system whose own *cognitive*
structure and processes of *semiosis* will constitute identifiable entities
(bottom) of some structure.

way it is (or is not). Obviously, such representational formats do not model the *processes* and the emergence of *structures* that constitute word meaning and/or world
knowledge, but merely make use of them[3]. As it is agreed that cognition is (among
other commitments) responsible for, if not even identifiable with, the processes of
how a previously unstructured surrounding for a cognitive system may be devided
into some identifiable portions the structural results of which are open to permanent
revision according to the system's capabilities, there are still considerable difficulties
to understand how by such a hypothetical structuring of the unstructured an (at
least) twofold purpose can be served namely

▷   to let such identifiable portions—as a sort of prerequisit to entity formation—
acquire *situational significance* (*Fig. 4*) for a system, and

▷   to let some of these entities by way of their particular situational significance be
recognized as *signs* whose interpretations may vary according to the *language
games* (*Fig. 5*) these entities are employed to be elements of.

---

[3]For illustrative examples and a detailed discussion see Rieger 1985b; 1988; 1989, Chapter 5:
pp.103–132.

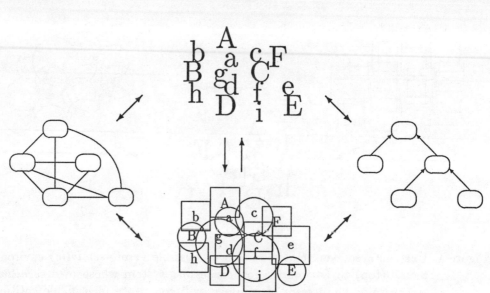

Figure 5: *Language games* employ, modify and/or create signs and entities by establishing entity-sign-relationships via recurrent types of correlated structures and items in processes of *semiosis*.

It should therefore be tried to reconstruct both, the *significance* of entities and the meanings of *signs* as a function of a first and second order semiotic embedding relation of *situations* (or contexts) and of *language games* (or cotexts) a cognitive information processing system (CIPS) is part of (*Fig.* 6). There is some chance for doing so because human beings as CIPSs with symbol manipulation and understanding capabilities of highest performance have *language* at their disposal. This is a powerful cognitive means not only to represent entities and their very complex relations but also to experiment with and to test hypothetical structures and models of entities by way of natural language texts. As their ontological status is again that of very complex structured entities whose *first order* situational significance appears to be identical with their being signs, aggregates, and structures thereof, their *second order* situational significance which allows for their semantic interpretatibility is constituted by their being an instatiation of some language game. Therefore,

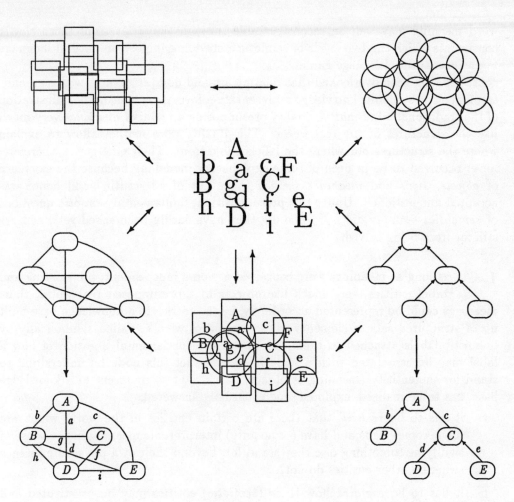

Figure 6: *World knowledge* and *word semantics* hinge on *language games* (center quadrangle) whose cognitive capacity to structure the unstructured is based upon *semiosis* (top triangle) and whose results are being mediated by language discourse or *texts*, providing a performative basis for and an empirical access to the constitution of differing *meaning* representations (bottom line).

word meaning may well be reconstructable through the analyses of those elastic constraints which the two levels of semiotic embedding impose on natural language texts constituting language games.

It has long been overlooked that relating arc-and-node structures with sign-and-term labels in symbolic knowledge representation formats is but another illustration of the traditional *mind-matter*-duality presupposing a realm of *meanings* very much like the structures of the *real world*. This duality does neither allow to explain where the structures nor where the labels come from. Their emergence, therefore, never occurred to be in need of some explanatory modelling because the existence of *objects, signs* and *meanings* seemed to be out of all scrutiny and hence was accepted unquestioned. Under this presupposition, fundamental *semiotic* questions of *semantics*—simply did not come up, they have hardly been asked yet[4], and are still far from being solved.

1.3 As long as meanings were conceived as some independent, pre-existing and stabel entities, very much like objects in a presupposed real world, these meanings could be represented accordingly, i.e. as entries to a knowledge base built up of structured sets of elements whose semantics were signalled symbolically by linguistic labels attached to them. However, the fundamental question of how a label may be associated with a node in order to let this node be understood to stand for the entitity (meaning or object) it is meant to represent in a knowledge base, has to be realized, explored, and eventually answered:

▷ it has to be *realized* that there are certain entities in the world which are (or become) signs and have (or acquire) interpretable meaning in the sense of signifying something else they stand for, beyond their own physical existence (whereas other entities do not).

▷ it has to be *explored* how these (*semiotic*) entities may be constituted and how the meaning relation be established on the basis of which regularities of observables (*uniformities*), controlled by what constraints, and under which boundary conditions of *pragmatic* configuration of communicative interactions like *situations*.

▷ it has to be *answered* why some entities may signify others by serving as labels for them (or rather by the *meanings* these labels purport), instead of being signified semiotically by way of positions, load values and/or states distributed

---

[4]see however Rieger (1977)

over a system of *semiotic/non-semiotic* entities which allows for the distinction of different *distributional* patterns being made, not however, for representing these by different (*symbolic*) labels.

In doing so, a *semiotic paradigm* will have to be followed which hopefully may allow to avoid (if not to solve) a number of spin-off problems, which originate in the traditional distinction and/or the methodological separation of the meaning of a language's term from the way it is employed in discourse. It appears that failing to mediate between these two sides of natural language semantics, phenomena like *creativity, dynamism, efficiency, vagueness,* and *variability* of meaning—to name only the most salient—have fallen in between, stayed (or be kept) out of the focus of interest, or were being overlooked altogether, sofar. Moreover, the classical approach in formal theory of semantics which is confined to the sentence boundary of propositional constructions, is badly in want of operational tools to bridge the gap between formal theory of language description (*competence*) and empirical analysis of language usage (*performance*) that is increasingly felt to be responsible for the unwarranted abstractions of fundamental properties of natural languages.

2 What is it based upon? or the situational setting.

The enthusiasm which the advent of the '*electronic brains*' had triggered during the 1950s and early 1960s was met by promising learning machines of which the pattern-recognizing perceptron-type[5] was widely discussed. Processing of numerical vector values representing features loadings of a described entity consists in cycles of systematic change of weights of features according to the actual input data. Starting from a random set of values such systems—under certain boundary conditions— would converge in a finite number of cycles to the desired set of feature weightings whose *distribution*, instead of a single *symbol*, would represent the entity.

Due to the apparently essential incapabilities these architectures were criticized for[6], neural networking went out of fashion with the early 1970s. Justified or not, mainstream research turned to *symbolic* instead of *distributed* representational formats of knowledge and information processing with the investigation in decision-making and problem-solving tasks gaining importance over knowledge acquisition and learning which became second rate problems.

---

[5]Rosenblatt (1962)
[6]Minsky/Papert (1969)

Meanwhile, the hardware situation has changed, microelectronic circuitry is available to allow parallel computing devices to then unforeseeable extent, and a revival of the early connectionist approaches can be witnessed. The reasonable attraction, however, which advances in parallel distributed processing (*PDP*)[7] have gained recently in both, cognitive and computer sciences appear to be unwarranted in respect to some of the underlying presuppositions that these models share with more traditional, declarative and predicative formats of word meaning and/or world knowledge representation.

2.1   From a computational point-of-view, the so-called *local* representation of entities appears to be the most natural: in a given network of computing elements each of them will be identified with one of the entities to be represented, so that the properties and the relations of the original's elements are mirrored by the structure of the network representing them. Alternatively,

> given a parallel network, items can be represented [...] by a pattern of activity in a large set of units with each unit encoding a micrifeature of the item. Distributed representations are efficient whenever there are underlying regularities which can be captured by interactions among microfeatures. By encoding each piece of knowledge as a large set of interactions, it is possible to achieve useful properties like content-addressable memory and automatic generalization, and new items can be created without having to create new connections at the hardware level. In the domain of continuously varying spatial features it is relatively easy to provide a mathematical analysis of the advantages and drawbacks of using distributed representations.[8]

Only very recently, however, the underlying presuppositions have been addressed critically to set forth some fundamental questioning[9]. To let a given network of *distributed* representations perform the way it does will necessitate the—mostly

---

[7]Rumelhart/McClelland (1986)

[8]Hinton/McClelland/Rumelhart 1986, p.108

[9]Learning in structured connectionist systems has been studied directly. A major problem in this formulation is "recruiting" the compact representation for new concepts. It is all very well to show the advantages of representational schemes [... of networks of distributed structures], but how could they arise? This question is far from settled, but there are some encouraging results. The central question is how a system that grows essentially no new connections could recruit compact groups of units to capture new concepts and relations. (Feldman 1989, p.40)

implicid— introduction of foils and filters, at least during the learning phase. In these models of automatic generalizing or learning, initial or underlying structures have to be presupposed in order to combine constraints of different level to match specified patterns or parts of it, instead of inducing them[10]

Approaching the problem from a *cognitive* point-of-view, it can still be conceded that any identification and interpretation of external structures has to be conceived as some form of *information processing* which (natural/artificial) systems—due to their own structuredness—are (or ought to be) able to perform. These processes or the structures underlying them, however, ought to be derivable from rather than presupposed to procedural models of meaning. Other than in those approaches to cognitive tasks and natural language understanding available sofar in information processing systems that *AI* or *CL* have advanced, it is argued here that *meaning* need not be introduced as a presupposition of *semantics* but may instead be derived as a result of semiotic modelling. It will be based upon a phenomenological reinterpretation of the formal concept of *situation* and the analytical notion of *language game*. The combination of both lends itself easily to operational extensions in empirical analysis and procedural simulation of associative meaning constitution which may grasp essential parts of what PEIRCE named *semiosis*[11].

2.2 Revising some fundamental assumptions in model theory, BARWISE/PERRY have presented a new approach to formal semantics which, essentially, can be considered a mapping of the traditional duality, mediated though by their notion of *situation*. According to their view as expressed in *Situation Semantics*[12] (*SS*), any language expression is tied to reality in two ways: by the *discourse situation* allowing an expression's meaning being *interpreted* and by the *described situation* allowing its interpretation being *evaluated* truth-functionally. Within this relational model of semantics, *meaning* appears to be the derivative of information processing which (natural/artificial) systems—due to their own structuredness—perform by recognizing similarities or invariants between situations that structure their surrounding

---

[10]In brief, there are more problems than solutions. Although it is true that one may view Connectionism as a new *research programm*, expecting it to solve the difficult problems of language without wiring in more traditional symbolic theories by hand is a form of day-dreaming. (Braspenning 1989, p.173)

[11]By *semiosis* I mean [...] an action, or influence, which is, or involves, a coöperation of *three* subjects, such as sign, its object, and its interpretant, this tri-relative influence not being in any way resolvable into actions between pairs. (Peirce 1906, p.282)

[12]Barwise/Perry (1983)

realities (or fragments thereof).

By recognizing these invariants and by mapping them as *uniformities* across *situations*, cognitive systems properly *attuned* to them are able to identify and understand those bits of information which appear to be essential to form these systems' particular view of reality: a flow of *types of situations* related by *uniformities* like individuals, relations, and time-space-locations which constrain an external "world teaming with meaning"[13] to become fragments of persistent *courses of events* whose expectability renders them interpretable.

In semiotic sign systems like natural languages, such uniformities appear to be signalled by *word-types* whose employment as *word-tokens* in texts exhibit a special form of *structurally conditioned* constraints. Not only allows their use the speakers/ hearers to convey/understand meanings differently in different discourse situations (*efficiency*), but at the same time the discourses' total vocabulary and word usages also provide an empirically accessible basis for the analysis of *structural* (as opposed to *referencial*) aspects of *event-types* and how these are related by virtue of word-uniformities accross phrases, sentences, and texts uttered. Thus, as a means for the *intensional* (as opposed to the *extensional*) description of (abstract, real, and actual) *situations*, the regularities of word-usages may serve as an access to and a representational format for those elastic constraints which underly and condition any word's *linguistic meaning*, the *interpretations* it allows within possible contexts of use, and the *information* its actual employment on a particular occasion may convey.

Owing to BARWISE/PERRYs situational approach to semantics—and notwithstanding its (mis)conception as a duality (i.e. the *independent-sign-meaning* view) of an information-processing system on the one hand which is confronted on the other hand with an external reality whose accessible fragments are to be recognized as its environment—the notion of *situation* proves to be seminal. Not only can it be employed to devise a procedural model for the situational embeddedness of cognitive systems as their primary means of mutual accessability[14], but also does it allow to capture and specify the semiotic unity of the notion of *language games* (i.e. the *contextual-use-meaning* view) as introduced by WITTGENSTEIN:

> And here you have a case of the *use of words*. I shall in the future
> again and again draw your attention to what I shall call *language games*.
> There are ways of using signs simpler than those in which we use the

---

[13]Barwise/Perry (1983), p.16
[14]Rieger/Thiopoulos 1989

signs of our highly complicated everyday language. Language games
are the forms of language with which a child begins to make use of
words. The study of language games is the study of primitive forms of
language or primitive languages. If we want to study the problems of
truth and falsehood, of the agreement and disagreement of propositions
with reality, of the nature of assertion, assumption, and question, we shall
with great advantage look at primitive forms of language in which these
forms of thinking appear without the confusing bachground of highly
complicated processes of thought. [...] We are not, however, regarding
the *language games* which we describe as incomplete parts of a language,
but as languages complete in themselves, as complete systems of human
communication[15].

2.3   Trying to model *language game* performance along traditional lines of cybernetics by way of, say, an information processing *subject*, a set of objects
surrounding it to provide the informatory *environment*, and some positive and/or
negative *feedback* relations between them, would hardly be able to capture the cognitive dynamism that self-organizing systems of knowledge acquisition and meaning
understanding are capable of[16].

It is this dynamism of cognitive processing in natural systems which renders the
so-called *cognitive paradigm* of information processing of current artificial systems so
unsatisfactory. Modelling the meaning of an expression along reference-theoretical
lines has to presuppose the structured sets of entities to serve as range of a de-
notational function which will provide the expression's interpretation. Instead, it
appears feasible to have this very range be constituted as a result of exactly those
cognitive procedures by way of which understanding is produced. It will be modelled as a multi-level dynamic description which reconstructs the possible structural
connections of an expression towards cognitive systems (that may both intend/pro-
duce and realize/understand it) and in respect to their *situational* settings, being
specified by the expressions' pragmatics.

---

[15]Wittgenstein (1958), pp.17 and 81; *my italics*

[16][...] feedback is a method of controlling a system by reinserting into it the results of its past
performance. If these results are merely used as numerical data for the criticism of the system
and its regulations, we have the simple feedback of control engineers. If, however, the information
which proceeds backward from the performance is able to change the general method and pattern
of perfomance, we have a process which may well be called learning. (Wiener 1958, p.60)

In phenomenological terms, the set of structural constraints defines any cognitive (natural or artificial) system's possible range in constituting its schemata whose instantiations will determine the system's actual interpretations of what it perceives. As such, these cannot be characterized as a domain of objective entities, external to and standing in contrast with a system's internal, subjective domain; instead, the links between these two domains are to be thought of as *ontologically fundamental*[17] or pre-theoretical. They constitute—from a *semiotic* point-of-view—a system's primary means of access to and interpretation of what may be called its "world" as the system's particular apprehension of its environment. Being fundamental to any cognitive activity, this basal identification appears to provide the grounding framework which underlies the duality of categorial-type rationalistic mind-world or subject-object separation.

In order to get an idea of what is meant by the pre-theoretical *proto-duality* of *semiosis*, any two of the feedback-related operational components separated in system-and-environment, in subject-and-object, or in mind-and-matter distinctions are to be thought of as being merged to form an indecomposable model which bears the characteristics of a self-regulating, *autopoietic* system

> organized (defined as a unity) as a network of processes of production, transformation, and destruction of components that produces the components which: (i) through their interactions and transformations regenerate and realize the network of processes (relations) that produced them; and (ii) constitute it as a concrete unity in the space in which they exist by specifying the topological domain of its realization as such a network.[18]

Together, these approaches may allow for the development of a process-oriented system modelling cognitive experience and semiotic structuring procedurally. Implemented, this system will eventually lead to something like machine-simulated cognition, as an intelligent, dynamic perception of reality by an information processing system and its textual surroundings, accessible through and structured by world-revealing (linguistic) elements of communicative language use. For natural language semantics this is tantamount to (re)present a term's meaning potential by a *distributional pattern* of a modelled system's state changes rather than a *single symbol* whose structural relations are to represent the system's interpretation of its

---

[17]Heidegger (1927)
[18]Maturana/Varela (1980), p.135

environment. Whereas the latter has to *exclude*, the former will automatically *include* the (linguistically) structured, pragmatic components which the system will both, embody and employ as its (linguistic) import to identify and to interpret its environmental structures by means of its own structuredness.

Thus, the notion of *situation* allows for the formal identification of both, the (*internal*) structure of the cognitive subject with the (*external*) structure of its environment. Perceived as a situational fragment of the objective world, a n d exhibited as systematic *constraints* of those systems that are properly attuned, the common *persistency* of courses-of-events will be the means to *understand* a linguistically presented reality.

Based upon the fundamentals of *semiotics*, the philosophical concept of communicative *language games* as specified by the formal notion of *situations*, and tied to the observables of actual language performance, allows for an empirical approach to word semantics. What can formally been analyzed as *uniformities* in BARWISEian *discourse situations* may be specified by word-type regularities as determined by co-occurring word-tokens in pragmatically homogeneous samples of natural language discourse. Going back to the fundamentals of structuralistic descriptions of regularities of *syntagmatic* linearity and *paradigmatic* selectivity of language items, the correlational analyses of discourse will allow for a two-level word meaning and world knowledge representation whose dynamism is a direct function of elastic constraints established and/or modified in communicative interaction by use of linguistic signs in language performance.

3   How could it be achieved? or the linguistic solution.

The *representation* of knowledge, the *understanding* of meanings, and the *analysis* of texts, have become focal areas of mutual interest of various disciplines in cognitive science. In linguistic semantics, cognitive psychology, and knowledge representation most of the necessary data concerning lexical, semantic and external world information is still provided introspectively. Researchers are exploring (or make test-persons explore) their own linguistic or cognitive capacities and memory structures to depict their findings (or to let hypotheses about them be tested) in various representational formats. By definition, these approaches can map only what is already known to the analysts, not, however, what of the world's fragments under investigation might be conveyed in texts unknown to them. Being *interpretative* and unable of auto-modification, such knowledge representations will not only be restricted to predicative and propositional structures which can be mapped in well established (concept-hierarchical, logically deductive) formats, but they will also lack the flexibility and dynamics of more *re-constructive* model structures adapted to automatic meaning analysis and representation from input texts. These have meanwhile been recognized to be essential[19] for any simulative model capable to set up and modify a system's own knowledge structure, however shallow and vague such knowledge may appear compared to human understanding.

3.1   Other than introspective data acquisition and in contrast to classical formalisms for knowledge representation which have been conceived as depicting some of the (inter)subjective reflections of entities which an external, objective world and reality would provide, the present approach focusses on the *semiotic* structuredness which the communicative use of language in discourse by speakers/hearers will both, constitute and modify as a paradigm of cognition and a model of *semiosis*. It has been based on the algorithmic analysis of discourse that real speakers/writers produce in actual situations of performed or intended communication on a certain subject domain. Under the notion of *lexical relevance* and *semantic disposition*[20], a conceptual meaning representation system has operationally been defined which may empirically be reconstructed from natural language texts.

Operationalizing the WITTGENSTEINian notion of *language games* and drawing on his assumption that a great number of texts analysed for the terms' usage *regularities* will reveal essential parts of the concepts and hence the meanings conveyed[21], such an description turns out to be identical with a analytical procedure. Starting from the sets of possible combinations of language units, it captures and reformulates their *syntagmatic* and *paradigmatic* regularities (provided the units function as *signs*) via two consecutive processes of abstraction based upon constraints that can empirically be ascertained.

In terms of *autopoietic* systems, it is a mere presupposition of propositional level approaches to natural language semantics that linguistic entities which may be combined to form language expressions must also have independent meanings which are to be identified first in order to let their composite meanings in discourse be interpreted. This presupposition leads to the faulty assumption that word meanings are somewhat static entities instead of variable results of processes constituted via

---

[19]Winograd (1986)
[20]Rieger 1985a
[21]Wittgenstein (1969)

semiotically different levels of abstraction. Although structural linguistics offers some hints[22] towards how language items come about to be employed the way they are, these obviously have not been fully exploited yet for the reconstructive modelling of such abstractions which will have to be executed on different levels of description and analysis too.

Thus, complementing the independent *sign-meaning* view of information processing and the propositional approach in *situation semantics*, the contextual *usage-meaning* view in word semantics may open up new vistas in natural language processing and its semantic models[23].

3.2   Within the formal framework of situation semantics, lexical items (as *word-types*) appear to render basic uniformities (as *word-tokens*) in any discourse whose *syntagmatic* or *linear* a n d *paradigmatic* or *associative*[24] relatedness can not only be formalized in analogy to *topos theoretical* constructions[25] but also allows for the empirical analyses of these structures and their possible restrictions in order to devise mechanisms to model operational constraints.

These constraints may be formalized as a set of *fuzzy subsets*[26] of the vocabulary. Represented as a set-theoretical system of meaning points, they will depict the distributional character of word *meanings*. Being composed of a number of operationally defined elements whose varying contributions can be identified with values of the respective membership functions, these can be derived from and specified by the differing usage regularities that the corresponding lexical items have produced in discourse. This translates the WITTGENSTEINian notion of *meaning* into an operation that may be applied empirically to any corpus of *pragmatically homogeneous* texts constituting a *language game*.

Based upon the distinction of the *syntagmatic* and *paradigmatic* structuredness of language items in discourse, the core of the representational formalism can be

---

[22]In subscribing to the systems-view of natural languages, the distinction of *langue/parole* and *competence/performance* in modern linguistics allowes for different levels of language description. Being able to *segment* strings of language discourse and to *categorize* types of linguistic entities is to make analytical use of the *structural coupling* represented by natural languages as semiotic systems.

[23]Rieger (1989b)

[24]According to the terminology of early linguistic structuralism as well as recent connectionistic models in cognitive networking.

[25]For the mathematical concept of *topoi* see Goldblatt (1984); for its application to natural language semantics see Rieger/Thiopoulos (1989)

[26]Zadeh (1965)

captured by a two-level process of abstraction (called α- and δ-abstraction) providing the set of *usage regularities* and the set of *meaning points* of those word-types which are being instantiated by word-tokens as employed in natural language texts. The resultant structure of these constraints render the set of potential *interpretations* which are to be modelled in the sequel as the *semantic hyperspace structure (SHS)*.

It has been shown elsewhere[27] that in a sufficiently large sample of *pragmatically homogeneous* texts produced in sufficiently similar *situational* contexts, only a restricted vocabulary, i.e. a limited number of lexical items, will be used by the interlocutors, however comprehensive their personal vocabularies in general might be. Consequently, the words employed to convey information on a certain subject domain under consideration in the discourse concerned will be distributed according to their conventionalized communicative properties, constituting *usage regularities* which may be detected empirically from texts. These are considered primarily as structured sets of strings of linguistic elements, not of sentences.

3.3 The statistics used so far for the analysis of *syntagmatic* and *paradigmatic* relations on the level of *words* in discourse, is basically descriptive. Developed from and centred around a correlational measure to specify intensities of co-occurring lexical items, these analysing algorithms allow for the systematic modelling of a fragment of the lexical structure constituted by the vocabulary employed in the texts as part of the concomitantly conveyed world knowledge.

A modified correlation coefficient has been used as a *first* mapping function α. It allows to compute the relational interdependence of any two lexical items from their textual frequencies. For a text corpus $K = \{t\}, t = 1, \ldots, T$ of pragmatically homogeneous discourse, having an overall length $L = \sum_{t=1}^{T} l_t; 1_t \leq l_t \leq L$ measured by the number of word-tokens per text, and a vocabulary $V = \{x_n\}; n = 1, \ldots, i, j, k, \ldots, N$ of word-types n whose frequencies are denoted by $H_i = \sum_{t=1}^{T} h_{it}; 1_{it} \leq h_{it} \leq H_i$, the modified correlation-coefficient $\alpha_{i,j}$ allows to express pairwise relatedness of word-types $(x_i, x_j) \in V \times V$ in numerical values ranging from -1 to $+1$ by calculating co-occurring word-token frequencies in the following way

$$\alpha(x_i, x_j) = \frac{\sum_{t=1}^{T}(h_{it} - h_{it}^*)(h_{jt} - h_{jt}^*)}{\left(\sum_{t=1}^{T}(h_{it} - h_{it}^*)^2 \sum_{t=1}^{T}(h_{jt} - h_{jt}^*)^2\right)^{\frac{1}{2}}} ; \qquad (1)$$

$$-1 \leq \alpha(x_i, x_j) \leq +1$$

[27]Rieger (1981)

where $h_{it}^* - \frac{H_i}{L} l_i$ and $h_{jt}^* = \frac{H_i}{L} l_t$.

Evidently, pairs of items which frequently either co-occur in, or are both absent from, a number of texts will positively be correlated and hence called *affined*, those of which only one (and not the other) frequently occurs in a number of texts will negatively be correlated and hence called *repugnant*.

As a fuzzy binary relation, $\alpha : V \times V \to I$ can be conditioned on $x_n \in V$ which yields a crisp mapping

$$\alpha \mid x_n \; : \; V \to C; \; \{y_n\} := C \tag{2}$$

where the tupels $\langle (x_{n,1}, \alpha(n,1)), \ldots, (x_{n,N}, \alpha(n,N)) \rangle$ represent the numerically specified, *syntagmatic* usage-regularities that have been observed for each word-type x_i against all other $x_n \in V$ and can therefore be abstracted over one of the components in each ordered pair, thus defining an element

$$(x_i, \alpha(i,1), \ldots, \alpha(i,N)) := y_i \in C \tag{3}$$

Hence, the regularities of usage of any lexical item will be determined by the tupel of its *affinity/repugnancy*-values towards each other item of the vocabulary which— interpreted as coordinates— can be represented as points in a vector space C spanned by the number of axes each of which corresponds to an entry of the vocabulary.

3.4 Considering C as representational structure of abstract entities constituted by *syntagmatic* regularities of word-token occurrences in *pragmatically homogeneous* discourse, then the similarities and/or dissimilarities between these abstract entities will capture the *paradigmatic* regularities of the correspondent word-types. These can be modelled by the *δ-abstraction* which is based on a numerically specified evaluation of differences between any two of such points $y_i, y_j \in C$ They will be the more adjacent to each other, the less the usages (*tokens*) of their corresponding lexical items $x_i, x_j \in V$ (*types*) differ. These differences may be calculated by a distance measure δ of, say, EUCLEDian metric.

$$\delta(y_i, y_j) = \left(\sum_{n=1}^{N} (\alpha(x_i, x_n) - \alpha(x_j, x_n))^2 \right)^{\frac{1}{2}} ; \tag{4}$$

$$0 \le \delta(y_i, y_j) \le 2\sqrt{n}$$

Thus, δ serves as a *second* mapping function to represent any item's differences of usage regularities measured against those of all other items.

As a fuzzy binary relation, also $\delta : C \times C \to I$ can be conditioned on $y_n \in C$ which again yields a crisp mapping

$$\delta \,|\, y_n \;:\; C \to S; \; \{z_n\} := S \tag{5}$$

where the tupels $\langle (y_{n,1}, \delta(n,1)), \ldots, (y_{n,N}\delta(n,N)) \rangle$ represents the numerically specified *paradigmatic* structure that has been derived for each abstract *syntagmatic* usage-regularity y_j against all other $y_n \in C$. The distance values can therefore be abstracted again as in (2), this time, however, over the other of the components in each ordered pair, thus defining an element $z_j \in S$ called *meaning point* by

$$(y_j, \delta(j,1), \ldots, \delta(j,N)) := z_j \in S \tag{6}$$

By identifying $z_n \in S$ with the numerically specified elements of potential paradigms, the set of possible combinations $S \times S$ may structurally be constrained and evaluated without (direct or indirect) recourse to any pre-existent external world. Introducing a EUCLEDIAN metric

$$\partial \;:\; S \times S \to I \tag{7}$$

the hyperstructure $\langle S, \partial \rangle$ or *semantic hyperspace* (*SHS*) is constituted (*Tab.* 1) providing the *meaning points* according to which the *stereotypes* of associated lexical items may be generated as part of the semantic paradigms concerned.

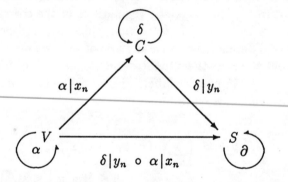

Figure 7: Mapping relations α and δ between the structured sets of vocabulary items $x_n \in V$, of corpus points $y_n \in C$, and of meaning points $z_n \in S$.

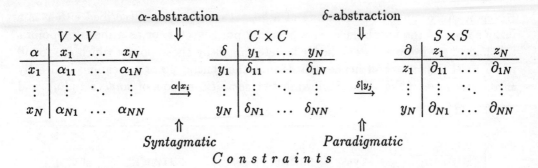

Table 1: Formalizing (*syntagmatic/paradigmatic*) constraints by consecutive (α- and δ-) abstractions over usage regularities of items x_i, y_j respectively.

As a result of the two *consecutive* mappings, any meaning point's position in SHS is determined by all the differences (δ- or distance-values) of all regularities of usage (α- or correlation-values) each lexical item shows against all others in the discourse analysed (*Fig. 7*). Thus, it is the basic analyzing algorithm which—by processing natural language texts—provides the processing system with the ability to recognize and represent and to employ and modify the structural information available to the system's performance constituting its *understanding*.

This answers the question where the labels in our representation come from: put into a discourse environment, the system's text analyzing algorithm provides the means how the topological position of any metrically specified meaning point $z \in \langle S, \partial \rangle$ is identified and labeled by a vocabulary item $x \in V$ according to the two consecutive mappings which can formally be stated as a composition of the two restricted relations $\delta \mid y$ and $\alpha \mid x$. It is achieved without recurring to any investigator's or his test-persons' word or world knowledge (*semantic competence*), but solely on the basis of usage regularities of lexical items in discourse which are produced by real speakers/hearers in actual or intended acts of communication (*communicative performance*).

3.5   Knowledge of word meanings (*lexical knowledge*) represented sofar is a relational data structure whose linguistically labeled elements (*meaning points*) and their mutual distances (*meaning differences*) form a system of potential *ste-*

reotypes. Accordingly, the *meaning* of a lexical item may be described either as a fuzzy subset of the vocabulary, as a meaning point vector, or as a meaning point's topological environment. The latter is determined by those points which are found to be most adjacent and hence will delimit the central point's meaning indirectly as its *stereotype* (*Tab.* 2). Following the *semiotic* notion of *understanding* and

WIRTSCHAFT/economy		0.000			
AUSLAND	3.785	BRITAIN	5.094	ENTWICKL	5.893
FOLGe	6.112	VERWALT	6.428	RAUM	6.903
EINSATZ	9.307	KONTAKT	9.934	HERRSCHen	10.163
GESCHÄFT	10.931	KRANK	11.732	VERKEHR	11.984
VERANTWORT	12.298	SPRACH	12.429	MÖGLICH	13.257
WEG	13.285	NEU	13.871	ZENTRAL	14.831
LEHR	15131	JUNG	15.550	ALLGEMEIN	15.796
MODE	15.850	AUFTRAG	15.952	MASCHINE	16.210
⋮	⋮	⋮	⋮	⋮	⋮

Table 2: Topological environment $E(z_i, r)$ of i = WIRTSCHAFT/economy listing points situated within the hypersphere of radius r in the *semantic hyperspace* $\langle S, \partial \rangle$ as computed from a text sample of the 1964 editions of the German daily DIE WELT (175 articles of approx. 7000 word tokens and 365 word types).

meaning constitution, the SHS-structure may be considered the core of a two-level conceptual knowledge representation system[28]. Essentially, it separates the format of a basic (stereotype) word meaning representation from its latent (dependency) relational concept organization. Whereas the former is a rather static, topologically structured (associative) memory, the latter can be characterized as a collection of dynamic and flexible structuring procedures to re-organize the memory data by semiotic principles under various aspects.

SHS being a distance-ralational data structure, well-known algorithmic search strategies cannot immediately be made to work. They are mostly based upon some non-symmetric relational structure as e.g. directed graphs in traditional meaning

---

[28]Rieger (1989b)

and knowledge representation formats. To convert the SHS-format into such a node-pointer-type structure, the SHS-model has to be considered as conceptual raw data or associative base structure which particular procedures may operate on to reorganize it. Thus, the *distributed* representational format of SHS which had appeared to be disadvantageous first, proved to be superior over more traditional formats of *symbolic* representation (as shown in *Fig.* 3 above). Other than in these pre-defined semantic network structures of predicative knowledge, non-predicative meaning relations of lexical *relevance* and semantic *dispositions* depend haevily on con- and cotextual constraints which will more adequately be defined procedurally, i.e. by generative algorithms that induce them on changing data only and whenever necessary. This is achieved by a recursively defined procedure that produces hierarchies of meaning points, tree-structured under given aspects according to and in dependence of their meanings' relevancy.

3.6 Unlike conceptual representations that link nodes to one another according to what cognitive scientists supposedly know about the way conceptual information is structured in memory[29], an algorithm has been devised which operates on the SHS-data to induce *dispositional dependency structures DDS* between its elements, i.e. among subsets of meaning points conceptually related. The procedure detects fragments from SHS according to different *perspectives* as specified by the meaning point it is started with, and it reorganizes *relevant* meaning points according to the constraints of semantic similarity encountered during operation . Stop-conditions may deliberately be formulated either qualitatively (i.e. naming a target point) or quantitatively (i.e. number of points, realm of distance or criteriality to be processed).

This so-called Δ-*operation* has been conceived as an *optimal spanning tree*-algorithm. The procedure is recursively defined to operate on the semantic space data $z_n \in \langle S, \partial \rangle$. Given one meaning point's position as a start, the algorithm will work its way through all labeled points—unless stopped under conditions of a given target node, number of nodes to be processed, or threshold of maximal distance—transforming prevailing similarities of paradigms as represented by adjacency of points to induce a binary, non-symmetric, and transitive relation of lexical *relevance* between them. This relation allows for the hierarchical reorganization of meaning points as nodes under a primed head in an n-ary tree called *dispositional dependency*

---

[29]Schank (1982)

structure (DDS)[30]. It is tantamount to a numerical assessment (*criterialty*) and a hierachical re-structuring (*tree*) of elements under a head point's aspect according to the dependency relation between descendant points along which activation might spread in case of the head point stimulation.

To illustrate the feasibility of the Δ-operation's generative procedure, a subset of the relevant, linguistic constraints triggered by the lexical item x_i, $i =$ WIRT-SCHAFT/economy is given in the format of a weighted semantic *dispositional dependency structure Fig.* 8. It has been generated by the procedure described from the SHS-data as computed from the corpus of German newspaper texts[31].

Weighted numerically as a function of an element's distance values and its associated node's level and position in the tree, $DDS(z_i)$ either is an expression of the head node's z_i meaning-dependencies on the daughter-nodes z_n or, inversely, expresses their meaning-criterialities adding up to an aspect's interpretation determined by that head[32]. For a wide range of purposes in processing DDS-trees, differing criterialities of nodes can be used to estimate which paths are more likely being taken against others being followed less likely under priming activated by certain meaning points.

4 What may it be used for? or the need for CIPS.

From the communicative point-of-view natural language texts, whether stored electronically or written conventionally, will in the foreseeable future provide the major source of scientifically, historically, and socially relevant information. Due to the new technologies, the amount of such textual information continues to grow beyond manageable quantities. Rapid access and availability of *data*, therefore, no longer serves to *solve* an assumed problem of lack of *information* to fill an obvious *knowledge* gap in a given instance, but is instead and will even more so in future *create* a new problem which arises from the abundance of information we are confronted with.

Thus, actual and potential (human) problemsolvers feel the increasing need to employ computers more effectively than hitherto for informational search through masses of natural language material. Although the demand is high for intelligent machinery to assist in or even provide speedy and reliable selection of relevant

---

[30]Rieger (1985)
[31]Randomly assembled from first two pages of the daily DIE WELT, Jg.1964, Berlin edition.
[32]Rieger (1989a)

267

Distributed Semantic Representation

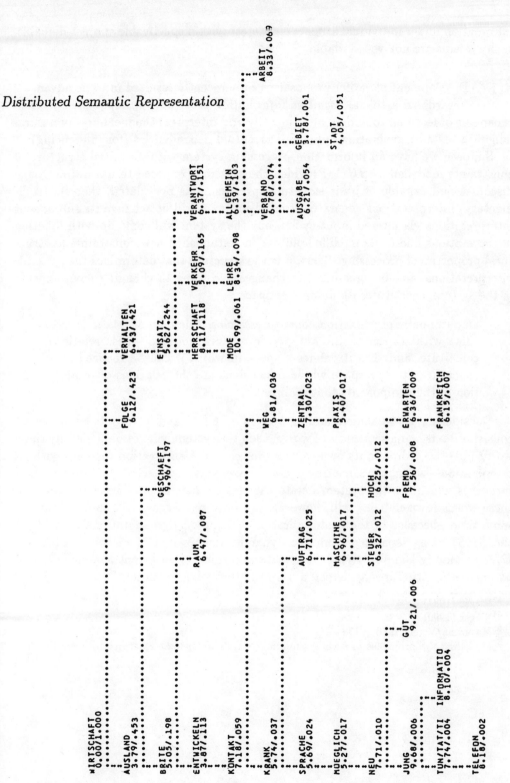

information under individual aspects of interest within specifyable subject domains, such systems are not yet available.

4.1 Development of earlier proposals[33], only recently resulted in some advance[34] towards an artificial *cognitive information processing system (CIPS)* which is capable of learning to understand (identify and interpret) the meanings in natural language texts by generating dynamic conceptual dependencies (for inferencing).

Suppose we have an information processing system with an initial structure of constraints modelled as *SHS*. Provided the system is exposed to natural language discourse and capable of basic structural processing as postulated, then its (rudimentary) interpretations generated from given texts will not change its subsequent interpretations via altered input-cycles, but the system will come up with differing interpretations due to its modified old and/or established new constraints as structural properties of processing. Thus, it is the structure that determines the system's interpretations, and being subject to changes according to changing environments of the system, constitutes its *autopoetic space*:

> an outopoetic organization constitutes a closed domain of relations specified with respect to the autopoetic organization that these relations constitute, and thus it defines a space in which it can be realized as a concrete system, a space whose dimensions are the relations of production of the components that realize it[35].

Considering a text understanding system as *CIPS* and letting its environment consist of texts being sequences of words, then the system will not only identify these words but—according to its own capacity for α- and δ-abstraction together with its Δ-operation—will at the same time realize the semantic connectedness between their *meanings* which are the system's state changes or *dispositional dependencies* that these words invoke. They will, however, not only be trigger *DDS* but will at the same time—because of the prototypical or distributed representational format (of the *SHS*) being separated from the dynamic organization of meaning points (in *DDS*)—modify the underlying *SHS*-data according to recurrent *syntagmatic* and *paradigmatic* structures as detected from the textual environment[36].

---

[33]Rieger (1984)

[34]Rieger (1990)

[35]Maturana/Varela 1980, p.135

[36]Modelling the principles of such a semiotic system's *autopoietic existence* by means of ma-

4.2   In view of a text skimming system under development[37], a basic *cognitive*
algorithm has been designed which detects from the textual environment
the system is exposed to, those strucural information which the system is able to
collect due to the two-level structure of its linguistic information processing and
knowledge acquisition mechanisms. These allow for the automatic generation of a
pre-predicative and formal representation of conceptual knowledge which the system will both, gather from and modify according to the input texts processed. The
system's internal knowledge representation is designed to be made accessible by a
front-end with dialog interface. This will allow system-users to make the system
skim masses of texts for them and display its acquired knowledge graphically in
dynamic structures of interdependently formed conceptualisations. These provide
variable constraints for the procedural modelling of conceptual connectedness and
non-propositional inferencing which both are based on the algorithmic induction of
an aspect-dependent relevance relation connecting concepts differently according to
differing conceptual perspektives in semantic *Dispositional Dependency Structures
(DDS)*. The display of *DDS*s or their resultant graphs may serve the user to acquire
an overall idea of what the texts processed are roughly about or deal with along what
general lines of conceptual dependencies. They may as well be employed in an knowledge processing environment to provide the user with relevant new keywords for
an optimized recall-precision ratio in intelligent retrieval tasks, helping for instance
to avoid unnecessary reading of irrelevant texts.

Dispositional dependencies appear to be a prerequisit not only to source-oriented, contents-driven *search* and *retrieval* procedures which may thus be performed
effectively on any *SHS*-structure. Due to its procedural definition, it also allows
to detect varying dependencies of identically labeled nodes under different aspects
which might change dynamically and could therefore be employed in conceptual,
pre-predicative, and *semantic* inferencing as opposed to propositional, predicative,
and *logic* deduction.

For this purpose a procedure was designed to operate simultaniously on two
(or more) *DDS*-trees by way of (simulated) parallel processing. The algorithm is
started by two (or more) meaning points which may be considered to represent
conceptual *premises*. Their *DDS* can be generated while the actual inferencing
procedure begins to work its way (breadth-first, depth-first, or according to highest
criteriality) through both (or more) trees, tagging each encountered node. When the

---

thematical *topoi* is the aim of a PhD-thesis (by C. Thiopoulos) presently being prepared at the
Deptartment of Computational Linguistics, University of Trier.

[37]Rieger (1988a)

first node is met that has previously been tagged by activation from another premise, the search procedure stops to activate the dependency paths from this *concluding* common node back to the *premises*, listing the intermediate nodes to mediate (as illustrated in *Tab.* 3) the *semantic inference paths* as part of the dispositional dependencies structures *DDS* concerned.

UNTERNEHM	0.0/1.000	⟸ ⟩ ⟹	0.0/1.000	WIRTSCHAFT
STADT/city	5.57/.428			
GEBIET/area	4.05/.239		6.43/.421	VERWALT/administrat
VERBAND/league	3.78/.144		5.62/.244	EINSATZ/effort/supply
ALLGEMEIN/general	6.78/.076			
Conclusion ⟹	6.39/.046	VERANTWORT	6.37/.151	⟸ ⟩

Table 3: *Semantic inference paths* from the premises UNTERNEHM/enterprise and WIRTSCHAFT/economy to the conclusion VERANTWORT/responsibility

4.3 It is hoped that our system will prove to provide a flexible, source-oriented, contents-driven method for the *multi-perspective* induction of dynamic conceptual dependencies among stereotypically represented concepts which—being linguistically conveyed by natural language discourse on specified subject domains— may empirically be detected, formally be presented, and continuously be modified in order to promote the learning and understanding of meaning by *cognitive information processing systems* (*CIPS*) for machine intelligence. Research is under way to emulate What sofar has been analysed as numerical constraints of correlational item distributions within a structural model of semantic usage regularities is presently being investigated to be remodelled in some connectionist architecture with the advantage of semiotically well established and linguistically well founded empirical data to provide numerical weights and grades of activation.

271

References

Barwise, J./Perry, J.(1983): Situations and Attitudes. Cambridge, MA (MIT)

Braspenning, P.J. (1989): "Out of Sight, Out of Mind → Blind Idiot. A review of Connectionism in the courtroom." *AiCommunications AICOM*, Vol.2,3/4, pp.168–176

Collins, A.M./Loftus, E.F. (1975): A spreading activation theory of semantic processing. *Psychological Review* 6(1975) 407–428

Feldman, J.A. (1989): "Connectionist Representation of Concepts" in: Pfeiffer/Schreter/Fogelman-Soulié/Steels, pp.25–45

Forsyth, R./Rada, R. (1986): Machine Learning. Chichester (Ellis Horwood)

Goldblatt, R. (1984): Topoi. The Categorial Analysis of Logic. (Studies in Logic and the Foundations of Mathematics 98), Amsterdam (North Holland)

Heidegger, M. (1927): Sein und Zeit. Tübingen (M.Niemeyer)

Hinton, G.E./McClelland, J.L./Rumelhart, D.E. (1986): "Distributed Representation" in: Rumelhart/McClelland, pp.77–109

Husserl, E. (1976): Ideen II (*Husserliana* III/1), DenHaag (M. Nijhoff)

Maturana, H. /Varela, F. (1980): Autopoiesis and Cognition. The Realization of the Living. Dordrecht (Reidel)

Minsky, M./Papert, S. (1969): Perceptrons. Cambridge, MA (MIT-Press)

Norvig, P. (1987): Unified Theory of Inference for Text Understanding. (EECS-Report UCB/CSD 87/339) University of California, Berkeley

Peirce, C.S. (1906): "Pragmatics in Retrospect: a last formulation" (CP 5.11 – 5.13), in: The Philosophical Writings of Peirce. Ed. by J. Buchler, NewYork (Dover), pp.269–289

Pfeiffer, R./Schreter, Z./Fogelman-Soulié/Steels, L. (1989)(Eds.): Connectionism in Perspective. Amsterdam/NewYork/Oxford/Tokyo (North-Holland)

Rieger, B. (1977): "Bedeutungskonstitution. Einige Bemerkungen zur semiotischen Problematik eines linguistischen Problems" *Zeitschrift für Literaturwissenschaft und Linguistik* 27/28, pp.55–68

Rieger, B. (1981): Feasible Fuzzy Semantics. In: Eikmeyer, H.J./Rieser, H. (Eds): Words, Worlds, and Contexts. New Approaches in Word Semantics. Berlin/NewYork (de Gruyter), pp.193–209

Rieger, B.B. (1984): "The Baseline Understanding Model. A Fuzzy Word Meaning Analysis and Representation System for Machine Comprehension of Natural Language." in: O'Shea, T.(Ed): Proceedings of the 6th European Conference on Artificial Intelligence (ECAI 84), NewYork/Amsterdam (Elsevier

Science), pp.748–749

Rieger, B.B. (1985): Lexical Relevance and Semantic Disposition. On stereotype word meaning representation in procedural semantics. In: Hoppenbrouwes, G./ Seuren. P./Weijters, T. (Eds.): Meaning and the Lexicon. Dordrecht (Foris), pp.387–400

Rieger, B. (1985b): "On Generating Semantic Dispositions in a Given Subject Domain" in: Agrawal, J.C./Zunde, P. (Eds.): Empirical Foundation of Information and Software Science. NewYork/London (Plenum Press), pp.273–291

Rieger, B.B. (1988a): TESKI – A natural language TExt-SKImmer for shallow understanding and conceptual structuring of textually conveyed knowledge. LDV/CL-Report 10/88, Dept. of Computational Linguistics, University of Trier

Rieger, B. (1988b): "Definition of Terms, Word Meaning, and Knowledge Structure. On some problems of semantics from a computational view of linguistics". in: Czap, H./Galinski, C. (Eds.): Terminology and Knowledge Engineering (Supplement). Frankfurt (Indeks Verlag), pp.25–41

Rieger, B. (1989a): "Situations and Dispositions. Some formal and empirical tools for semantic analysis" in: Bahner, W. (Ed.): Proceedings of the XIV. Intern.Congress of Linguists (CIPL), Berlin (Akademie) [in print]

Rieger, B. (1989b): Unscharfe Semantik. Die empirische Analyse, quantitative Beschreibung, formale Repräsentation und prozedurale Modellierung vager Wortbedeutungen in Texten. Frankfurt/Bern/NewYork (P. Lang)

Rieger, B. (1990): "Reconstructing Meaning from Texts. A Computational View of Natural Language Understanding" in: Proceedings of the 2nd German-Chinese-Electronic-Week (DCEW 90), Berlin (VDE) [in print]

Rieger, B.B./Thiopoulos, C. (1989): Situations, Topoi, and Dispositions. On the phenomenological modelling of meaning. in: Retti, J./Leidlmair, K. (Eds.): 5th Austrian Artificial Intelligence Conference. (ÖGAI 89) Innsbruck; (KI–Informatik-Fachberichte Bd.208) Berlin/Heidelberg/NewYork (Springer), pp.365–375

Rosenblatt, F. (1962): Principles of Neurodynamics. London (Spartan)

Rumelhart, D.E./McClelland, J.L (1986): Parallel Distributed Processing. Explorations in the Microstructure of Cognition. 2 Vols. Cambridge, MA (MIT)

Schank, R.C. (1982): Dynamic Memory. A Theory of Reminding and Learning in Computers and People. Cambridge/London/NewYork (Cambridge UP)

Sklansky, J./Wassel, G. (1981): Pattern Classifiers and Trainable Machines. Berlin/Heidelberg/NewYork (Springer)

Varela, F. (1979): Principles of Biological Autonomy. NewYork (North Holland)

Wiener, N. (1956): The Human Use of Human Beings. Cybernetics and Society. NewYork (Doubleday Anchor)

Winograd, T. (1983): Language as a Cognitive Process. Vol.1 Syntax. Reading, MA (Addison-Wesley)

Winograd,T./Flores, F. (1986): Understanding Computers and Cognition: A New Foundation for Design. Norwood, NJ (Ablex)

Wittgenstein, L. (1958): The Blue and Brown Books. Ed. by R. Rhees, Oxford (Blackwell)

Wittgenstein, L. (1969): Über Gewißheit - On Certainty. NewYork/San Francisco/London (Harper & Row), [No.61–65], p.10e

Zadeh,L.A. (1965): Fuzzy sets. *Information and Control* 8(1965), pp.338–353

LOGIC OF COGNITIVE REPRESENTATIONS
AND THEIR EVOLUTION

Henri WERMUS
Le Fort 4
CH – 1268 Begnins

> "Thought activity is creation of meanings step by step."
>
> J.P. PEIRCE

Abstract

"Thinking is processing of meanings of cognitive representations" (J. Piaget).

Cognitive representations combine to more complex semantic nets, among others, by means of connectives belonging to different levels of "natural logics" (= protologics) attained by a subject S. I intend to describe a system of these protologics and analyse briefly the resulting inferential competences of the mind. In contrast to formal logic, the protological connectives are only partially defined functions of their valuation tables.

Following a particular principle of extension – the reflection functor – one can express the transitions from lower level logics to higher ones: these different protologics form a lattice with growing power of "discernability". Higher protologics allow much better reasoning abilities and consequently induce an import increase of cognitive competences. The classical formal logic may be seen as the unique maximal element of the lattice; experiences show that "every–day" argumentations very seldom reach the level of classical logic.

Further study of other reflection functors may be of interest: they allow formalisation of a kind of "transactional logic" and of a system of modal logic.

The models and concepts presented here were suggested and supported by reports of numerous experiments within the framework of the Piaget'ean School of psychology in Geneva.

Key concepts: "Natural thinking" (NT), protologic ("natural logic") λ^{nj} of level n and "strength" j, formal logic λ^F, occultation (= undefined value), reflection (or reflexive) functor of a cognitive representation, cognitive meaning Mc.

1. INTRODUCTION

I intend to present here some ideas about cognitive processes suggested by experimental psychology. The main concepts and models used here like: representations, cognitive inferences, levels of cognitive competence, use of conditional and negation, intentionality ect. are derived from reports of numerous experiments and the theoretical framework of Piaget's work. Of course I tried to reformulate several terms of this work in a way which is better suitable to a logical investigation.

I have no competence in neuro–physiological processes, parallel, sequential or embedded, which are the "causa efficiens" of the conscious or tacit thinking activity of the mind. Gerhard Dalenoort noticed the unavoidable mystery which may exist between the neuronal activity and the emergence of semantic properties like meaning, beliefs, intentionality, truth–valuations etc.. I agree, there seems to be a mystery, but I am strongly interested in a kind of a converse of this mysterious relation between mind an brain: how can the brain machinery be activated in order to realise processes which achieve goals or intentional states, the "causa finalis" bared by the mind [1]). Remember that semantics cannot be constructive, cf. the well known metatheorems about (higher) formal systems; neither there exist atoms of meanings and despite Leibniz, no Caracteristica Universalis, no Ars Inveniendi etc. Also Boolean Algebra does not reflect the "laws of thought", as they appear in every–day reasoning.

These laws have to be investigated using cautiously observations of behaviours of subjects during psychological experiments. I intend to present a small sample of these experiments which may give an insight into behaviours which I call "natural thinking" (NT) as opposed to valid procedures of formal logic or scientific argumentation. Most of these experiences are well known to psychologists and I apologise for not always quoting their first autor. They are reported and extensively discussed in Piaget EEG (36 volumes !) [2]), Gruber and Vonèche, Bideau. Many examples and exercises in reasoning (use of logical connectives, inferences, properties of relations etc.) are described in Wermus 1975, cf. bibliography.

I have already described a representational system of cognition in different previous papers (cf. Wermus 1977, 1978, 1982, 1986, 1989) using such concepts as amalgameted predicates, partially defined connectives of "natural logic " [3], cognitive constructs (= cognitive propositions), and semantic nets as well as transformations of several kinds and intentional factors. An outline of a lattice of various "natural logics" (= protologics) is the core of this paper.

The fallacies of children reasoning in problem solving may be caused by wrong theories about the world or incomplete world knowledge, but they are mainly caused by a lack of logical and operational competence and the use of child–specific mental processes.

Acknowledgement

I owe great thanks to Dr. Branka Zei and to my colleage Professor Hermine Sinclair for their comments and their help in writing a first draft in english of my manuscripts about cognitive processes. I am also most grateful to Dr. Jörg Becker, organizer of Wopplot 89, who made ready for printing and corrected the english version of the crude manuscript of my talk at this workshop. Without his kind encouragements this manuscript, like many others, would wait a long time to be worked out for a publication.

2. PRELIMINARY REMARKS

a) It must be strongly emphasized that cognitive processes are not to be confused with their verbal expression. I shall not deal here with the difficult problem of the relation between cognition and linguistic competence, and shall try to maintain a clear distinction between the concept of (mental) representation and the meaning related to it on the one hand and verbal expressions on the other. Thinking is to be viewed as a silent meaningful game.

I adopt Piaget's dogma: "Thinking is transformation of meanings"; cognitive meanings are of course closely related to mental (i.e. cognitive) representations and consequently depend on the different transformations of these representations. The processing of representations and corresponding meanings is determinded by the level of cognitive development of the subject and particularly by his level of logical compentence. This development, according to Piaget, is guided by the fundamental tendency to equilibration and adaptation of the subject to his environment. Thus, in general, intelligence is seen as a highly elaborated tool of biological adaptation.

b)   The restriction of the study to cognitive processes in children does not mean that it would not be applicable to adult cognition. In the first place "normal" adult natural logic does not differ all that much from that of children in their later stages of development, although world knowledge does differ. Secondly, the diachronic dimensions of cognitive activity are easier to observe during development in childhood than in adult behaviour. Thirdly, the psychological relevance of certain behaviours is much clearer when they are observed in children than in adults: verbal habits, introspection, strong interaction with observers etc.... introduce a bias into the analysis of the "real" cognitive processes of adults.

3. A SAMPLE OF INTRODUCTORY EXAMPLES OF NATURAL THINKING

In this section we give three examples of Natural Thinking. The first experiment refers to cognitive transfers, i.e. the context dependency of a transitive relation. This example is due to Piaget et al (1972):

```
|    |    |
|    |    |
     a    b    c
```

We have three sticks of different lengths a, b, and c. The observer makes the children notice that a<b and b<c. When subsequently the sticks are hidden children under the age of seven cannot reach a conclusion about the relation between a and c; this means that the concept of transitivity is still missing at this age (in this context).

However, as Botson and Deliege (1975) have shown transitivity can be "induced" through colors: if e.g. stick a is yellow, stick b is red, and stick c is blue children already at the age of 4 or 5 can conclude a<c after the sticks have been hidden

Extensive comments on these and similar experiments on transitivity can be found in Bideau (p. 273 ff).

The second experiment has been reported by Globerson (1986) [4]) and refers to the rotation of an object in a toy plane. In the first situation the nose of the toy plane heads towards the tangent of the circle of rotation. When asked what would happen if one would cut the string a significant part of schoolchildren gave the physically correct answer (other answers were v_1 or v_3). If, however, the nose of the toy plane is heading towards the centre of the circle an overwhelming part of the subjects indicated (and tried to jusitfy) the trajectory v_4 – i.e. that the toy plane would fall towards the centre. As Globerson comments it is the direction of the nose which is important for the cognitive representation of the trajectory, but not the physical law. These inferences are induced by perceptive salience.

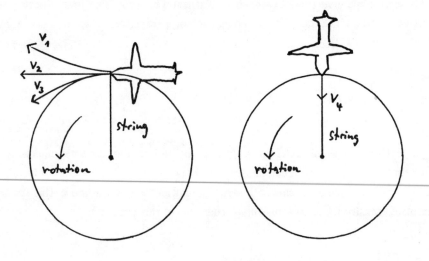

Fig. 1:   Rotation of a toy plane:
           Which direction will it go when we cut the string?

The third experiment, due to Fluckiger (1972), is similar in nature. Children are asked at what point a ball rotated on a circle should leave the circle to hit a given target. Most children indicate the point L_C of the circle which is closest to the target. The interpretation of this answer is as follows: L_C is situated on the straight line from the centre to the target as well as on the circle. To be on a circle and on a straight line are two strongly salient perceptive properties which in NT induce this inference for the indicated conclusion.

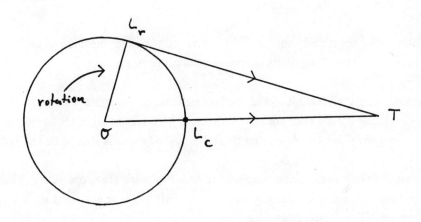

Fig. 2:   The "slingshot ball" :
                Where to leave the sling in order to hit the target

Formally: $\alpha_1 : L_C \in OT$

             $\alpha_2 : L_C \in$ Circle

             $\gamma : L_C$ is the leaving point in NT

L_C is the common element in the three cognitive propositions of this inference (cf. below the requirement of common context condition (c.c.c.) in natural logics.)

4. EXAMPLES OF NATURAL REASONING; PROTOLOGICS

The usual meanings of <u>negations</u> in NT are:

neg(neg A) <≠> A :   double negation of a proposition is not equivalent
                            to that proposition

neg A   is the contrary to A, if it exists
              i.e. neg(white) = black

neg A   rejection of A; A is not processed further in cognition

neg A   may also express (or represent) an opposition of a subject (in a dialog)
              to A.

There are also well–known problems for NT to compute the exact meaning of negations of quantified propositions or of negations of opaque (modal) contexts.

Examples: neg ("all inhabitants of the house choose one delegate")
                 neg ("John does not distribute all of his books to all of his friends")
                 neg ("John believes that there is somebody who is able to do the job").

Verbal expressions often need special emphasis in order to mark the scope of a negation of a complex proposition, i.e. neg ("John is married and rich"). These examples show some of the difficulties for NT in interpreting negations.

There are other problems in handlung negations, e.g. the "fallacy ad ignorantiam": a proposition not (yet) proven to be true is believed to be false in NT, and similar fallacies.

The problems of negation joined to the "vagaries in use of the conditional" – to quote Johnson–Laird – show the human weakness in dealing with logical processes.

I list a sample of examples of experiments which exhibit reasoning habits of NT. If not specified otherwise, these examples are listed in Wermus 1975 and have been used as exercises for students. It is important to note that logical competences of adults on the average show little progress in comparison with 12/13 years old youngsters. They exhibit also an important problem in teaching, namely: formal simplicity does not imply representational and understanding easiness and also conversely.

The following experiment shows differences in the syntax of natural language (NL) and in the formal logic (λ^F). Take a sequence of items characterised by two properties: size S and color C. The sequence is as follows:

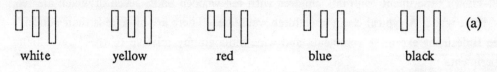

| white | yellow | red | blue | black |

The task is to construct a formula (a predicate) Bxy which is true iff the object x comes before the object y in the sequence (a). The usual linguistic expression in NL for Bxy is:

"If x is clearer than y, **then** x comes before y, **and** if the colors are the same, **then** x is smaller than y."

In contrast, the correct logical definition of the predicate Bxy in λ^F is:

$$Bxy \quad <=>_{def} \quad Cxy \lor (E_1 xy \land Sxy)$$

where Cxy means x is clearer than y, $E_1 xy$: x has the same color as y, Sxy: x is smaller then y;

and $\qquad Cxy \quad <=> \quad \neg E_1 xy.$

It appears that the NL expression (containing two conditionals and a conjunction) is **very different** from the formula in λ^F, and may be misleading if it is active in NT as a premise in an inference.

Experiments in class inclusion problems demonstrate the lack of a correct conditional in NT (Piaget, Matalon, Bideau). For example, let be

$\quad$ a := set of daisies (D)

$\quad$ b := set of flowers (Fl)

(1) $\qquad$ D implies Fl: $\qquad$ D $\subset$ Fl $\quad <=> \quad$ a is subset of b.

However, the question "Are there more daisies than flowers?" is a difficult one for children up to the age of 7 or 8; and the question (Bideau), "What can you do in order to have more daisies than flowers?" does not yield a clear answer before the age of 10. Another well–known experiment confronts children with ten wooden balls, seven of which are red, and three white. A typical claim of children would be, "There are more reds than wooden". (See Bideau for extensive comments and wide bibliography relative to these and similar experiments.)

I should like to add that inclusion depends strongly on the kind of conditional used in cognition (see formula (1) above). The truth table looks like this:

X	Y	$X \supset Y$ def. in λ^F	$X \supset_n Y$ in NT
1) +	+	+	+
2) +	−	−	−
3) −	+	+	*
4) −	−	+	*

where "+" stands for TRUE, "−" for FALSE, and "*" for UNDEFINED. Thus, there are value gaps in NT. This is a well–experienced fact (e.g. in teaching): A false antecedens (X) of the conditional doesn't generate any (clear) meaning; lines 3) and 4) in the truth table are inactive in cognition.

Also Wason and Johnson–Laird (1972) investigated the use of conditional in NT. Given a set of cards,

(1) $\boxed{B}$ $\boxed{2}$ $\boxed{E}$ $\boxed{7}$ "abstract condition"

people should test the assertion

α: "If vowel on one side, then even number on the other side."

So the task is to see which of the cards you must turn over in order to decide whether α is true or false, and to use a minimal number of turns. The success in this experiment was below 20%, which is very little.

A logically equivalent test,

β: "**If** envelope closed, **then** stamps on the other side",

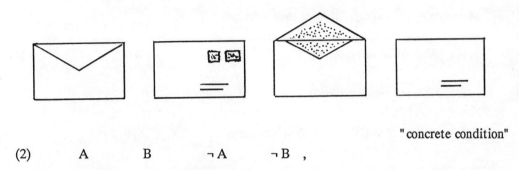

"concrete condition"

(2)   A   B   ¬ A   ¬ B ,

resulted in a much better success (> 70%).

Johnson–Laird commented about the above experiments: "There was obviously no transfer between the two forms of the task (...the "abstract" one 1) and the "concrete" condition 2) ...), and only two subjects (of 22) subsequently acknowledged any logical relation between them".

One has to pay attention to the fact, which I shall discuss below, that in task (1) – called "abstract" by the authors – there is only a very loose content connection between vowels and even numbers. In contrast the "common content condition" (c.c.c) between envelopes and stamps is common in NT.

Johnson–Laird [5]) has noted that composition of comparisons depend on the converse operations and negations involved. In the case
    "A better than B, C worse than B"
for a comparison between A and C a converse operation is needed; the success is average. In the cases
    "A better than B, B not as bad as C"
    "A better than B, C not as good as B"
negations and converse operations have to be used; this is much more difficult, takes much time, and success is low. The case
    "B not as bad as C, A not as bad as B"
involves converse operations, negations, and a permutation; it turns out to be very difficult and to require a very long time.

The conditional "If A, then B" in NT may have various meanings, such as

cause $\to$ effect

symptom $\to$ cause (e.g., "if fever then sickness)

A is a justification or reason for B

A is a sufficient condition for B

A is a subclass of B ("if square then rectangle)

B follows A in time

...

In contrast, in (interpreted) mathematical logic λ^F the only facts valid in general are:

If "A implies B" is true, then extension (A) $\subset$ extension (B);
and, in addition, if A is also true, then B is true: (A entails B).

5. EXAMPLES OF SOME OTHER DEVIANT PHEMOMENA IN λ^F

a)   The expression
$$\text{"to be or not to be"}$$
is a tautology in λ^F and as such containing no information about reality (is true in every interpretation), but may be highly significant in the mind (of a non–logican).

b)   The common content condition (c.c.c) constraint in NT:
$$\text{"if } 2 + 2 = 5 \text{, then Geneva is in Switzerland"},$$
is a well formed formula (even true) in λ^F, but not common in NT.
(see the precise definition of the c.c.c. in Wermus 1978).

c) The enthymema phenomenon: A part of an intended meaning is not expressed.
Let α be the sentence

$$R => \neg C$$
"if it is raining, then I don't come"

The intended meaning in NT is most likely an equivalence: α and it's converse α^{-1}:
$$"R => (\neg C) \text{ and } (\neg R) => C"$$
"If it's raining then I don't come, and if it does not rain then I come."

d) Amalgamated representations of objects in cognition:
In cognition there doesn't exist a representation of exactly <u>one</u> property of an object, say the color of an eye of a given person; there is in the same time a great number of other "contextual" properties: caracteristics of eyeball, the location in the face, properties of the look etc. This is in contrast to a treatment of an explicity stated predicate in λ^F.

This fact is most important in the cognitive processing of representations (Wermus 1978, 1982).

Some laws (or theorems) of formal logic λ^F (in propositional calculus) seem paradoxical from the point of view of NT; such as

L_1: $\vdash (A => B) \lor (B => A)$

L_2 $\vdash (A => B) \lor (A => \neg B)$

L_3 $\vdash A => (\neg A => B)$ ("ex falso quodlibet")

L_4 $\vdash (A => B) \lor (\neg A => B)$

...

Here "$\vdash A$" means "A is a theorem".

These laws (tautologies), and many others, are irrespective to the common content condition (c.c.c.); substitutions of contents may give rise to unusual, for NT strange, instancies (assertions); e.g.,

L_{10}: "If there is life in other galaxies, then Paris is the capital of France,

   or if Paris is the capital of France, then there is life in other galaxies."

In λ^F, L_{10} is a true, well–formed stetament, but highly unusual in NT.

Some fallacies of reasoning in NT are due to invalid inference schemes. Consider, for instance, this example:

Inference I_1: α_1:   if $3 = 2$, then $2 = 1$

            α_2:   if $2 = 1$, then $1 = 0$

            α_3:   but $3 \neq 2$

            _____

Conclusion γ: therefore, $1 \neq 0$.

All premises in the example as well as the conclusion γ are true, but the inference I_1 is invalid, as can be seen from the following counterexample:

Inference I_2: β_1:   If A is a citizen of Geneva, then he is Swiss.

            β_2:   If A is Swiss then he is European.

            β_3:   But A is not a citizen of Geneva.

            _____

            γ':   A is not European ???

Both I_1 and I_2 are instances of the same formal inference scheme:

         $x => y$
         $y => z$
         $\neg x$
         _____
         $\neg z$

which is known to be invalid. However, I$_1$ is usually accepted in NT, in spite of existing counterexamples, like I$_2$. This demonstrates the low competence of NT (of non–professionals) to discriminate between valid and invalid inferences.

The following example of reasoning, which is taken from a test for 1st year students, regards a concrete argument in NT (Wermus 1975):

α_1: If John is telling the truth **then** Paul is guilty.

α_2: If Dan is telling the truth **then** John lies.

α_3: But investigation has shown that Dan is right.

_____

Conclusion γ : What about Paul?

o γ_a: guilty

o γ_b: not guilty

o γ_c: H = { α_1, α_2, α_3 } is inconsistent

o γ_d: no conclusion about Paul is possible

o   don't know

The formal scheme of these arguments is

$$x => y$$
$$z => \neg x$$
$$z$$
_____

γ : $z \wedge \neg x$

This example was one of the questions in a test of reasoning competences of <u>students</u> with entered our Faculty of Psychology. In 1976 an overhelming part, 75 %, spontaneously gave the γ_b conclusion ("Paul not quilty"), the rest was more or less uniformly distributed over other conclusions. The correct conclusion is γ_d. The predicate "is guilty" was replaced (next year) by "left to Zürich" which was affectively (intentionally) neutral, the logical form of the inference remaining exactly the same. "Only" 60 % of students indicated the conclusion γ_b' ("has not left") which would correspond to γ_b. There were also other significant justifications of the responses to the test which suggested several requirements for a realistic modelisation of human reasoning habits.[6]) It appeared also clearly that intentional or affective components like "guilty" work as attractors or inhibitors in most natural inferences.

The next example suggests another important feature of protologics namely the wide use by NT of only partially defined logical connenctives (cf. also the protoconditional in class inclusion problems mentioned above). Consider the proposition:

α: "If this substance is an (efficient) remedy (R) and has no side effects (S), then it should be applied (A)".

The formal meaning of α as given in Propositional Calculus (PC) is defined by the thruth table for the 8 possible combinations of truth values of the three relevant parameters R, S. A. In cognition however, the valuation "R is false" (R is not an (efficient) remedy) is usually not further processed and consequently activates no meaning components in mind. The resulting meanings in λ^F and in cognition are different. The valuations considered relevant in NT and consequently processed are in general a subset of the complete truth values table of a given proposition.

These and many other experiments and teaching experiences, exhibit several properties of protologics which may be used as requirements for a model of natural logic λ^n:

-   There is no "logical omniscience" in NT (apart from a small subset, there is no correct use of other valid inferential rules)
-   No use of paradoxical laws of formal λ^F
-   Particular meanings of negation and conditional
-   Common content condition for logical links
-   Connectives of λ^n have only partially defined valuations
-   Valuations for compound propositions may occur in spite of value gaps of its parts
-   Strong dependency of inferences upon (tacit) modal functors. Desire preserving inferences.
-   Reluctance to use variables, or complex deductions.
-   Most of the processing of representations and of inferences are tacit, not (necessarily) expressed in NL (natural language).

6. INTRODUCTION TO THE PROTOLOGICS

<u>Remark</u>: In order to distinguish cognitive representations from the formal predicates and propositions, I use in the sequel, if not specified otherwise, $\hat{P}$, $\hat{\alpha}$ (with a "$\wedge$") for cognitive predicates and "constructs" (= compound propositions); see Wermus 1978 for more details.

Several $\hat{P}$'s may combine into a cognitive construct $\hat{\alpha}$ by means of protological links belonging to a kind of <u>"natural logic"</u> or protologic λ_s^n of the subject s. In contrast to formal logic connectives ("and", "or", "if ..., then ...", etc....) the protological links are only <u>partially defined</u> by means of their truth value table (VT).

In early stages of development valuations of constructs are not alethic (i.e. truth, falsity, necessity, etc....) but <u>attributive</u> (as I call it).

Let $V = \{+,-,*\}$ be the set of attributive "values".

$Val_s \hat{P}\alpha = +$ means for s: $\hat{P}$ matches a given object α, or abstracts (selects) $\hat{P}$ from α <u>and</u> accepts $\hat{P}$ for further processing. $Val_s \hat{P}\alpha = -$ means inconvenience or rejection of $\hat{P}$, even if this rejection may be further processed. $Val_s \hat{P}\alpha = *$ roughly means that no valuations of $\hat{P}$ occurs in the mind of s: we call the "*" an <u>occultation</u> [7]) of valuation or an undefined value (value gap). There may be different reasons for such an "occulation": no correspondence, no endeavour to match $\hat{P}$ to an object α takes place, $\hat{P}$ is not active in the mind of s at t (s does no "see" $\hat{P}$), no decision about the match or mismatch is reached (lack of decision for a valuation), nor about whether or not $\hat{P}$ applies to a given object α. One can even imagine that an s conceives a $\hat{P}$ without an existing or a possible object α on which a valuation could be tried. This latter case of distinction between cognitive conceivability and possibility is left here as an open question.

The connectives of the λ^2 and λ^3 protologics are defined by the following value tables. To simplify the notations, the index s, the "$\wedge$" above the P's are omitted and val P is written instead of $val_s (\hat{P}\alpha)$, unless it is specifically required otherwise. Parantheses are also omited where confusion should not appear. Subscripts as in Λ_2, Λ_3, $\rightarrow_3$, etc....indicate the corresponding level of protologic, to which the connective belongs.

Value tables for λ^2, λ^3, λ^4:

P	Q	PΛ_2Q	P$\rightarrow_2$Q	PV_2Q	PΛ_3Q	P$\rightarrow_3$Q	PV_3Q	PV_{3e}Q	PΛ_4Q	P$\rightarrow_4$Q	PV_4Q
+	+	+	+	*	+	+	+	−	+	+	+
+	−	−	−	+	−	−	+	+	−	−	+
−	+	*	*	+	−	+	+	+	−	+	+
−	−	*	*	*	*	*	*	*	−	+	−

(VT) of negation:

P	N(P)
+	−
−	*
*	*

Cases where val($P\alpha$) or val($Q\alpha$) are undefined (*) will be discussed below.

Λ_4, $\rightarrow_4$, V_4 are the usual connectives of formal logic.

It is a peculiar feature of natural negation NP that it remains the same in all the λ^n, except for a particular subclass of λ^4 which will be specified later.

Note that Λ_2 is not commutative: the rejection of the first conjunct P makes the whole representation of the conjunction inactive. Consequently, no further meaning is generated in such a case. In general two conjuncts do not have the same importance for a subject. This fact is very common in the (natural) thinking of children. Note further that in λ^2 the connectives Λ_2 und $\rightarrow_2$ are indistinguishable. This distinction appears only at the Λ^3 level. The lack of a correct conditional in earlier stages may explain the problem children have with class inclusions; cf. the well–known experiments, carried out in many countries, when young children are asked if "There are more roses than flowers" and other similar problems (cf. examples above).

There may be also an uncertainty in the definition of the "V_2, V_3" protodisjunction connectives, particularly in λ^3; some other possibilities could be considered, but changes in the partial definitions of the protoconnectives do not entail significant changes in the general features of the protologics.

The attributive valuations (acceptance, rejection) dominate usually the alethic ones (truth, falsity, necessity). The correct use of alethic valuations and the emergence of a classical negation functor appear later in the development i.e. in the so called hypothetico–deductive stage [8]). In the λ^n we do not have $N(N\hat{P})$ equivalent to $\hat{P}$, except special cases specified below.

Note that the (*) has not the same status as the determined values "+" or "–". (*) is a value gap which may become determined by a new information or by solving "backwards" a kind of logical equation when a desired value of a compound construct activates values of its parts; see also the comments about "strong protologics", which I am going to define soon. The value gaps may also be determined by representations of higher order by means of so called "reflexive functors", which will be described in Nr. 9.

<u>Definition of weak an strong protologics:</u> In dealing with the (*), I shall distinguish for each λ^n, n = 2,3,4, the <u>weak protologic</u> λ^{n1} and the <u>strong protologic</u> λ^{n2} according to the following definitions: Let in the so called "polish notation" $C_n(\epsilon_1, \epsilon_2)$ stand for the valuation function corresponding to the conjunction in λ^n, $I_n(\epsilon_1, \epsilon_2)$ the function for the conditional, $D_n(\epsilon_1, \epsilon_2)$ the function for the inclusive disjunction where $\epsilon_k \in \{+,-,*\}$.

Let $f_n^j(\epsilon_1, \epsilon_2)$ stand for one of these valuation functions and j = 1,2. Then for determined values of $\epsilon_1, \epsilon_2 \in \{+,-\}$ with

$$f_n^j(\epsilon_1, \epsilon_2) = \text{as defined in the (VT),}$$

$f_n^1(\epsilon_1, \epsilon_2)$ belongs to the <u>weak</u> λ^{n1} iff:

$$f_n^1(*, \epsilon_2) = f_n^1(\epsilon_1, *) = *$$

$f_n^2(\epsilon_1, \epsilon_2)$ belongs to the <u>strong</u> λ^{n2} iff:

$$f_n^2(*, \epsilon_2) = \epsilon_2, f_n^2(\epsilon_1, *) = \epsilon_1$$

Many examples of children's behaviour may be interpreted as an use of the weak (or strong) protologic. In problem solving procedures, the desired value of a construct α is often assumed without knowing the values of all of the parts of α. Having in mind the desired value of the construct α, the subject may try to solve a kind of protological equation in order to determine previously undefined values (*) of its parts (goal–driven valuations). In contrast, weak logic shows dominance of an undefined value (*) over a defined one; this often appears in exploring the consequences of a hypothesis, when, in the processing of one of its assumed values, an undefined value is encountered which blocks the whole process and leads to the abandoning of the hypothesis.

One can study different laws of each of these types of protologic λ^{nj} (j=1,2) (cf. Wermus, 1977 and 1978). Proto–equivalence, prototautology, proto–contradiction may be defined in ways analogous to the corresponding concepts in usual propositional calculus. I.e. construct α is a prototautology if its VT contains only " + " and " * " and similarly for the other concepts. A construct is totally occulted if its VT contains only " * ".

Note that

$$PV_3(NP) \text{ and } P\Lambda_3(NP)$$

are in λ^{3j} only a prototautology and a proto–contradiction, so that classical analytical expressions may acquire other possible (i.e. not analytical) interpretations in the λ^{nj} logics; e.g. "to be or not to be" (example above, No. 5). The law of the excluded middle and the principle of non–contradiction are not (fully) valid in these λ^n.

7. SOME PROPERTIES OF PROTOLOGICAL SYSTEMS λ^{nj}. LEVELS OF DISCERNABILITY

The λ^{nj} protologics are of course poor types of logic but any of the so–called paradoxical laws of classical logic are valid for them, as shall be verified in some examples.

It is easy to see that in a weak logic a valuation of a construct $\alpha(P,Q,...)$ becomes undefined as soon as one atom, or one of its parts, is undefined. In building complex constructs or inferences in a weak λ^n, where several protoconnectives occur, the undefined values tend to proliferate. Thus the whole construct or inference becomes finally undefined (respect. conclusionless) and stops beeing active in the mind. This fact is in accordance with many data from natural reasoning.

In a strong logic the undefined values of parts of a construct α are ignored for val(α): the whole construct α is valuated in computing suitably the remaining determined valuations of his parts.

Let us say $\alpha(P,Q,...)$ is more discernible in λ^{mj} than λ^{nk} if the valuation function of α in λ^{mj} is an extension of the one in λ^{nk} i.e.

(i) the defined values of α in λ^{nk} are all the same as in λ^{mj}

(ii) some undefined value in λ^{nk} becomes determined in λ^{mj}; $n,m=2,3,4$; $j,k=1,2$.

Note that there are constructs α and λ^{mj} and λ^{nk} such that α is neither more discernible in λ^{mj} than λ^{nk} nor conversely. This is the case for example for PVQ in λ^{22} and λ^{31}: condition (i) is in this case not satisfied. The relation D (α, mj, nk) as defined by (i), (ii) is also only a partial ordering.

Note that $P\Lambda_n Q$, $PV_n Q$ are more discernible in strong λ^3 than in all λ^2 and λ^3 weak. Consequently it is easy to see that all constructs are more discernible in λ^3 strong than in all the other protologics, except the λ^4 ones (see below).

In case where D(α, mj, nk) holds we say λ^{nk} is lower than λ^{mj} for α. It may also be easiliy shown that λ^{21} is lower than other protologics λ^{nj} for all α. The discernibility comparison fails only for λ^{22} and λ^{31} (respectively for λ^{32} and λ^{41}).

Using VT it is also easy to verify that the so-called paradoxical laws of classical propositonal calculus (CPC) are not valid even in the strong λ^3.

I.e. $\alpha \to_n (\beta \to \alpha)$, $\alpha \to_n (\beta \to_n \beta)$, $(\alpha \to_n \beta) V_n (N\alpha \to_n \beta)$, $(\alpha\Lambda_n\beta) \to_n \alpha$, $(\alpha\Lambda_n\beta) \to_n \beta$

are all contingent, even in the λ^3 strong; and for example,

$\alpha \to_n (N\alpha \to_n \beta)$, $(\alpha \to_n \beta) V(\sigma \to_n N\beta)$, $(\alpha \to_n \beta) V_n (\beta \to_n \alpha)$ are only prototautologies.

If $D(\alpha, \lambda^{mj}, \lambda^{nk})$ holds for all construct α, then I say λ^{mj} is of higher level discernability than λ^{nk}, or in short λ^{mj} is higher than λ^{nk}. The discernability relation defined as

$$D(\lambda^{mj}, \lambda^{nk}) <=>_{def} \forall \alpha\, D(\alpha, \lambda^{mj}, \lambda^{nk})$$

induces a partial ordering on the set of protologics; we get the lattice shown in the following figure:

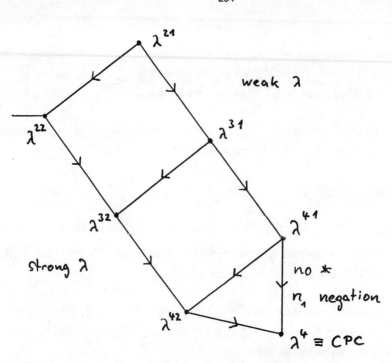

Fig. 3:   Lattice of protologics with growing discernability relations

The classical propositional calculus (PC) is attained from λ^{41} or λ^{42} by using

    (i)   only the determined values " + " or " – "

    (ii) the formal negation neg(neg α) < = > α.

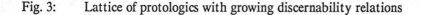

It is easy to verify that these protologics meet most of the requirements for a description of natural logics.

The inference rules resulting from the definitions of the λ^{mj} can be investigated in an analogous strategy as in the formal logic. For example the transitive inference rule:

$$P \to_n Q, \; Q \to_n R \; \|{\to}_n \; P \to_n R$$

is valid in the λ^{21} protologic only in the trivial case where all the atoms, P,Q,R have the value " + " and is not conclusive otherwise, i.e. the conclusion is in all these other cases undefined. This rule beccomes conclusive in other cases in the higher level protologics. The rule

$$P \vee_n Q, NP \parallel\!\!\rightarrow Q$$

– a case of Modus Tollens – becomes valid only for the λ^{32} and higher protologic. This is in accordance with the well–known problems children have in dealing with this rule.

8. FURTHER SIGNIFICANT FACTS OF PROTOLOGICAL SYSTEMS

Neither $P \wedge_n P$ nor $P \vee_n P$ are protoequivalent to P in any of the λ^{nj}, (exept a particular case of λ^{41} isomorphic to <u>classical propositional calculus (CPC)</u>, see below.

Most interestingly $(P \wedge_n Q) \wedge_n R$ is associative in the weak λ^{21} and λ^{31} protologics, but ceases to be associative in the strong λ^{32}. Analogous evolution takes place with the distributive and some other basic rules of CPC.

The numerous value gaps of such protoequivalences in the weak λ^{21} and λ^{31} protologics prevent the appearance of values which destroy these protoequivalences in strong λ^{32}. The validity of these rules is recovered in λ^4 binary logic which is in fact the CPC.

It is interesting to note that the well–known confusion in natural language of (NP) $\wedge_n Q$ with $N(P \wedge_n Q)$ has its counterpart in the protoequivalence of these constructs in the λ^3 weak protologic; this protoequivalence disappears in the strong λ^3. Most important (and in accordance with the meanings occuring in natural languages) (NP)$\vee_n Q$ and $P \rightarrow_n Q$ constructs are not protoequivalent in any of the λ^{nj} protologics, except the $\lambda^4 = CPC$.

The lack of useful inferential rules in the λ^{mj} protologics may partly explain the fact that NT uses other means instead of inference rules in order to understand, argue or take decisions. There exist also a reluctance in "everyday" dialogues to argue by means of deductions; NT uses mainly examples, descriptions, comparisons, metaphors, persuasion, quick induction etc. in order to justify or to convince.

9. TRANSITIONS FROM LOWER TO HIGHER LEVEL PROTOLOGICS; REFLEXIVE FUNCTORS

As developmental psychology shows, there may exist in mind representations of other representations: second order representations, reflections, introspection etc. One can think about one's own (lack of) decision. These representations undergo also analogous operations and valuations as the previous ones. Let

$$\Theta \hat{P} = \hat{\hat{P}} = \hat{P} \ (2)$$

be a higher level representation of the fact represented by $\hat{P}$. The symbol Θ abreviates a particular case of "reflecting abstraction" [9]), which, following Piaget, construct higher cognitive stages from lower ones.

I define several valuation functions: u_0, n_0, r_0, ..., u_1, ... etc. from the set of valuations of $\hat{P}$, i.e. from $\{+, -, *\}$, to valuations of $\hat{\hat{P}}$ by the following valuation tables (VT_1):

Val $\hat{P}$	$U_0\hat{\hat{P}}$	$U_1\hat{\hat{P}}$	$U_2\hat{\hat{P}}$	$n_0\hat{\hat{P}}$	$r_0\hat{\hat{P}}$
+	+	+	+	−	*
−	*	−	+	*	*
*	*	*	*	*	+

(VT_1)

I use a slightly incorrect abreviation in writing $u_0\hat{\hat{P}}$, $u_1\hat{\hat{P}}$ etc., instead of $u_0(\text{val } \hat{P})$, $u_1(\text{val } \hat{P})$ etc. Let me call the u_0, u_1, ..., functions and their compositions "reflexive functors". Some of them particularly u_0, n_0 and r_0 have an interesting interpretation as peculiar processes in cognition. Instead of $\hat{\hat{P}}$ we shall now write $\hat{P}^{(2)}$.

u_0 may be interpreted as a functor which accepts (retains) $\hat{P}^{(2)}$ only if $\hat{P}$ itself is accepted and is inactive in the other cases. n_0 rejects $\hat{P}^{(2)}$ in case of the matching (or acceptance) of $\hat{P}$ with an object α. The latter can mean that the case val $\hat{P} = +$ is for some other reason rejected (valued as unfavourable for the s). The most interesting is the r_0 functor which accepts to process (to investigate) $\hat{P}$ if the val $\hat{P}$ is undefined.

Different compositions of these functors allow the analysis of transitions from <u>one level of protologic to higher</u> ones (i.e. from λ^{2j} to λ^{3k}, j, k = 1 or 2). One has to accept a kind of reduction principle which allows to combine valuations of representations of different levels. This principle is, to some extent, natural because all these operations are performed in the same cognition. The mind <u>may simultaneously process several representations.</u> It can also build representations of representations and conversely use chains or references of these iterated representations.

In the following formulae, parantheses will be omitted as well as the "^" above P when no ambiguity occurs; the term "conjunctions" C_{ik} instead of "valuation functions" C_{ik} etc. will be used. Let me denote

$$u_2 (P) =_{def} {}^r{}_0({}^r{}_0 P), \quad \text{and further}$$
$$n_1 (P) =_{def} C_{22} (u_2 P, n_0 P)$$

Note that C_{22} is the strong conjunction of λ^{22}. The $n_1 P$ then becomes the <u>formal</u> <u>negation</u>, as can be easiliy verified by using the $VT_2 : n_1 (n_1 (val\ P)) = val\ P$.
In fact the value table of $n_1 P$ is

	val P	val(n_1P)
(VT$_2$)	+	−
	−	+
	*	*

One can easily generate further reflexive functors using compositions of the already defined and the C_{22} connective of λ^{22}. There is a conjecture that the $3^3 = 27$ reflective functors form a semi–group with u_1 as unit element, but I will not investigate this issue here.

I mention other functors which occur essentially in the formulas espressing the formation of connectives of higher protologics by means of the lower ones:

val P	r_1 P	r_2 P	r_3 P	r_4 P
+	*	*	+	+
−	*	+	−	−
*	−	+	+	−

(VT$_3$)

It's easy to verify the following equations using the already established value tables:

$$r_1 = n_o \cdot r_o; \quad r_2 = r_o \cdot u_o$$
$$r_3 = C_{22}\,(u_1\,P, r_o\,P); \quad r_4 = C_{22}\,(n_1 P, r_1 P)$$

10. FORMULAS EXPRESSING TRANSITIONS FROM LOWER TO HIGHER PROTOLOGICS

Let ϵ_1 = val P and ϵ_2 = val Q, where as usual $\epsilon_i \in \{+, -, *\}$.

$C_{nk}\,(\epsilon_1.\,\epsilon_2)$ denote valuation functions of conjuctions, $\text{Imp}_{nk}\,(\epsilon_1.\epsilon_2)$ of conditionals, $\text{DE}_{nk}\,(\epsilon_1,\,\epsilon_2)$ of exclusive disjunctions, $D_{nk}\,(\epsilon_1,\,\epsilon_2)$ of inclusive disjunctions in the λ^{nk} protologics.

The following formular express transitions from λ^{3j} to λ^{31}, j = 1, 2:

$C_{31}\,(\epsilon_1,\,\epsilon_2) = C_{22}\,[C_{21}\,(\epsilon_1,\,\epsilon_2),\,C_{21}\,(\epsilon_2,\,\epsilon_1)]$
$\text{Imp}_{31}\,(\epsilon_1,\,\epsilon_2) = C_{22}\,[C_{21}\,(\epsilon_1.\,\epsilon_2),\,U_2\,C_{21}\,(\epsilon_2,\,\epsilon_1)]$
$\text{DE}_{31}\,(\epsilon_1,\,\epsilon_2) = C_{22}\,[r_o\,C_{31}\,(\epsilon_1,\,\epsilon_2),\,D_2\,(\epsilon_1,\,\epsilon_2)]$

From λ^{3j} to λ^{41}:

$C_{41}\,(\epsilon_1,\,\epsilon_2) = r_4\,C_{31}\,(\epsilon_1,\,\epsilon_2)$
$\text{Imp}_{41}\,(\epsilon_1,\,\epsilon_2) = r_3\,\text{Imp}_{31}\,(\epsilon_1,\,\epsilon_2)$
$\text{DE}_{41}\,(\epsilon_1,\,\epsilon_2) = r_4\,\text{DE}_{\,8}\epsilon_1,\,\epsilon_2)$
$D_{41}\,(\epsilon_1,\,\epsilon_2) = r_3\,\text{DE}\,(\epsilon_1,\,\epsilon_2)$

The strong valuation functions can be defined as follows:

Let $F_{n2}(\epsilon_1, \epsilon_2)$ denote one of the strong valuation functions of conjunctions, conditionals, disjunctions in λ^{n2}, for n = 2, 3, 4.
Then

$$F_{n2}(\epsilon1, \epsilon_2) = \begin{cases} F_{n1}(\epsilon_1, \epsilon_2) \text{ for } \epsilon_1, \epsilon_2 \in \{+, -\} \\ \\ \epsilon_1 \text{ if } \epsilon_2 \text{ is } * \\ \\ \epsilon_2 \text{ if } \epsilon_1 \text{ is } * \end{cases}$$

As mentioned before, the classical Propositional Calculus (PC) appears as the λ^4 logic where the negation n_1 is used and no undefined values (*) for atoms occur.

The given transition formulas are not unique, but every such a formula require an occurance of at least one reflexive functor, except the above formula for C_{22}.

Many of the reflexive functors may have intersting interpretations as representations of "higher" thinking processes i.e. concept formation, introspection, or when a sudden insight take place (an unknown value (*) or a hidden property becomes known) etc.

The lattice of protologics as in fig. 3 can be refined in considering the possibility of using the n_1 formal negation also at lower levels, instead of the n_0 alone. In this case we get 12 different protologics and the 13th as the classical PC which is the unique maximal element of this extended lattice, see the figure on the following page.

11. A CONCLUDING REMARK

These researches in NT and particularly in protologics rise several important questions in (genetic) epistemology, didactics, argumentation etc.. To end up, I mention only one of them.

This is the <u>accessibility</u> relation between two cognitions, for example between a teacher t and a student s and concerning the understanding of an expression α. Note that this accessibility relation Acs $(t, s; \alpha)$ has an interesting analogy to the same concept used in the "possible world semantics" [10]. It appears that, if the teacher uses in expressing α (or in an argument) the protologic λ_t and a student conceives α, or the argument, say, in a lower

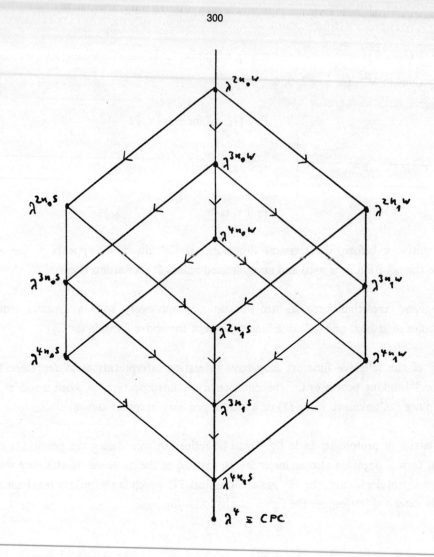

Fig. 4: Lattice of protologics refined (see preceding page):

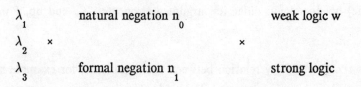

λ_1	natural negation n_0	weak logic w
λ_2	×	×
λ_3	formal negation n_1	strong logic

protologic λ_s, then the cognitive meaning generated by s is poorer than the meaning intended by t; the activated semantic nets differ and the representations of s have more occultations (or indetermination values), than these of t.

The natural maturation processes of the minds of adults are in general very slow. This rises the problem of existence and of the kind of didactic means able to improve the protological abilities and consequently to render more efficient the accessibility relations.

BIBILOGRAPHY

Anderson, AR. & Belnap, N.D. Entailment, vol. I, Princeton Univ. Pr. 1975

Bideau J. Logique et bricolage, Presses Univ. de Lille, 1988

Blair, J.A. & Johnson, R.H. (Eds) Informal logic, Iverness, California Edgepress, 1980

Botson C., Deluge M. Le développement intellectuel de l'enfant, II, Andenne Magermans (Eds) 1975.

FodorJ. The language of thought, Sussex Harvester Press, 1976.

Gruber H & Vonèche, J. The essential Piaget. New York, Basis Books 1977.

Hughes G.E. & Creswell M.J. An introduction to modal logic, Methuen, London, 1972.

Johnson–Laird P.N. Models of deduction, in R. Falmage (Ed):
        Reasoning, Representation and Process, Erlbaum, Hillsdale 1975

Johnson–Laird, P.N. & Wason P.C. Thinking (reading in Cognitive Science), Cambridge Univ. Press, 1977.

Kitchener R.F. Genetic epistemology, History of Science and Genetic Pschyology, in Synthèse vol 65, 1985 devoted to Jean Piaget (B. Richards & H. Wermus, guest eds.), Reidel publ. Co.

Matalon B. Etude génétique de l'implication, in Beth W. et al. Implication, Formalisation et logique naturelle, Etudes d'Epistémologie Génétique, vol. XVI, Paris, PUF 1962.

Piaget J. <u>La génèse des structures logiques élémentaires: classifications et sériations,</u> Neuchatel, Delachaux et Niestlé, 1972.

Piaget, J. <u>Genetic epistemology,</u> W.W. Norton, New York, 1971

Piaget, J. <u>L'équilibration des structures cognitives, problème central du développement,</u> RUF, Paris, 1975.

Piaget, J. & Garcia R. <u>Vers une logique des significations,</u> Murionde éd. (Science nouvelle), Genève, 1987

Searle, J. <u>Intentionality, an essay in philosophy of mind,</u> Cambride Univ. Press, 1983

Wason P.C. & Johnson–Laird <u>Psychology of Reasoning,</u> Univ. Harvard P., 1972

Wermus H. <u>Introduction à la logique formelle, (cahier d'exercises)</u> Centrale universitaire des polycopiés, Genève, 1975

Wermus H. <u>Essai de développement d'une prélogique à partir des foncteurs partiels,</u> Archives de Psychologie, No 174, Genève 1977

Wermus H. <u>Esquisse d'un modèle des activités cognitives,</u> in Dialectica, vol 32, Nos 3/4, Bienne 1978

Wermus H. <u>Procedures de la pensée naturelle et schèmes formels,</u> in Cahiers de la fondation Archives Jean Piaget, Nos 2/3, Genève 1982.

Wermus H. <u>Formation des opérations nouvelles à l'aide des operateurs réflexifs de la pensée,</u> in Archives de Psychologie, 51, Grnève 1983

Wermus H. <u>Des connaissances et des croyances,</u> in Cptes rendus des VIIIèmes journées d'éducation sscientifique, Giordan & Martinand eds. UER Paris VII, Paris 1986

Wermus H. <u>Représentations du monde dans l'esprit des enfants,</u> in Comment parler avec un enfant aujourd'hui, Cptes rendus Colloque J. Korczak à l'Uni. Genève, éd. DelVal, Fribourg 1989

Zimmermann B.J. & Blom D.E. <u>On resolving conflicting views of cognitive conflicts,</u> in Developmental Review, 1983 b No 3

Cf. also volumes <u>Etudes d'Epistémologie Génétique,</u> directed by Jean Piaget, 36 volumes, Presses Univ. de France, 1955–80, for detailed descriptions and comments of numerous experiments with children.

FOOTNOTES

1) I do not intend now to discuss this hard issue.

2) EEG = Etudes d'Epistémologie Génétique, published under the direction of Jean Piaget, 36 volumes, Presses Universitaires de France (PUF), Paris 1955–80. These volumes include proceedings of annual colloquia organised by the "Centre", chairman Jean Piaget, meet every weak to discuss talks on epistemology and reports in experimental psychology. Some of the examples given in this paper are abstracts of my records of these reports.

3) "Natural logic" in this paper must not be confused with Gentzens theory which deals with strongly formal logical rules.

4) Experience related by Professor Tamara Globerson of Tel–Aviv University in a talk at our Faculty of Psychology in Geneva, 1986.

5) Similar experiments with analogous results are reported in several EEG by Piaget et al.

6) Tests given to first year students (with some changes) from 1972 to 1978. The figures of 75 % and 60 % represent a mean value of the responses during these years. The shift from γ_b to γ_b' is significant of people which have had logic courses in colleges.

7) "Occultation" comes from ophtalmology and means something intervenes blocking the vision: it is an inability to perceive or to be aware of a property (at a given moment t).

8) In his famous "stage theory" Piaget uses the term "formal stage"; in fact this term seems to me inappropriate as the mind almost never processes uninterpreted (formal) symbol. The label "Hypothetico–deductive stage", sometimes used also by Piaget, appears to me more appropriate.

9) My translation of the french term "abstraction refléchisante" which is for Piaget one of the most important transition factors leading from one psychogenetic level to a higher one.

10) See i.e. Hughes & Creswell.

A MODAL PROPOSITIONAL CALCULUS
FOR QUANTUM FACTS AND DYNAMICAL THEORIES

H. ATMANSPACHER[1], F. R. KRUEGER[2], AND H. SCHEINGRABER[3]

1. Facts and theories

An interdisciplinary meeting like WOPPLOT invites to look across boundaries between different scientific disciplines, and it challenges understanding from a viewpoint superior to that of the individual disciplines. Hence it contains a "transdisciplinary" aspect in the sense of a unifying or "meta" - view on different disciplines. A contribution to such a view concerning facts and theories and their mutual interrelations is attempted in this article.

A fact is generally not to be considered as an independent isolated object in an "objective world" (*noumenon*), but as a contextual object (*phenomenon*). Contextual is here understood with regard to both observation and interpretation of the fact. The contextual aspect emphasizes the relevance of pragmatic and semantic influences onto the statement of a fact. This means that noumena would be relevant only in a situation without any observational and interpretational context. Strictly speaking, noumena can therefore be considered as a (artificial) limiting case of phenomena. Being contextual objects, facts are phenomena: they are pragmatically and semantically contaminated.

A theory provides order relations among facts, i.e. it provides an interpretation of facts. (Note that the latin origin of "interpretation" is *interpres*, the mediator – in our regard among facts.) As facts (considered as phenomena) are contextual objects, theories (considered as interpretations) are contextually subjective constructions designed as compressed descriptions of facts. All one can ultimately get from a theory is thus consistency with facts concerning their contextual interpretation.[1] There is no genuine semantic difference between facts and theories.

Figure 1 shows a schematic representation of this situation. Interfactual relations in systems

---

[1]Max-Planck-Institut für Physik und Astrophysik, D-8046 Garching, FRG,
   ICAS - Institut für Cybernetische Anthropologie Starnberg, D-8130 Starnberg, FRG

[2]D-6100 Darmstadt 12; Guest at Max-Planck-Institut für Kernphysik, D-6900 Heidelberg, FRG

[3]Max-Planck-Institut für extraterrestrische Physik, D-8046 Garching, FRG

S_1 and S_2 constitute two types f_1 and f_2 of interfactual syntax. The theories T_1 and T_2 to which S_1 and S_2 are related constitute an intertheoretical syntax t. The main purpose of the present article concentrates on studying a specific syntax f_1 and a specific syntax t. We shall speak of a system S_1 as the support of interfactual relations f_1, and we shall speak of a metatheory MT (which is identically an intertheory) as the support of intertheoretical relations t.

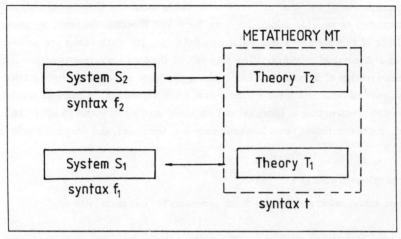

Figure 1: Schematic illustration of relations between facts and theories. Interfactual syntax (as denoted by f_1 and f_2) in systems S_1 and S_2 is related to theories T_1 and T_2. The relation between these theories constitutes an intertheoretical syntax (denoted by t) within a metatheory MT.

Our specific example for facts and theories refers to quantum facts on the level of a system S_1 (syntax f_1) and to dynamical theories T_1 and T_2 with an intertheoretical syntax t. Quantum facts are phenomena in the domain of quantum systems. Quantum systems are systems where Planck's action h is not negligible with respect to the relevant action corresponding to the Hamiltonian of the system.[4] Typical quantum facts are position and momentum. Dynamical theories are theories providing temporal order relations among facts. As specific examples refering to t we shall use Liouvillean dynamics and information dynamics.

Quantum facts and dynamical theories are formalized in the frame of functional analysis, i.e. we have operators P, Q for momentum and position (as usual in quantum theory), and we have operators L, M for Liouvillean and information dynamics. The formalism giving rise to L and M has originally been introduced by the Brussels school of Prigogine and coworkers.[2,3,4,5,6] A reinterpretation of this formalism in terms of information processing in dynamical systems has previously been proposed and investigated to some detail.[7,8,9] In Sec.2 we shall briefly present the main features required for the application of quantum facts and dynamical theories to the scheme shown in Fig.1. In particular we shall compare the incommensurabilities resulting from the commutation relations for the operators P, Q and L, M.

---

[4]This is a weak characterization since no boundary distinguishing a system from its environment is specified. This weakness is intentional since we do not want to exclude open quantum systems.

These incommensurabilities are further explored using a lattice theoretical approach. For quantum facts this has already been started in the seminal work of Birkhoff and von Neumann.[10] The lattice structure of dynamical theories has been studied recently.[9] Motivated by quite general arguments[11] the lattice structure corresponding to both facts *and* theories can be unified by the *dual* properties of *one* single lattice. This unification is the goal of Sec.3.

As a consequence of the dual properties concerning facts and theories, the obtained lattice can be interpreted to provide a logic for both facts and theories. In Sec.4 we show a formal correspondence of this logic to a four–valued modal logic. Its truth values are pairwise relevant to factual and theoretical relations. (The two–valued Boolean logic results as the limiting case for strict conservation of information, i.e., tautology.) Since the duality of the lattice reflects a basic equivalence of facts and theories, those four truth values can be pairwise combined into a new two–valued construction of (positive and negative) contingency and (positive and negative) definiteness. Some concluding remarks about meaning, semantics, and pragmatic information are given in Sec.5.

2. Incommensurabilities

The basic analytical expressions to define operators for quantum facts are

$$P\rho = ih\frac{\partial}{\partial q}\rho \tag{1}$$

$$Q\rho = q\rho \tag{2}$$

where P is the momentum operator, Q is the position operator, ρ is a distribution function in quantum mechanical phase space, h is Planck's action, and $i = \sqrt{-1}$.

An algebraic formulation of (1) and (2) would be advantageous to illustrate correlations, non–separability, and symmetry breakings in composed and open quantum systems.[3,12,13] Since our purpose in the present article will be focussed on incommensurabilities and their consequences we prefer the analytical representation. It immediately provides the commutation relation:

$$i[P,Q] = h\,\mathbf{I} \tag{3}$$

with $\mathbf{I}$ as the identity operator.

Quite analogous in spirit, Liouvillean dynamics and information dynamics can be formalized by corresponding operators L and M:

$$L\rho = i\frac{\partial}{\partial t}\rho \tag{4}$$

$$M\rho = (M_o + Kt)\rho \tag{5}$$

with the commutation relation:

$$i[L,M] = K\,\mathbf{I} \tag{6}$$

Relations (1) — (6) have been extensively discussed earlier,[3,7,8] so that we only repeat the most essential points. The Liouvillean operator L gives rise to a unitary evolution of a system as characterized by its phase space distribution ρ. This evolution is temporally reversible in accordance with its information conserving character. Since L is given by canonical quantities,

$$L = \sum \left(-i\frac{\partial H}{\partial p}\frac{\partial}{\partial q} + i\frac{\partial H}{\partial q}\frac{\partial}{\partial p} \right) \tag{7}$$

it refers to systems of type S_1 with an interfactual relation f_1 as expressed by (3). However, this reference is rigorous only if L can be shown to be applicable to quantum systems. Much work of the Brussels school has been devoted to this issue in recent years,[3,12] and a considerable degree of evidence for a quantum mechanical L has been provided. Since it is difficult to estimate whether this evidence is sufficient and since the present status of the question does not restrict the validity of our subsequent argumentation, we may content ourselves with a solid confidence concerning the quantum relevance of L.

The information operator M has been constructed to account for classical systems processing information. There are different classes of systems providing a non–vanishing information flow as a function of time. In the conservative domain, these are K–flows, and corresponding dissipative systems are characterized by chaotic attractors. The central quantity for these systems is given by the Kolmogorov – Sinai entropy K.[14,15] Since K can be related to the sum of positive Ljapunov exponents of the system,[16] K is positive semidefinite. A positive K–entropy is a necessary and sufficient condition for the existence of K–flows or chaotic attractors. Interpreting K as an information production rate,[17,18] it is possible to formulate an information operator M according to (5).[7] Here M_o is the informational content of a solution of a system in phase space with respect to a given resolution at a time $t = 0$. The dynamical entropy K provides the linear approximation to the generally nonlinear information increase as a function of time and with the same resolution as at $t = 0$. This information increase is a consequence of the sensitive dependence of the state of the system on initial conditions which are fundamentally uncertain. For a detailed discussion of Eq.(5) we again refer to earlier publications.[7,8,9] In contrast to L the information operator provides a non–unitary evolution whenever $K > 0$. It is therefore capable of describing the temporal evolution of information processing ($K > 0$) as well as information conserving ($K = 0$) systems.

Although the commutation relations (3) and (6) are structurally equivalent, they show a significant difference between the commutators. While h is a universal constant, K is an invariant characterizing a particular system. In this difference the objective aspect of facts and the subjective aspect of theories as mentioned in the introduction appears again. This ambivalence will be resolved on the basis of a unifying lattice theoretical view to be developed in the subsequent section.

To summarize, we have facts given by (1) and (2) in a system S_1 constituting a syntax f_1 which contains the incommensurability (3). We also have theories T_1 and T_2 given by (4) and (5) where T_1 relates to S_1 via (7). The incommensurability (6) belonging to the syntax t is settled on the level of a metatheory MT.

3. Propositional lattice

With respect to quantum systems, Birkhoff and von Neumann[10] have initiated a particularly appropriate mathematical view toward incommensurabilities. It is based on propositional lattices, and contains elements of logics, set theory, and lattice theory. A comprehensive documentation of the history of quantum logics is given in Jammer's book.[19] An analogous approach concerning dynamical theories has been made recently.[9] In the present section we shall review, discuss, and compare both approaches.

In quantum systems one has to deal with propositions $a, b, \ldots$ about facts which are typically of the form:

$$a : \quad \text{"}p = p_o\text{"}$$
$$b : \quad \text{"}q = q_o\text{"}$$

In principle, each pair of incommensurable observables provides such an example. We use p and q since their incommensurability is most commonly known. However, it should be mentioned that the example of spin components is better suited to illustrate certain logical relations to be discussed below.

An investigation of the truth values of these propositions in a given system is to be imagined as a dialogue between the system and an observer (or observing system).[5] The resulting propositions are required to be affirmative.[21] They are object to a dialogical (proponent – opponent) demonstration of their validity. In this picture a measurement is a message from a system (proponent) to an observer (opponent). A possible dialogue could be as follows:

System: "$q = q_o$" (i.e. a measurement provides q_o as the position of an object).
Observer: "$q \neq q_o$" (i.e. the observer, representing the dialogical role of the opponent, has to doubt the proponent's proposition).
System: "$q = q_o$" (i.e. the repetition of the measurement, if prepared exactly as before, yields the same result).

A dialogue of this type implies that the proposition "$q = q_o$" has to be considered true. If there were repetitions of the measurement providing $q \neq q_o$ (which is possible for quantum mechanical incommensurable quantities) then the proposition "$q = q_o$" has at least to be considered as doubtful: the position may or may not be q_o.

A set G of propositions $(a, b, \ldots)$ with ordering relations $(\geq, \leq)$ is called a partially ordered set (poset). A poset is called a lattice V if it satisfies the basic properties of idempotency, commutativity, associativity, and absorption. They can easily be shown for propositions as indicated above. In the context of incommensurabilities the most interesting additional properties are complementation and distributivity. It has been shown[10] that in the context of quantum systems a

[5]This viewpoint has already been advocated in a series of articles by Cyranski. He suggests an axiomatic formulation[20] of a conceptual basis which seems to be very similar to ours. However, our approach represents a step beyond Cyranski's arguments in considering the duality of lattices as a crucial property. This duality will be shown to imply the ultimate resolution of the dichotomy of facts and theories.

complement a' exists for each a (and b) such that:

$$a \wedge a' = 0 \tag{8a}$$
$$a \vee a' = I \tag{8b}$$

Here, 0 is the null set, and I is the set containing all propositions possible within G. In particular, the complement exists for propositions about facts which are incommensurable with other ones. However, for incommensurable quantum facts this complement is not unique. This non–uniqueness is formally derivable through the distributive inequalities:

$$a > (a \wedge b) \vee (a \wedge b') \tag{9a}$$
$$\wedge \quad b > (b \wedge a) \vee (b \wedge a') \tag{9b}$$

It is generally not easy to understand the content of these inequalities. A dialogue as above would be most helpful in this respect. Since such a dialogue were too lengthy to be reproduced here, we refer the reader interested in a corresponding exercise to the example sketched by Jammer (page 349, footnote 31).[19] Formally speaking, the appearance of two inequalities in (9a) and (9b) proves that the lattice is non–distributive. This result is equivalent with the non–uniqueness of the complement.

As it has been indicated before, the most important general property of lattices which we want to make use of is their duality. Dual properties of a lattice are obtained by simply interchanging $\wedge$ and $\vee$ (as in (8a) and (8b)). If inequalities are considered, this exchange has to be accompanied by the exchange of $<$ and $>$. Using these rules the relations dual to (9a) and (9b) can easily be written down. However, it turns out that propositions about incommensurable quantum facts do not make any sense for these dual relations. On the other hand, if (9a) and (9b) are satisfied by propositions about quantum facts then another class of propositions *must* exist in order to provide a reasonable interpretation for the dual relations.

Guided by general arguments,[11] it has recently been shown[9] that propositions about incommensurable theories can fill this gap. As an example the dynamical theories of Liouvillean dynamics and information dynamics have been used. Relevant propositions concerning these theories are for instance given by:

$$a : \quad ``K = 0"$$
$$b : \quad ``K = K_s > 0"$$

Proposition a means that a state $\rho(t)$ of a system can be predicted with the same accuracy as an initial state $\rho(0)$ is given. Such a system is said to conserve information. Proposition b means that the time interval of prediction is limited by K^{-1}. The prediction uncertainty increases, and the information about the state of the system decreases. The system accordingly produces information. Both, propositions a as well as b refer to predicitons about the temporal evolution of a system.

The truth values of these propositions can be checked in a dialogue between an observer (proponent) and a meta–observer (opponent). For instance, the observer as a proponent claims:

Observer: "If I know the dynamical laws of a system, then I can predict its state $\rho(t)$ with the same accuracy as the initial conditions at time $t = 0$."

Meta–Observer: "You are thinking of systems with $K_s = 0$. If your system is information producing, the prediction uncertainty corresponding to $\rho(t)$ increases with time, and $K_s > 0$. I claim that $K_s > 0$ such that the state of your system will not be predictable as you think."

Observer: "Now that t is over, I have measured $\rho(t)$. The uncertainty of $\rho(t)$ is indeed greater than it was at $t = 0$. My prediction was incorrect."

In this case $K_s = 0$ is false. On the other hand, if the accuracy of $\rho(t)$ had remained constant, $K_s = 0$ could not be classified to be a true theoretical proposition, since only one particular system has been checked. This case allows to speak of a corroborated proposition in the sense that is has not yet been falsified.

As has been shown[11] two types of complements may exist for dynamical systems. If the complement is an element of G we have an inner complement, otherwise we speak of an outer complement. The example of dynamical theories[9] contains both kinds. It shows that the inner complement is unique, but it is a pseudocomplement, i.e. $a'' \neq a$. On the contrary the outer complement is non – unique and satisfies one of the distributive inequalities dual to (9a) and (9b):

$$a < (a \vee b) \wedge (a \vee b') \tag{10a}$$
$$\vee \quad b < (b \vee a) \wedge (b \vee a') \tag{10b}$$

We have thus identified propositions concerning theories to provide the distributive inequalities dual to (9a) and (9b). From a mathematical point of view this is necessary since otherwise the lattice relating to incommensurable quantum facts would be incomplete with respect to the class of relevant propositions. However, from the same point of view the duality of the non – distributivities concerning facts and theories indicates that both are inseparably related to each other. Facts and theories have formally turned out to be nothing but two aspects of one and the same universe of discourse. This is so because dual properties of a lattice are tautological. They do not hide any additional degree of freedom. No semantic distance exists between facts of a system and theories about a system based on these facts.

4. Modalities

Since the propositional lattice discussed in the preceding section is complemented and non – distributive, the corresponding propositional logic is non – Boolean. A non – Boolean logic requires at least one truth value additional to the Boolean truth values "true" and "false". The dialogical examples considered above indicate which additional truth values have to be introduced in order to account for incommensurabilities of facts and theories.

For incommensurable facts as intensive "objective" properties of a system, we have to refer to a dialogue between a system and an observer, usually called an experiment. If the validity of a fact, for instance a position $q = q_0$, can affirmatively be demonstrated in an experiment, then the proposition (a : "$q = q_0$") is necessarily true. Otherwise, if an affirmative demonstration of a's validity cannot be performed, then a is not necessarily true (*vulgo* doubtful). In our example, this situation occurs if an experiment does not permanently reproduce $q = q_0$.

For incommensurable theories as extensive "subjective" properties of a system the situation is different. A theoretical proposition is falsified if the system under consideration does not behave according to a prediction of the proponent. In this case the proposition is necessarily not true, i.e. it is necessarily false. Otherwise if an affirmative procedure does not falsify a proposition, then the proposition is not necessarily not true, i.e. not necessarily false (*vulgo* corroborated).

The truth values corresponding to the levels of facts and of theories can be formalized as follows.[22]

$$\Box a : \quad a \text{ is necessarily true (true)}$$
$$\neg \Box a : \quad a \text{ is not necessarily true (doubtful)}$$
$$\Box \neg a : \quad a \text{ is necessarily not true (false)}$$
$$\neg \Box \neg a : \quad a \text{ is not necessarily not true (corroborated)}$$

The symbols on the left are the monadic modal functors, or modal operators. Acting on a proposition they generate the basic modalities of a four – valued modal propositional logic. With respect to our context, the four modalities refer to two levels, namely that of facts ($\Box a$ and $\neg \Box a$) and that of theories ($\Box \neg a$ and $\neg \Box \neg a$). Hence the Boolean truth values true ($\Box$) and false ($\Box \neg$) are relevant to different levels, on each of which they are supplemented by an additional modality ($\neg \Box$ and $\neg \Box \neg$). The resulting structure of modal truth values is illustrated in Fig.2.

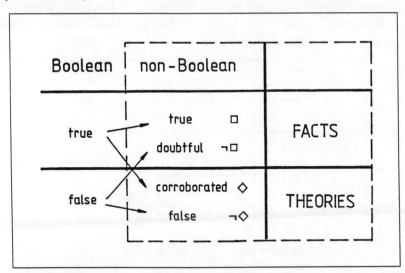

Figure 2: Schematic illustration of the splitted Boolean truth values into the mutually nested truth values "true" and "doubtful" (for facts) and "false" and "corroborated" (for theories). The formal modal operators as described in the text are indicated.

Since the levels of facts and theories have been identified as aspects of one underlying lattice structure, a corresponding reduction of the four basic modalities into two groups referring to each level seems desirable. This can formally be achieved using a possibility operator with:

$$\Diamond a : \quad a \text{ is possibly true}$$

$$\neg \Diamond a : \quad a \text{ is not possibly true}$$

It is defined by $\Diamond := \neg\Box\neg$ so that $\neg\Diamond = \Box\neg$. (Note that all four basic modalities can be constructed identically with $\Diamond$ and $\Box$.) Applying the modalities in this modified manner, one immediately recognizes that $(\Box, \neg\Box)$ and $(\Diamond, \neg\Diamond)$, each containing a basic operator and its negation, provide the same levels as those corresponding to the duality of facts and theories (see Fig.2). While the pair of operators $(\Box, \neg\Box)$ refers to necessarily true (true) and not necessarily true (doubtful) facts, the pair of operators $(\Diamond, \neg\Diamond)$ refers to possibly true (corroborated) and not possibly true (false) theories. Hence the levels of facts and of theories are dual aspects concerning the non – distributivity of the corresponding propositional lattice *and* concerning the basic modalities of the corresponding propositional logic. As facts and theories coincide as dual aspects of a propositional lattice, the operators $\Box$ and $\Diamond$ coincide as aspects of a fundamental two–valued logic, based on the antinomy of positive and negative. As a consequence, the two–valuedness anchored in the negation has to be considered as a very basic feature, exceeding the significance of the duality of $\Box$ and $\Diamond$ or facts and theories.

From a different point of view, $\Box$ and $\neg\Diamond$ could be grouped together in order to classify the definite Boolean truth values. Accordingly, $\Diamond$ and $\neg\Box$ would then represent the class of contingent truth values. The duality of definiteness and contingency can thus be illustrated orthogonal to that of positive and negative modalities and hence orthogonal to the duality of facts and theories (see Fig.3). This relation indicates that the levels of facts and theories, once they are separated, cannot be arranged in an arbitrary manner.

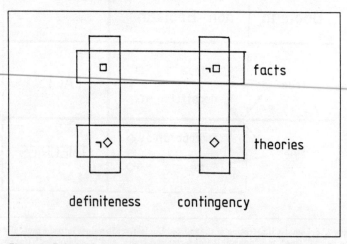

Figure 3: Relation between the duality of facts and theories and the duality of definite and contingent modalities. Note that the antinomy of positive and negative modalities introduces a further type of duality since it connects definite modalities of facts with contingent modalities of theories and *vice versa*.

5. Semantics

The central points of the present article culminate in the principle of duality. Here, duality is understood in the well–defined mathematical sense of two properties turning out as aspects of the same structure. In this sense, the following dualities have been discussed:

- Duality of incommensurabilities between quantum facts and dynamical theories (Sec.2).
- Duality of distributive inequalities concerning these incommensurabilities (Sec.3).
- Duality of pairwise different modalities as truth values of propositions about facts and theories (Sec.4).

The structure which unifies these dual aspects is a bicomplemented non–distributive lattice providing a non–Boolean type of logic. This unification implies that facts of a system and theories about a system based on these facts are merely required to be consistent. The notion of facts as an *explanandum* and of theories as an *explanans* has formally turned out as an illusion. The amalgamation of facts and theories corresponds to the semantic (as well as pragmatic) inseparability of both. Since no clearcut demarcation of facts and theories is possible, both have to be considered as an inseparable whole.

This result raises an important question. If semantic distances do not apply to a difference between facts and theories, to what sort of difference do they apply? With respect to this question, we propose to look for meaning and semantics in the evolution of systems and theories. This viewpoint arises very naturally if one considers systems and theories as dual aspects related to the level of interfactual and intertheoretical syntax. From Fig.1 it can be seen that this is obvious for systems. In order to explain the substitution of the corresponding metatheory by a theory, we use the argument that a metatheory of evolution must describe the evolution of theories as well. Hence a theory describing evolution is a metatheory.

The evolution of systems and of theories again provides a duality, which can be illustrated in terms of symmetry breakings and symmetry recoverings. The evolution of systems ("objective" aspect) leads to an increasing diversity by means of consecutive symmetry breakings, whereas the corresponding evolution of theories ("subjective" aspect) recovers symmetries in the sense of an increasing degree of generality.

The relation between an advanced theory T' and a retarded theory T can be expressed as:

$$T = T' + \{\text{contingent conditions}\}$$

so that T' indeed represents an extension (generalization) of T. An "evolved" system S' and a "degenerated" system S are related dually to the theoretical relation:

$$S' = S + \{\text{contingent conditions}\}$$

The contingent conditions contain the symmetries which are dually broken and recovered, respectively. We suggest to consider them as the origin of meaning, thus providing a proper starting point toward a formalization of semantics.

On the level of theories, a recovering of a symmetry can be described according to an inductive inference, referring to an outer complement in the lattice theoretical formulation. This is equivalent to an extension of a theory: the set of propositions relevant in the framework of a theory is extended. In contrast, purely deductive inference would refer to an inner complement. A theory developed on a purely deductive basis is therefore tautological to its predecessor. It cannot provide any generalization, since no symmetry is recovered.

On the level of systems, an increasing diversity by symmetry breaking might have to do with an increasing complexity. Consistent with the remarks above one might speculate that a complexity increase caused by a symmetry breaking generates meaning which can be measured by well–defined quantities of the system. A corresponding and promising attempt has been carried out for dynamical instabilities in a multimode laser system.[23] The key quantities in this approach are efficiency and pragmatic information. It is most interesting that they are general enough to overcome restrictions to particular domains of applicability.

ACKNOWLEDGEMENTS

It is a great pleasure to thank F.Mündemann and J.Becker for the warm and creative atmosphere during WOPPLOT 89. The written version of the contribution to the workshop includes ideas which have emerged in discussions with W. v. Lucadou, B. Rieger, and H. Wermus.

REFERENCES

1. P. Mittelstaedt, *Sprache und Realität in der modernen Physik* (BI Verlag, Mannheim, 1986)
2. I. Prigogine, *Non – Equilibrium Statistical Mechanics* (Wiley, New York, 1962)
3. I. Prigogine, *From Being to Becoming* (Freeman, San Francisco, 1980)
4. B. Misra, *Proc. Ntl. Acad. Sci. USA* **75**, 1627 (1978)
5. B. Misra, I. Prigogine, and M. Courbage, *Proc. Ntl. Acad. Sci. USA* **76**, 4768 (1979)
6. Y. Elskens and I. Prigogine, *Proc. Ntl. Acad. Sci. USA* **83**, 5756 (1986)
7. H. Atmanspacher and H. Scheingraber, *Found. Phys.* **17**, 939 (1987)
8. H. Atmanspacher, in *Evolutionary Models and Strategies*, eds. F. Mündemann and J. Becker (Springer, Berlin, 1990)
9. H. Atmanspacher, *Found. Phys.* .., ... (1990)
10. G. Birkhoff and J. von Neumann, *Ann. Math.* **37**, 823 (1936)
11. F. R. Krueger, *Physik und Evolution* (Parey, Berlin, 1984)
12. I. Prigogine and T. Petrofsky, *Physica* **147 A**, 33, 439, 461 (1987)
13. H. Primas, *Chemistry, Quantum Mechanics, and Reductionism* (Springer, Berlin, 1983)
14. A. N. Kolmogorov, *Dokl. Akad. Nauk.* **119**, 861 (1959)
15. Y. Sinai, *Dokl. Akad. Nauk.* **124**, 768 (1959)
16. J. B. Pesin, *Russ. Math. Survey* **32**, 455 (1977) (*Usp. Math. Nauk* **32**, 55 (1977))
17. S. Goldstein, *Israel J. Math.* **38**, 241 (1981)
18. R. Shaw, *Z. Naturforsch.* **36 a**, 80 (1981)
19. M. Jammer, *The Philosophy of Quantum Mechanics* (Wiley, New York, 1974)
20. J. F. Cyranski, *Found. Phys.* **15**, 753 (1985)
21. P. Lorenzen, *Metamathematik* (BI Verlag, Mannheim, 1962)
22. T. Bucher, *Einführung in die angewandte Logik* (deGruyter, Berlin, 1987)
23. H. Atmanspacher and H. Scheingraber, *Can. J. Phys.* .., ... (1990)

Self-Organization in Computational Systems:

Advance of a New Computational Paradigm?

Friedhelm W. Mündemann

University of the Federal Armed Forces Munich
Faculty for Computer Science
D-8014 Neubiberg

and

ICAS - Institute for Cybernetic Anthropology Starnberg
D-8130 Starnberg

Introduction

Computer paradigms have long life cycles, due to the reluctance of
people to learn a new one, and due to existing software that has
been in use for many years. An example is the never-dying language
FORTRAN which has been in use in science for three decades now.

However, FORTRAN is only a suboptimal solution, speaking in evolutionary terms, and computer science has developed many more paradigms since then. In particular we want to consider Neural Nets
and Evolutionary Strategies / Genetic Algorithms in this paper.
These approaches have their roots in biology.

Thus, several ideas which are used in Neural Networks have their
origin in neurophysiology, such as massively parallel processing
with simple processing elements and plasticity of the connections.
But do we have the right to compare "Neural" Networks to the brain
just because of some rudimentary biological ideas woven into the
wires of our artificial nets?

Here we do not want to argue about the biological basis, about the
much larger complexity (in structure and in function) of natural
neural nets, as compared to the artificial ones, let alone the in-
timate connection of the neuronal and the hormonal systems in living beings little of which is understood. It will suffice to compare Neural Nets to other paradigms that have been developed in
computer science in recent years in order to demystify the Neural
Network approach.

We will first look at the embedding of Neural Networks into computation paradigms in the second chapter. In the third chapter we
consider their constituents and operational principles. The fourth
chapter deals with self-organization and Neural Networks whereas
the fifth chapter introduces concepts and meanings of evolution.
The sixth chapter brings together Neural Networks and the approach
of evolution. Some issues for further discussion conclude the contribution.

Computation Paradigms and Constituents of Neural Networks

If we want to embed Neural Nets into other computer science paradigms, it is useful to look at general models of computation.

Usually we may divide a computational problem into a task domain and a data domain; algorithms are used to describe the "active part" and data structures serve as the "passive part" of the solution in terms of instructions to be performed using operand values. The operational model determines the way how all this is executed on a multi-processor system after an allocation step that assigns instruction (blocks) to processing elements.

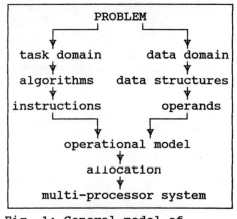

Fig. 1: General model of computation

Operational models may be classified according to the following scheme [TRELEAVEN 82]:

operand supply	instruction model	synchronization strategy
by literal	flow model	sequential (control driven)
by value	state model	parallel (data driven)
by reference		recursive (demand driven)

Table 1: Elements of operational models

Operand supply for instructions may be done by inserting values known at compile time directly (by literal), by inserting copies of values (by value) or by following up a pointer to the value stored elsewhere (by reference).

Instruction models determine whether information is kept from one execution of an instance of an instruction to the next one (state model) or not (flow model).

Synchronization strategies determine the sequence in which instructions are executed.

To give two examples, we may look at the von Neumann machine program execution model

(literal/reference, state, sequential)

and the data flow model [DENNIS 74]

(literal/value, flow, parallel).

A further classification of parallel processes may be achieved by considering the granularity of tasks:

granularity of task	"sub-"paradigm
programs	multiprogramming
processes	concurrent processes
methods	object-oriented programming
statements	macro-dataflow
basic operations	dataflow / reduction
Boolean functions	microprogramming (register-transfer level, gates, ...)
primitive functions	neural networks

Table 2: Classification of parallel processing concepts from granularity

In this contribution we may understand as execution subparadigms

multiprogramming: the execution of program(segments) that is controlled by an operating system working in time-multiplex,

concurrent processes: the execution of programs that have been decomposed into cooperating sequential parts (CSP [HOARE 78], OCCAM [MAY 83], [HULL 86]),

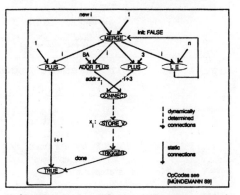

Fig.2: Data Flow program graph

objectoriented programming: the execution of programs that have been decomposed into so-called agents, which interact through messages in time ([AGHA 85], OOP,SMALLTALK [GOLDBERG 83]),

macro-dataflow: the execution of programs that have been decomposed into independent statements which are executed asynchronously ([BABB 84]),

dataflow and reduction: the execution of programs that have been decomposed into independent basic operations which are executed asynchronously ([DENNIS 74]-[DENNIS 84], s. [TRELEAVEN 82]), using static and/ or dynamic links [MÜNDEMANN 89].

microprogramming on the regis-ter-transfer level: the execution of programs that have been decomposed into boolean operations and their control and that are executed asynchronously,

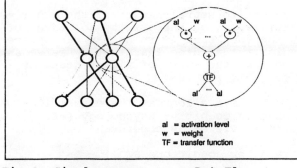

al = activation level
w = weight
TF = transfer function

Fig.3: Single neuron as a DataFlow subgraph

neural networks: the execution of programs that have been decomposed from boolean functions to functions composed of primitive functions and that are executed asynchronously ([WIDROW 88]).

As such, the Neural Net approach appears as the most basic processing approach in parallel computation.

But, on the other hand it is possible to view a single neuron as a data flow subgraph as can be seen from fig. 2; a general outline of the processing element's architecture of a neuron is given in fig. 3 (taken from [NEURALWARE 88]).

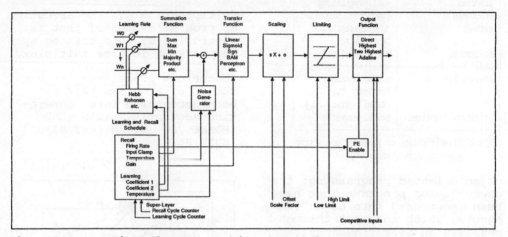

Fig. 4: Processing element architecture of a neuron

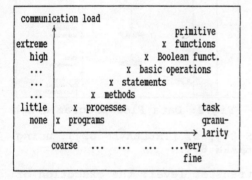

As granularity gets finer and finer the communication overhead rises, which means that connectivity has to be increased (see fig.5).

Fig.5: Correlation between task granularity and communication load in computing systems

Constituents and operational principles of neural networks

Granularity is only one ingredient to Neural Networks; the other one is the adaptivity or plasticity, i.e. the ability to learn. Of course we should be aware of the fact that learning may take place also at other levels of granularity; machine learning may serve as an example. Yet it may well be that the level of finest granularity is particularly suitable for this purpose.

At this point it is worthwhile to have a look at the classification of problems as it is common in cognitive psychology (cf. [OSTERLOH 83]). Here, a problem is viewed as consisting of

- a (set of) initial state(s) or conditions S_α ,

- a (set of) final state(s) or conditions S_Ω , and

- a (perhaps composed) transformation $T_{\alpha\Omega}$,

that transformes the initial state into the final state. We may
now classify problems according to the bits and pieces of a problem which are known, and to those which are unknown. This classification apparantly corresponds partly to paradigms known from
computer science (Table 3).

S_α	$T_{\alpha\Omega}$	S_Ω	class
known	known	?	classic: compute the output
known	partly known	known	rule-based programming
known	?	known	Neural Networks

Table 3: Problem classification

This table is by no means
exhaustive, and the reader may
easily form other problem
classes, which, however, have
not yet been treated in
computer science. (Evolutionary
strategies or genetic
algorithms may well be suitable
to attack even more complicated
classes; see below.)

In the case of Neural Nets, the solution $T_{\alpha\Omega}$ is represented by the
net architecture. This is unknown a priory, but the net may adjust
its connections and couplings such that it transforms the input
(stimulus, e.g. picture data) into the desired output (reaction,
e.g. classification); this may be called learning. It may be
achieved by presenting historical data to the net during a learning phase, but there are also nets which learn continuously. We
may further supervise the learning procedure from the outside

(external teacher), or we may design a net which adjusts its couplings according to correlations in
the data. It must be pointed out
that learning is restricted to the
connections, whereas the functionality of the "neurons" (e.g., the
transfer functions) is usually
kept fixed.

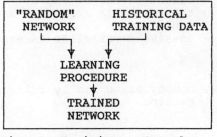

Fig. 5: Training a Neural
Network

To summarize some of the operational principles of Neural Networks, we have

constituents:
- simple processing elements with an interconnection structure
- nets initially preconnected
- direction of the connections forward or forward/backward

plasticity:
- proper modification/insertion/deletion of interconnections
- training data and strategies
  (supervised, unsupervised, self-supervised)

operation:
- according to the connections and the functionality of the
  neurons

We may now compare these principles to those of other devices and
operational principles (Table 4).

connections				
				neural nets
variable		event flow(?)	communication networks	adaptation phase
fixed	multi-programming (CSP, OOP, ...)	data flow systems	hardware circuits	neural nets production phase
"functionality" of nodes	mighty	simple	mighty	simple
	variable		fixed	

Table 4: Connections and nodes in various operation principles

"Functionality" of nodes means the computational power of nodes in terms of the granularity of computation performed.

Ideally, communication networks as well as Neural Networks (adaptation phase) have variable connections although in their technical realization they might have fixed connections. While multi-programming as well as data flow systems use variable computational nodes and are usually executed on general purpose machines, hardware circuits and some Neural Networks (production phase) use fixed nodes realized by special purpose hardware.

To summarize some distinguishing aspects of the Neural Net approach, too, we may say that

- data flow may serve as an operational model;
- it uses "meta-rules" for parameter estimation and operand value determination;
- the result is a general mapping instead of partial recursive functions;
- massive parallelism is exploitable - in principle,

but:

- nobody has a really existing Neural Net as a computing machine.

Self-Organization and Neural Networks

The term "self-organization" has become fashionable. But what does it mean: organization   - of itself,
    - by itself,
    - of its self,
    - by its self?

In [DARPA 88, p. 275] it is stated: "It is not known, which parts of the brain are hardwired and which are fully self-organized". There self-organization is defined as "the autonomous modification of the dynamics of a complete neural network via learning in some or all of its processing elements to achieve a specified result" [p. 389]. It must be asked, however, how and by whom those results are specified.

Other examples of the use of the term "self-organization" are:

- clustering or (the ability for) category formation on input data samples;
- unsupervised training or operating without a teacher;
- self-adaptation or self-regulation, in the sense of fine-tuning circuits in response to the environment.

Furthermore, it is distinguished between small-scale self-organization (= organization of the neural nodes or layers as receptive fields cf. [LINSKER 88]) and global self-organization (= system adjustment to environmental conditions).

By system we mean (as a working definition only) a totality of components, between which relations of some kind exist or may be constructed. Complexity may arise from a great variety of the components (kind, diversity of function), from a great connectivity between the components (kind, diversity of transmission characteristics), from a great number of elements and connections, or from basic nonlinearities.

However, the term "complexity" may have various meanings, such as complexity of behaviour, as we see it in chaotic systems; or computational complexity, which we encounter for instance in the travelling salesman problem, where the time to get the exact solution rises exponentially with the number of cities to be visited; or algorithmic complexity, which deals with the shortest algorithm to solve a given problem: such algorithms may themselves be infinitely long, as in the case of a (true) random sequence, where the "algorithm" is the sequence itself - just to quote a few out of many possible meanings of the term.

Frequently it is necessary to "reduce" the complexity of a system to keep it manageable. Three techniques are commonly used: First, choose only a few aspects of the systems; second, divide the system into sub-structures; and third, introduce levels of abstraction (sometimes also for the sub-systems). It is clear that "reducing" refers to the models of a system; as we must be aware of the fact that terms like structure, goals, meaning, etc. are already interpretations, or models, of our observations (cf. also the contributions by CAIANIELLO, by QUADRANTI, and by DALENOORT to this volume).

Thus we may say that the term "self-organization" is used to indicate that a system shows behaviour on a higher level of abstraction with respect to some optimization function (... also this subjected to some model, of course!). Hence, self-organization refers to a "swap" from a micro-level of operation to a macro-level, based upon a set of (perhaps unknown) "inner rules" of the system under consideration. An example is provided by [LINSKER 83], who has shown that feature-analyzing functions of a neural net may originate ("emerge") from a design rule based on the principle of maximal information transfer: A single neuron does not "know" about the purpose of the net, so the best thing it can do is to pass on as much of the incoming information as it possibly can.

It is commonly believed that such effects of self-organization can also occur in groups of humans (cf. the contribution of MOLZBERGER to this volume). We may define it as a goal-oriented cooperation and use of resources, where the goal is defined and accepted by all members of the group, either consciously or unconsciously.

Evolution: Meanings and Concepts

Natural evolution (or rather a model of it!) may be viewed as a system of rules governing the process of self-organization to optimally adapt living systems to their changing environment. Appearently we have to consider living beings as a part of their environment such that adaptation of individuals or populations leads to a change of the environment which, in turn, leads to further adaptation etc. Thus it may be doubted that natural evolution is optimization, at all. Such considerations are usually left out when mimicking evolution in technical systems.

What ingredients are necessary for evolution to work?

We need

- an environment,
- a population of individuals, where
  - individuals are characterized by attributes,
    - individuals possess genes,
  - genes are factors that determine attributes,
  - the possible values of the genes are called alleles,
  - mutation arises from an accidental, but persisting variation of genes.

Here, accidental means that a possible benefit has no influence on the frequency of certain variations. A simplified set of rules for the neo-Darwinistic evolution model (for higher organisms) then reads:

1. Individuals reproduce sexually; the offsprings bear new combinations of the parental genes (heredity); casual mutation, crossing over etc. may occur; genes may be extended, reduced, doubled, or eliminated.

2. Well-adapted individuals survive and are fertile.

3. GOTO step 1.

	individual techniques	collective techniques
without use of experience	trial & error	Monte Carlo methods
with use of experience	learning, heuristics	evolution

Table 5: Classification of optimization techniques

Thus, evolution is a perpetual process of natural selection of heredity changes that incidentally occur in every generation. An important fact is that "knowledge" about the environment is distributed over the population, and recombined in sexual reproduction. (Sexual reproduction is not an essential ingredient, because also monocellular living beings which propagate asexually have means of exchanging genetic information with other individuals, either by direct contact, or by means of a virus-postman.) Hence, evolution is a collective optimization strategy. In a nutshell we may classify various optimization methods as shown in table 5.

But, one has to remember that evolution does not take care of each
single individual, but only concerns kinds of individuals as a
whole.

Bridging the Gap: Evolution, Self-Organization and Neural Nets

There are but a few ideas as how Neural Nets should be designed.
This ignorance concerns all elements and levels of Neural Nets:
neurons, connections, structure, dynamics, emergent properties,
higher level functions, learning strategies, etc. In principle,
any neural net of a sufficient "complexity" should be able to solve a given problem (out of a suitable complexity class); this follows from equivalence arguments. Hence, additional criteria have
to be provided that, together with the problem, defines "optimality" (i.e., the "evaluation function") for a Neural Net.

Such criteria could be:

- look for a net in a given class of net architectures;
- look for an efficient net, i.e. a net that does the job with a
  minimal number of elements and/or connections;
- look for a net that may be implemented in a specific hardware
  technology; etc.;

or a combination of such requirements.

It is an appealing idea to use evolution as a tool for optimizing
Neural Nets according to a given set of criteria. (For a first attempt, cf. LOHMANN's contribution to this volume.) Hence, we may
use evolution as a self-organization tool to develop a Neural Net,
only by applying the rules of mutation, propagation, and selection. It is known that evolution is an efficient strategy to
search a multidimensional parameter space with a richly structured
evaluation function; this is precisely the case for Neural Net design. Evolutionary strategies may especially be useful to shorten
the training phase of a Neural Network.

The first step is the coding of the problem in evolutionary terms.
This means that either the "state" of the system is coded directly
as the quantity to be treated by evolution ("evolutionary strategies"), or that it is encoded in a set of numbers ("genes") from
which the system state ("phenotype") is to be determined ("genetic
algorithms").

Again the term "state" may refer to some complex description of
the system under consideration; for a Neural Net, it would include
the transfer function of the neurons, the density of connections,
the net structure, the couplings, the time scales involved, the
learning rules, etc.

Next, the desired properties have to be described by some kind of
evaluation function of the system state. As a matter of principle,
the computation of the system state from the genes in the case of
genetic algorithms could be absorbed in the evaluation function,
such that there is no real difference between evolutionary strategies and genetic algorithms; but for practical matters it may be

convenient to keep the distinction. In some cases the computation
of the evaluation function is the bottleneck of evolution anyway;
but this may well be true for natural or cultural evolution, too.

Once the problem has been coded we may generate a first population; we may let the individuals happily reproduce, with sexual
recombination, mutation, and crossing over involved; we may then
evaluate the system states of the offsprings and select some of
them for further reproduction. There are many examples in this
volume, therefore it will not be necessary to go into many details
here.

Instead we shall briefly reflect on the difference between learning and evolution in Neural Nets. Learning means adaptation of
the net couplings with respect to a given task; evolution means
adaptation of the net architecture with respect to a task, a class
of tasks, or a task changing over of time.

It should be pointed out that evolution is quite a general tool
for system design, and that it works on any level of abstraction,
for the whole system, for any subsystem alone, for any system aspect alone; it may even be applied to the level of algorithms.
Whether or not such an approach will really lead to the "emergence" of "higher" functions, and thus to self-organization, is still
a matter of dispute.

Summary

To summarize, we should like to give some (doubtful?) answers to
some (doubtful?) questions:

What are Neural Networks?

- the omnipotent medicine for solving AI-problems or
- just a new paradigm for information processing

What is evolution?

- the omnipotent medicine for curing neural network shortcomings
  or diseases or
- just a new paradigm for tuning system parameters according to
  some optimization function or
- discovering new mechanisms or "frames of effect",
  i.e. a set of meta-rules for meta-programming, hoping that the
  cardinality of that set of meta-rules is less than the cardinality of the set of common program writing rules.

What is the governing principle behind evolution?

- selection of the fittest in a hostile environment or
- an optimistic force driven from within living beings ("new biology"; cf. MÜHLENBEIN's contribution to this volume).

What is self-organization?

- just a new gap-filler for unexplicable phenomena in systems.

References

[AGHA 85] Agha, G., Hewitt, C.: Concurrent Programming Using Actors-Exploiting Large-Scale Parallelism, Fifth Conference on Foundations of Software Technology and Theoretical Computer Science, Springer Verlag 1985, S.19 Zit. nach: Informatik-Spektrum, Bd.10, Heft 1, Febr. 1987, S.43

[BABB 84] Babb II, R.G.: Parallel Processing with Large-Grain Data Flow Techniques, IEEE Computer, July 1984, S.55-61

[DARPA 88] DARPA Neural Network Study (October 1987-February 1988), AFCEA International Press

[DENNIS 74] Dennis, J.B., Misunas, D.P.: A Computer Architecture for Highly Parallel Signal Processing, ACM Proceedings of the 1974 National Conference, New York, Nov. 1974, S.402-409

[DENNIS 79] Dennis, J.B.: The Varieties of Data Flow Computers, 1th International Conference on Distributed Computing Systems, Huntsville/Alabama, Oct. 1979, S.430-439

[DENNIS 80] Dennis, J.B.: Data Flow Supercomputers, Computer, Nov.1980, S.48-56

[DENNIS 84] Dennis, J.B., Gao, G.-R., Todd, K.W.: Modeling the Weather with a Data Flow Supercomputer, IEEE Transactions on Computers, Vol.C-33, No.7, July 1984, S.592-603

[GOLDBERG 83] Goldberg, A., Robson, D.,Ingalls, D.: SMALLTALK-80: The Language and its Interpretation SMALLTALK-80: The Interactive Programming Environment, Edison-Wesley, Reading/Mass., 1983

[HOARE 78] Hoare, C.A.R.: Communicating Sequential Processes Communications of the ACM, Vol.21, Aug. 1978, S.666-677

[HULL 86] Hull, M.E.: Implementations of the CSP Notation for Concurrent Systems, The Computer Journal, Vol.29, No.6, 1986 S.500-505

[LINSKER 83] Linsker, R.: From basic Network Principles to Neural Architecture, Proc. Natl. Acad. Sci. USA, Vol.83, 3 Parts: Emergence of spatial-opponent cells, p.7508-7512, Emergence of orientation-selective cells, p.8390-8394, Emergence of orientation columns, p.8779-8783

[LINSKER 88] Linsker, R.: Self-Organization in a Perceptual Network, IEEE Computer, March 1988, p. 105-117

[MAY 83] May, D.: OCCAM, SIGPLAN Notices, Vol.18, April 1983, S.69-79

[MÜNDEMANN 89] Mündemann, F.W.: Konzepte zur Verwaltung n-dimensionaler Felder als Operanden in Datenfluß-synchronisierten Programmen, Ph.D. Thesis, Universität der Bundeswehr München, Fakultät für Informatik, Juni 1989 (in German)

[NEURALWARE 88] Neural Works: An Introduction to Neural Computing, NeuralWorks User's Guide, NeuralWare Inc.; Sewickley, USA 1988

[OSTERLOH 83] Osterloh, M.: Handlungsspielräume und Informationsverarbeitung, Huber 1983

[TRELEAVEN 82] Treleaven, P.C., Hopkins, R.P., Rautenbach, P.W.: Combining Data Flow and Control Flow Computing, The Computer Journal, Vol.25, Nr.2, Febr.1982, S.207-217

[WIDROW 88] Widrow, B., Winter, R.: Neural Nets for Adaptive Filtering and Adaptive Pattern Recognition, IEEE Computer, March 1988, p.25-39

Adaptation and Extension: Learning in Rigid Systems

Jörg D. Becker

ET/Physik, Universität der Bundeswehr München, D–8014 Neubiberg; and
ICAS – Institut für Cybernetische Anthropologie Starnberg eV, D–8130 Starnberg

Neural Nets and Nonlinear Dynamics have much in common; one may even say that the difference is rather a matter of point of view than of principle [1]. In both fields the question of diversity and structure, as well as the question of information gain and learning, must be considered as open. In this paper we shall discuss some ideas which may be relevant for these issues.

When studying a system we may use different representations. For instance, we may represent system elements and relations between elements by means of one of the following descriptions:

- □ semantic nets

- □ inference systems

- □ interaction patterns

- □ system dynamics (system elements = variables)

- □ distributed system dynamics (system elements = distributions)

Of course, we may also use various descriptions concurrently. These descriptions form a hierarchy, in the sense that one may derive higher levels from deeper levels (e.g., interaction patterns from system dynamics), whereas the converse procedure is ambiguous. The relationships between the levels are not well established. (The list given above might also not be complete.) One must also be aware of the fact that the choice of a level of description has an influence on the kind of questions one may ask. For instance, there is no superposition of assertions in a semantic net, whereas superposition is something natural in the language of distributions. (This fact may be important when one wants to superimpose conflicting opinions in an expert system; cf. [2].)

Learning, or evolution, may take place at any level of description. However, also here effects depend on the level of description. We assume that for neural nets and genetic algorithms the language of distributions is most adequate since the information is distributed over the net resp. population. As Caianiello's contribution to this volume shows [3] there exist uncertainty relations for information changes in systems the elements of which are described by distributions. In such a representation we may derive an uncertainty relation for the change of information S with respect to time [4]:

$$\tau^2 \left(\partial S / \partial t \right)^2 = F(Z) \geq 1 ;$$

where $F(Z)$ denotes a functional of the system state, and τ is a typical time constant. As in quantum physics, this is a formal relationship which may have various interpretations. An obvious interpretation is that τ is the time horizon for predictions; as such, the relation is a generalization of an equation known from chaos theory [5] :

$$\tau \left(\partial S / \partial t \right) = 1 \text{ with } \tau = 1 / \lambda ;$$

λ the Lyapunov exponent. Alternatively we may interprete τ as an internal constant of the system which tells us something about the learning speed of the system. This sounds trivial, but nevertheless it is interesting to have an uncertainty relation for the common phenomenon that some systems learn faster than others. As we shall see below, cultural systems may have very large time constants.

Consequences of such uncertainty relations for neural nets or for genetic algorithms still have to be discussed. Here we shall concentrate on the question how systems actually do learn. In neural nets it is generally assumed that learning is achieved by a change of the couplings, usually on a larger time scale than the one for the operation of the net. Take, for example, Grossberg's ART net [6] the equations of operation of which read

$$dz_k / dt = - b_k + \sum_1 A_{kl} z_l ;$$

here z_k denotes the activity of neuron (k) at the time t, b_k leads to a relaxation when there is no input, and the last term describes the input from other neurons (l) to neuron (k) with the coupling A_{kl}. (More generally, we may use nonlinear coupling functions $A_{kl} (z_l)$.)

The learning eqations read

$$dA_{kl} / dt = - c_{kl} + L_{kl} z_k z_l ;$$

the latter equation is a Hebb type learning rule: the couplings A_{kl} are increased iff both neurons (k) and (l) are frequently active at the same time (expressed by the product $z_k z_l$).

If a system learns by a change in the couplings we shall speak of **adaptation**. Now the question arises whether there are other types of learnig in natural systems. Consider, for instance, systems which may be described by a set of equations of the logistic type. Let us start with a simple system with just one state variable (in the language of system dynamics), z(t) . The logistic equation reads

$$dz / dt = a z - b z^2 ;$$

it describes the growth of a system (bacteria in a Petri dish, sunflower seeds) limited by the available ressources (crowding, competition) or by the genetic code. This equation allows for an analytic solution:

$$z (t) = z^* \left(1 - e^{-(bt+s)} \right)^{-1} ;$$

here, $z^* = a/b$ is the saturation point, b is the parameter which sets the scale for the effective speed of growth, and s is determined by the initial condition.

It has been known since over a century that many biological systems follow this law. However, as Marchetti [7] has shown, also many economical and cultural phenomena can be described by the logistic equation. It is common to take as the effective growth time the interval between $z = 0.1 z^*$ and $z = 0.9 z^*$, i. e. from the time z has reached 10 % of its saturation until the time when it has reached 90 %.

We just give a list of the effective growth times for various phenomena, as analysed by Marchetti:

PHENOMENON	GROWTH PEROID (years)
Gothic cathedrales in Europe (first stones)	200
Witch processes, Northern France	170
Witch processes, Scotland	70
Red brigades, number of victims	6.5
CDC, mainframe computers produced	22
IBM, mainframe computer s produced	56
World air traffic	32
Registered cars in Italy	22

This is not the place to discuss these phenomena in detail. The interesting fact for us is that obviously there is some information conserved in these systems (cultural genes?), because otherwise we could not describe them by means of a deterministic differential equation. (There are more complicated cases, like the competition of primary energy sources for the energy market, which can be described by a set of coupled logistic equations.)

Let us now see whether we can detect any sign for learning in these systems. For instance, one might expect that systems adjust themselves to external forces by changing there parameters, i. e. there growth rates or there saturation points. One prominent event certainly was the last oil crises in 1973, where one would have guessed that systems like the number of cars registered, the world air traffic, or the share of crude oil among its competitors for the primary energy market would change their parameters; but, to our big surprise, we do not see the slightest deviation in the logistic parameters! (Both phenomena, oil crisis and logistic growth curves, might have some connection with yet another deterministic phenomenon in economics and culture: the Kondratjew cycles.)

So if cultural systems are able to learn they must have a different mechanism. Appearently, information changes by means of innovation (new technologies, new styles), which is a kind of **extension** rather than **adaptation**.

Let us consider another class of systems: the class known as Hierarchical Modular Systems (HMS) in the sense of Caianiello [8]. Such systems have a finite number N of elements, as well as a finite number L+1 hierarchical levels. Each element is uniquely assigned to a specific hierarchy level k, k = 0 ... L. The population of level k is called n_k.

Furthermore there is a value v_k assigned to each level. From the two requirements

    □ maximal efficiency (= maximal number of value states with N fixed)

    □ mean value $<v> = N^{-1} \Sigma\, n_k\, v_k$ invariant under structural changes

it can be shown to follow that

    □ values and populations must be modular, and

    □ $v_{k+1} = M\, v_k$; $n_{k+1} = M^{-1/2}\, n_k$.

M, which is called the modulus, is independent of k. Several systems have been identified as being in the class of HMS, such as the monetary system, the distribution of settlements in a country (levels: capital, big cities, cities, ... , villages), the officers of the Federal Armed Forces, natural language (levels: letters, syllables, words, ...), as well as several parallel processing architectures, like pyramid or hypercube architecture.

In our context mainly two findings are interesting. The first is that the value of an information is proportional to the square of its length p (measured in bits or digits),

    $v \sim p^2$;

a relation that follows from the study of natural language. This suggests that it is rather the correlations between pieces of information that count than the information itself. Language, and sense, is a matter not of vocabulary but of combining words. The second finding is that the modulus M has a typical value for specific systems, such as listed in the following table.

SYSTEM	MODULUS
monetary system	2.18 ($= 10^{1/3}$)
army officers	2.2
natural language	4.5
distribution of settlements	4.5
hypercube	4
bin pyramid	4
quad pyramid	16

Thus, the modulus for a specific system seems to be constant – and systems react to changes in their environment not by a change of the modulus but by adding new levels, i. e. by changing L: another example of **extension**.

More examples of **extension** are provided by natural evolution. We may see **adaptation** within the genes of a species, i. e. mutations and crossover in genes of a fixed length; but the growth of the lenghts of the genomes in the course of evolution, as well as the emergence of higher forms of organization, as the Hypercycle [9], are clear examples for **extensions**.

Let us now turn to technology. If our goal was to cast neural nets into hardware we have few realistic possibilities. Disregarding futuristic stuff like molecular electronics [10], we have the choice between VLSI technology, optical computers, or hybrid devices of these two. Let us stay with silicon planar technology, which is the most developed one. If we want to represent learning by variable couplings we may choose floating gates or MNOS structures. In these types of devices electrons may be stored in insulated conducting zones in the bulk of silicon or at O–N–interfaces. Learning may consist in applying strong fields that make electrons tunnel into the floating gate resp. into the interface. These charges may then provide fields for switching the channels of field effect transistors underneath.

Both solutions are used in digital silicon technology. However, the electrons will not stay in their cages forever: they will diffuse away. The point is that for the digital mode of operation a 10 % change in the charge does not matter. Things are different for neural nets: a change in a coupling of 10 % may change the behaviour of the whole net! So which are the ways out?

One possible solution might be solid state chemistry. If electrons leave their floating gate cages we could design structures that bind electrons more tightly. Biology teaches us how to do that: make use of redox systems! A prominent example from living systems is iron:

$$Fe^{2+} \leftrightarrow Fe^{3+} + e^- \; ;$$

but the atomic radius of iron is much smaller than the one of silicon, and thus iron diffuses through silicon like hell. But – do not give up: there is another well known redox system – cerium:

$$Ce^{3+} \leftrightarrow Ce^{4+} + e^- \; ;$$

and cerium has an atomic radius of the same size as silicon. So which technologist would be prepared to contaminate his equipment with cerium?

However, as we have seen, there is also an alternative to adaptation: it's extension. So let us turn to another example from non–linear dynamics: to the well–known sequence [5]

$$z(t+\tau) = \left[z(t) \right]^2 - c \; ;$$

We may interprete z as the state variable and c as the control parameter. This sequence develops a very complex bahaviour, depending on the control parameter c, from a simple stable fixed point up to chaos. The range of c which is of interest for us is the Feigenbaum scenario of bifurcation – which we may interprete as a repeated extension controlled by c. Such a series of bifurcations, controlled by a parameter, may well be an alternative to adaptation in VLSI nets, even if our simple example will certainly be too simple-minded.

333

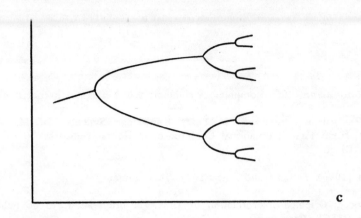

z

c

Feigenbaum bifurcation scenario:
attractors for the system state z as a function of the control parameter c

To summarize we could say that **adaptation** (= change in the coupling constants) is certainly one way of learning; but there is an alternative: **extension** (= enlargement of the state space). The second mechanism should be particularly interesting for systems that cannot (or do not want to) change their coupling constants.

There is a well known example for a task where extension is needed: how can you make four triangles out of six matches? By shifting around the matches in the plane (adaptation !) you will never be able to solve this problem. You have to go to three-dimensional space (extension !) to do it.

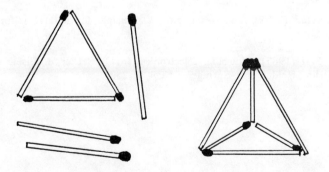

Four triangles from six matches:
adaptation (left, 2D) and extension (right, 3D)

References

[1] J. Becker, F. Mündemann, Neurocomputer. Weinheim: vch Verlag (in preparation)

[2] B. Kosko (1987), Adaptive Inference in Fuzzy Knowledge Systems. In: M Caudill, Ch. Butzler (Eds), IEEE First International Conference on Neural Networks. Vol. II, 261. San Diego

[3] E. R. Caianiello (1990), Systems and Uncertainty. This volume

[4] J. Becker, A Process-Oriented Interpretation of Time. (in preparation; to be published in: Journal of the ESSCS, Groningen)

[5] H. G. Schuster (1984), Deterministic Chaos. Weinheim: Physik-Verlag

[6] S. Grossberg (1987), Recent Developments in the Design of Real-Time Non-Linear Neural Network Architecture. In: M. Caudill, Ch. Butzler (Eds), IEEE First International Conference on Neural Networks. Vol. I, 1. San Diego

[7] C. Marchetti (1988), The Future. In: G. Caglioti, H. Haken, L. Lugiato (Eds), Sinergetica ed instabilità dinamiche. Rendiconti della scuola internazionale di fisica "Enrico Fermi", IC corso. Società Italiana di Fisica, Bologna

[8] E. R. Caianiello (1977), Some Remarks on Organization and Structure. Biol. Cybernetics 26, 151

J. Becker (1987), Structural aspects of organizing parallel processing machines. In: K. Ecker (Ed), Proc. ICoLe '87. Informatik-Bericht 88/3, Inst. f. Informatik, Universität Clausthal

J. Becker (1988), Structure, Justice, and Efficiency. To be published in: Proc. of a Workshop "Modeling processes of structural chenge in social systems". Beiträge zur Sicherheitspolitik Nr. 3. Forschungsprojekt Stabilitätsorientierte Sicherheitspolitik, Max-Planck-Institut Starnberg

[9] M. Eigen, P. Schuster (1979), The hypercycle: A principle for natural selforganization. Berlin, Heidelberg: Springer

[10] M. Mehring, H. Sixl (1987), Molecular Electronics. In: J. Becker, I. Eisele (Eds), WOPPLOT 86. Berlin, Heidelberg: Springer LNCS # 253

CHAOTIC DYNAMICS OF GENERATING MARKOV PARTITIONS,

AND LINGUISTIC SEQUENCES MIMICKING ZIPF'S LAW.

ANASTASSIS A. KATSIKAS
Department of Mathematics, School of Sciences.
University of Patras, Patras 26110, Greece.

JOHN S. NICOLIS.
Department of Electrical Engineering, School of Engineering.
University of Patras, Patras 26110, Greece.

Abstract: A chaotic dynamics model of creating Markovian strings of symbols as well as sequences of words is presented, and its possible relevance to Zipf's law is discussed.Comparison from contemporary Greek prose as well as nucleotide sequences (strings) of mRNA Polymerase III and embryonic cDNA displaying a quasi-Zipf's law behavior is made.The order of generated markovian strings as well as the information transferred between two symbols t steps apart (the "transinformation") is also estimated.

Key Words : Chaotic dynamics, Markov chain, information processing, dynamic linguistics, transinformation, Zipf's law.

Introduction

It has been long believed that a complex functional (Software) repertoire - such as e.g. language, be it natural, genetic or artificial - requires a structure (Hardware) of comparable complexity. Chaotic dynamics of dissipative systems provides a par excellence example of very complicated symbolic functions springing out from very simple, low-dimensional, non-linear systems.

In this brief communication we treat a specific application of chaotic dynamics to linguistics at the syntactical level. Chaotic dynamics as applied to information processing has been recently subjected to intensive investigations [1-6] and the results so far look promising. On the other hand, in fields of science such as physics, biology and sociology dynamics displaying inverse power law distributions have also been used as models with considerable success [6]. In linguistics in particular, there is Zipf's empirical law relating the probability of the appearance of words in long text with their rank order [7]. From the point of view of information processing in biology, any natural language can be considered as a partition that the processor applies to the external word with the aim of splitting this "word" into a set of descrete categories (attractors).

A Markov process may be thought of as an information source that with restricted randomness generates each of N symbols at discrete timesteps. These symbols can be thought of, also as the outcome of sequential measurements. If the occurence of a given symbol depends on N-1 precending symbols, then the Markov procces is of Nth order. In sec.1 we examine the Zipf's law of contemporary Greek prose. In sec.2 we investigate the Zipf's like law behavior of two different genetic texts - that of mRNA Polymerase III and embryonic cDNA. In sec. 3 an artificial "language" created by a Markov partition (5th order Markov chain) by the logistic map at the chaotic regime (R=4) is examined.

The cumulative distribution of "words" of 3 symbols separated by a fourth symbol taken as the "pause" is calculated, and the comparison with the Zipf's law manifest a rather satisfying similarity. From the same map the transinformation I(t) relating the information transfer between two symbols t steps apart, and its non-monotonic behavior is commented upon. The same two macroparameters (Zipf's law and the transinformation) are also calculated for the genetic texts. The reader is invited to judge for himself, the extent to which a Markovian partition on a simple recursive chaotic map, - although by no means it models on a quantitative fashion - emulates a complex phenomenon such as written natural or "genetic" language at the syntactical level.

1. Zipf's law and Greek language

Zipf's law is an empirical law in Linguistics, relating the probability $P(r)$ of words in a natural language, with their rank r:

$$P(r) \sim \frac{A}{r^\lambda} \qquad (1)$$

Where A and λ are empirically determined parameters. So if we test a long text of a common language e.g. (English), and then we make a log-log plot of $P(r)$ vs. rank order r, the result is in some cases a straight line, with slope -1. This means that the second most probable word in text has probability $P(r_2) = P(r_1)/2$, the third most probable word in the text has probability $P(r_3) = P(r_1)/3$ and generally we have:

$$P(r_n) = \frac{P(r_1)}{n} \qquad (2)$$

For the Greek language the analysed text consisted of 50 pages, from the book: "A N T I L E G O M E N A" by I. M. Panayiotopoulos with 8253 words totally, and 2615 different words. Out of these 2615 different words there exist 51 with discriminable probabilities. The result is presented in Fig.1. Some changes have been brought up in the document due to grammar peculiarities of the Greek language. Specifically we treated all different declinations of nouns and adjectives in each case as one word. The renewed version now has 2518 words, 50 of them with discriminable probabilities. The result is presented in Fig.2. In Fig.3 the comparison between document1 and document2 is presented for $r \leq 12$ because after this value the two diagrams are identical. In Table 1 the probabilities of words for the two documents are presented with their rank r.

2. Zipf's law and genetic language.

In a previous paper [5] the Zipf's law has been deduced theoritically from a variational principle with a cost function associated with the maximization of information carried from average word length. If we consider many candidate proto - languages (natural or genetic) competing as it were for predominance in a given population, it is not surprising that the winner(s) should be the ones mimicking to some degree Zipf's law. Hence the rough similarity between the curves of fig.4. For the genetic language two strings are examined (nucleotide sequences); with length $l_1=4854$ or 1618 words for the the first sequence (RNA Polymerase III) and length $l_2=8109$ (Embryonic cDNA) or 2703 words for the second sequence. The results are presented in fig.5,6. In this case the alphabet contains four letters A, G, C, T and all the "words" have the same length ($l=3$). There are 64 possible "words" in the examined sequences.

3. Chaotic dynamics of Markov partitions, and memory persistence (transinformation)

There are two distinct "dynamical" ways of producing either strings of symbols or sequences of words. Either by a dissipative non-linear chaotic flow $x=f(x;p)$, $\nabla.f<0$, or by a non-linear chaotic map . We use the second mechanism, a very simple chaotic model in order to simulate strings of symbols and "text" e.g. sequences of symbols (words) separated by the pause. Specifically we take a non-linear chaotic map $X_{n+1}=F(X_n, \xi)$ with a smooth invariant density $\mu(x)$. In our case the non-linear model is the classical logistic map onto which we apply a partition of 4 (A,B,C,D) equal subintervals (symbols).

$$X_{n+1} = 4X_n(1 - X_n/4) \tag{3}$$

$$\text{with } \mu(\bar{x})=(1/\pi)\left[\frac{x}{4}\left(1-\frac{x}{4}\right) \right]^{-1/2} \tag{4}$$

Consider in general a finite partition {Bi}, of M non-overlapping subintervals.The restriction of non-overlapping cells is needed because transition points between sequential elements must be mapped on transition points.In any other case of non-Markovian partition,the generated "alphabet" (by the partition on the x - axis) is distorted by the flow of the system or the sequence of iterations. It is possible to generate a Markov process via cascades of iterations giving rise to strings of symbols. For a M-cell with M = 4 we consider one cell as the pause.The M-1 remaining symbols are organized into words, interrupted by the pause.It is obvious, that these words have unequal length.The whole "alphabet" and the evolution of the words depend from the position of the pause in the partition.Two cases were examined for the non-linear map (3).In case 1 the pause symbol for the alphabet is "D" with "D"=(3,4).In case 2 the pause symbol is "C"=(2,3). We can see from fig.7 and fig.8, that our model in both cases presents a similarity with Zipf's law.In table 2 the words characterized with different discriminable probabilities from both cases are presented with their rank order r.

There exist some parameters, which can give us "bulk" information about the "grammar" and "syntax" of the "language" namely,

a) The set of $M \times M$ elements of transition probabilities

$$p^t(B_i/B_j) = \frac{\bar{\mu}\left[F^{-t}(B_i) \bigcap (B_j)\right]}{\bar{\mu}(B_j)} \qquad (5)$$

of jumping from the symbol ("letter") B_j to the symbol B_i in t discrete steps (iterations).In the present case simple calculations lead to the matrix

$$
\begin{array}{c|cccc}
 & A & B & C & D \\
\hline
A & a & 0 & 0 & a \\
B & b & 0 & 0 & b \\
C & c & 0 & 0 & c \\
D & 0 & 1 & 1 & 0 \\
\end{array}
\qquad
\text{where}
\quad
\begin{array}{l}
a = 2 - \sqrt{3} \\
b = \sqrt{3} - \sqrt{2} \\
c = \sqrt{2} - 1
\end{array}
\quad \text{for } t=1
$$

b) The (Shannonian) Entropy of the partition

$$S = -\sum_{i=1}^{M} \bar{\mu}(B_i) \log_2 \bar{\mu}(B_i) \text{ bits} \qquad (6)$$

gives a measure of the degree of monitoring of the process that is the information needed to locate one symbol-subinterval amongst the M.
In our case S=1.20032 bits

c) The Kolmogorov - Sinai Entropy.

$$S_0 = -\sum_{i}^{M}\sum_{j}^{M} \mu_i P_{ij} \log_2 \bar{P}_{ij} \text{ bits} \tag{7}$$

where P_{ij} is given by (5). The K–S Entropy offers a measure of our ability of forcasting the process namely it gives the uncertainty when a single jump takes place from one symbol to another. We can accept the K–S Entropy, as measuring the extra amount of knowledge gained about an initial condition with each new measurement. In our case $S_0 = 0.96607$ bits.

d) The Block Entropy.

$$S_I = -\sum_{i=1}^{n} p(C_L) \ln p(C_L) \text{ bits} \tag{8}$$

where $p(C_L)$ is the probability of formation of words C_L with length L. The Block Entropy can give us enough information about the richness of the repertoire of the language. In our case $S_I = 1.49725$ bits.

e) The Transinformation or Mutual Information

$$I(t) = \sum_{i=1}^{M}\sum_{j=1}^{M} \bar{\mu}\left[F^t(B_i)\cap B_j\right] \log_2 \left\{\frac{\bar{\mu}\left[F^t(B_i)\cap B_j\right]}{\bar{\mu}\left[F^t(B_i)\right]\bar{\mu}(B_j)}\right\} \text{ bits} \tag{9}$$

I(t) is the information contained in an initial condition about the state of the system t time units, or iteration steps, or number of symbols, into the future. The slope of I(t) is a measure of the rate of loss of the initial information and under certain conditions equals the Kolmogorov-Sinai Entropy. I(t) may decrease monotonically but it is also possible as we see in fig.9 that I(t) is not decreasing monotonically. Fig.10,11 displays the transinformation for the two examined nucleotide sequences of RNA Polymerase III and embryonic cDNA. We observe a non monotonic decay. We may suggest then that the existance of more than one time scales in the sequence of symbols, indicates a regeneration or persistence of memory. It is possible to examine the statistical properties of symbols of eq. (3). Using 10000 iterations we may generate 1500 "words" aprroximately. The results are recorded starting at t=500 in order to have all transients die out. The number of observed singlets, doublets, or triplets are counted. The conditional probabilities are then deduced from such relations as:

$$P(B/A) = \frac{P(AB)}{P(A)} \qquad (10)$$

$$P(C/AB) = \frac{P(ABC)}{P(AB)} \qquad (11)$$

The fact that doublets like DD or CC never occur shows that our sequence is not completely random. As it is shown in table 3, good agreement with the numerically computed frequencies of septuplets is reached by assuming a 5th order Markov chain:

$$P(A_1....A_7) = P(A_7/A_2...A_6) * P(A_6/A_1...A_5) \qquad (12)$$

where $P(A_7/A_2...A_6)$ represents the numerically computed conditional probability of the last symbol given the quinitplet $A_2...A_6$. In table III we display the probability of words deduced from counting vs. the probability inferred from a fifth order Markov chain. A Chi-square test has been deviced to verify the order of the Markovian character of the artificial text. The result is displayed in table IV. Moreover we see that out of 4^7 (16384) possible words only 253 appear in the text, thereby corroborating our hypothesis that the simple chaotic generator we have been using acts essentially, as a "linguistic filter" allowing the emergence of a "language" with enough variety but also good reliability - a mixture in short, of innovation and order. Finally we check the extent to which the "language" created by our chaotic generator displays a broken symmetry in one dimensional physical space - thereby avoiding "palindroms". We found that amongst 1500 words only 7 palindroms exists.

Aknowlegment

We want to thank Mrs. Aglaia Athanasiadou - Gika for her help in the study of the genetic strings.

References

1) J. S. NICOLIS "Dynamics of Hierarcical Systems" Vol. 25 SYNERGETICS Series. Springer-Verlag ,Berlin (1986)

2) J. S. NICOLIS "Chaotic dynamics applied to information processing" Rep. Prog. Phys. 1986,49, p. 1109-1196

3) J. S. NICOLIS and I. TSUDA "On the parallel between Zipf's law and 1/f processes in systems possesing multiple attractors" Progress in Theoritical Physics (to appear)

4) J. S. NICOLIS "Chaotic Dynamics in Biological information Processing: A Heuristic Outline" Nuovo Cimento Vol. 9D pp.1359-1388

5) G. NICOLIS, C. NICOLIS & J.S. NICOLIS "Chaotic Dynamics, Markov partitions and Zipf's law" J. of Stat. Phys. (N° 3|4, February 1989, p. 915)

6) G. NICOLIS, G. RAO, J. RAO & C. NICOLIS "Generation of spatially asymmetric, information-rich structures in far from equilibrium systems" preprint 1988

7) G.K. ZIPF "Human behavior and the principle of least effort" Addison-Wesley , Cambridge, Mass (1949)

Figure   Captions

Figure 1. Zipf's law for contemporary Greek prose [text:"ANTILEGOMENA"
by I.M. Panayiotopoulos]
Case 1: Original text.

Figure 2. Zipf's law for contemporary Greek prose [text:"ANTILEGOMENA"
by I.M. Panayiotopoulos]
Case 2: Modified text.

Figure 3. Comparison of words from case 1,2 with rank order r<12.

Figure 4. Comparison of words from all the examined languages.

Figure 5. Zipf's law for the mRNA Polymerase III string.

Figure 6. Zipf's law for the Embryonic cDNA string.

Figure 7. Num. Simulation of Zipf's law in equation $4X(1-X/4)$
Case 1: Blank symbol for the alphabet is "D"=(3,4)

Figure 8. Num. Simulation of Zipf's law in equation $4X(1-X/4)$
Case 2: Blank symbol for the alphabet is "C"=(2,3)

Figure 9. Mutual Information for the equation $X=4X(1-X/4)$

Figure 10. Mutual Information for mRNA Polymerase III string.

Figure 11. Mutual Information for Embryonic cDNA string.

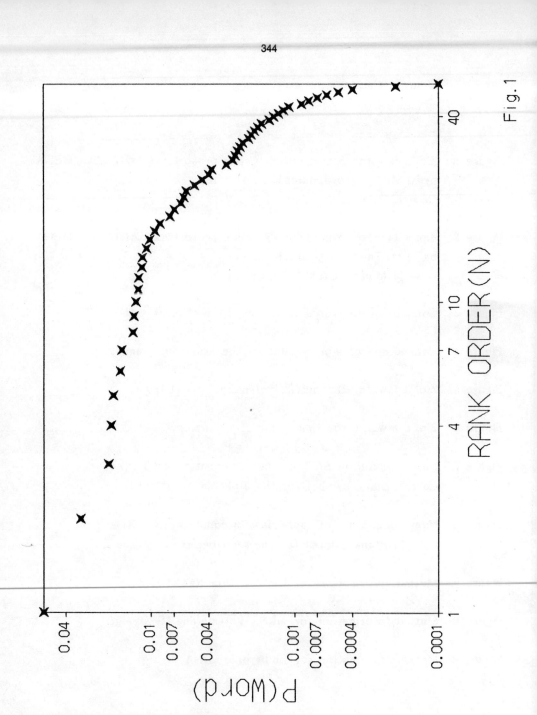

Fig. 1

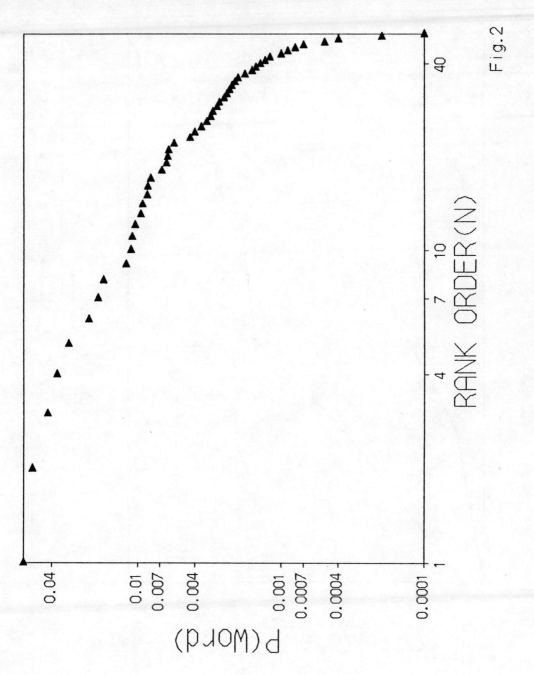

Fig. 2

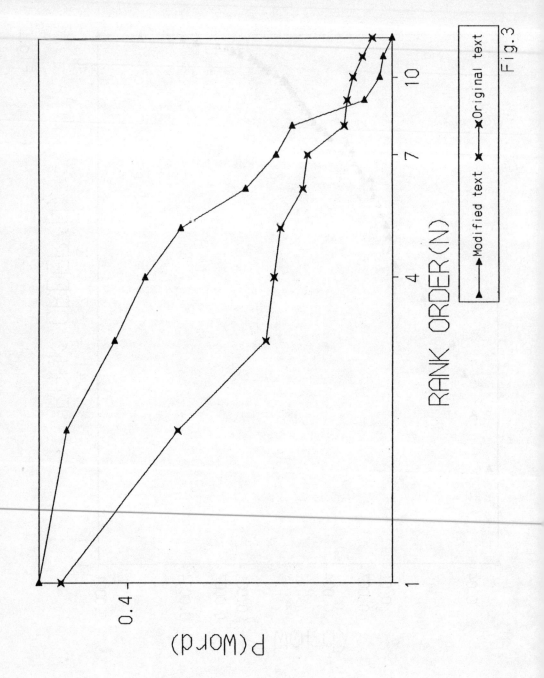

Fig.3

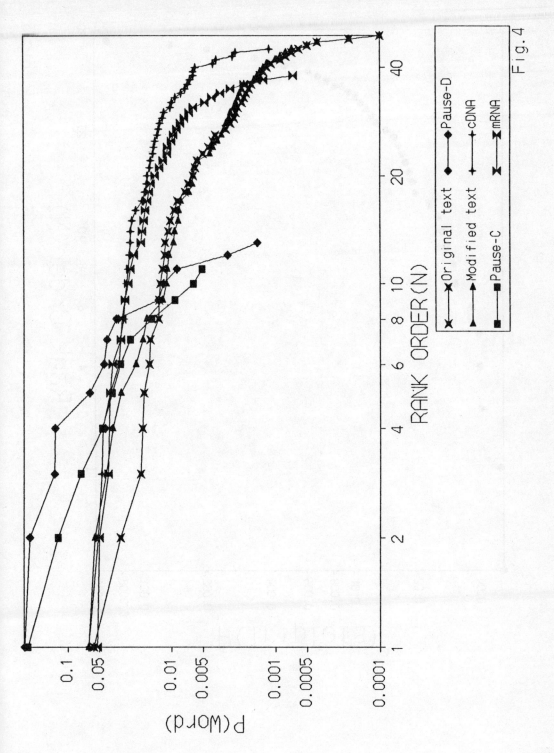

Fig. 4

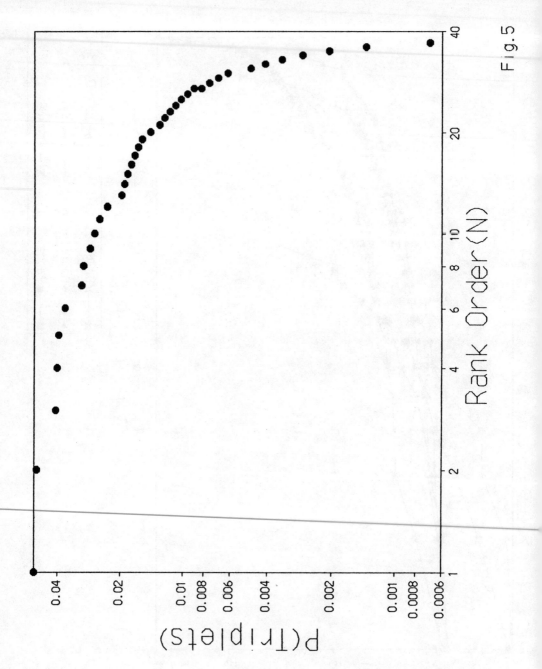

Fig.5

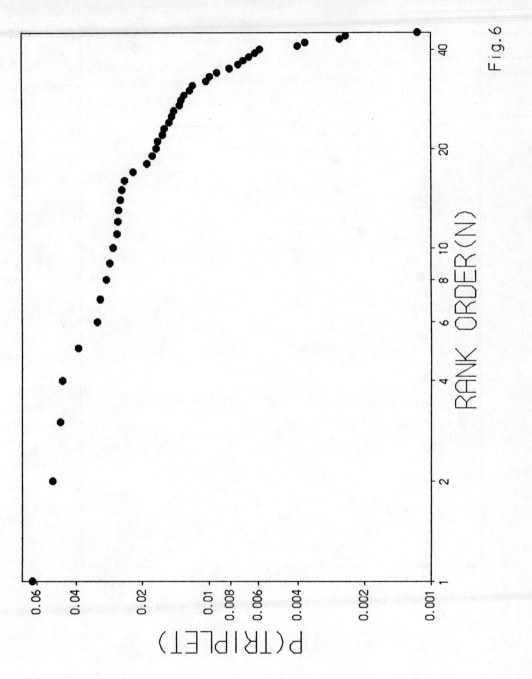

Fig. 6

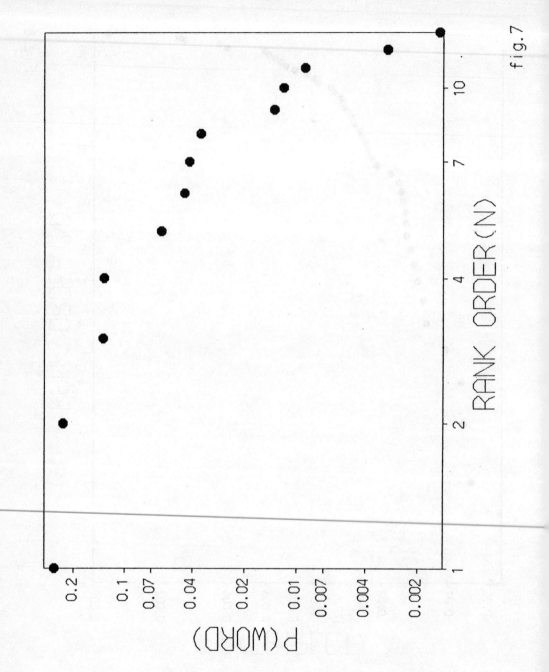

fig.7

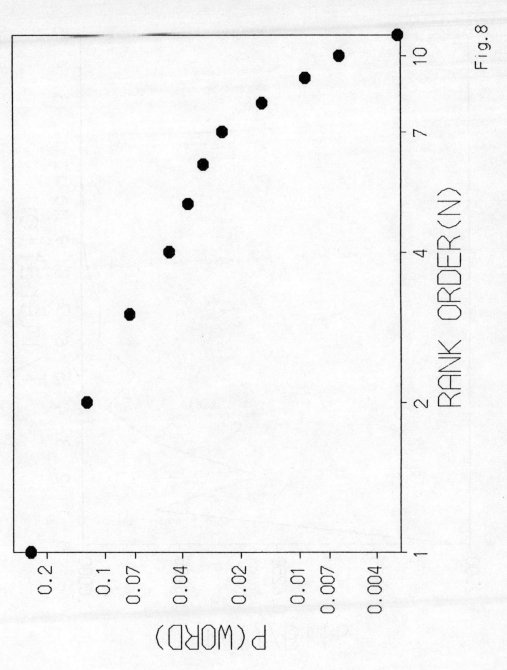

Fig. 8

352

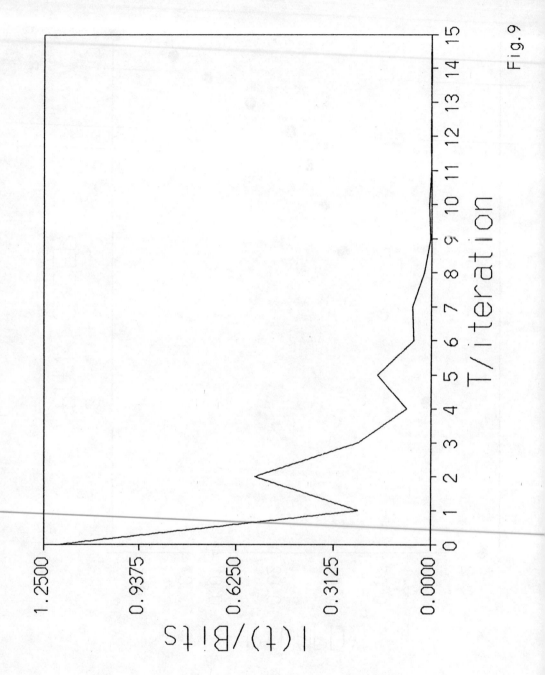

Fig.9

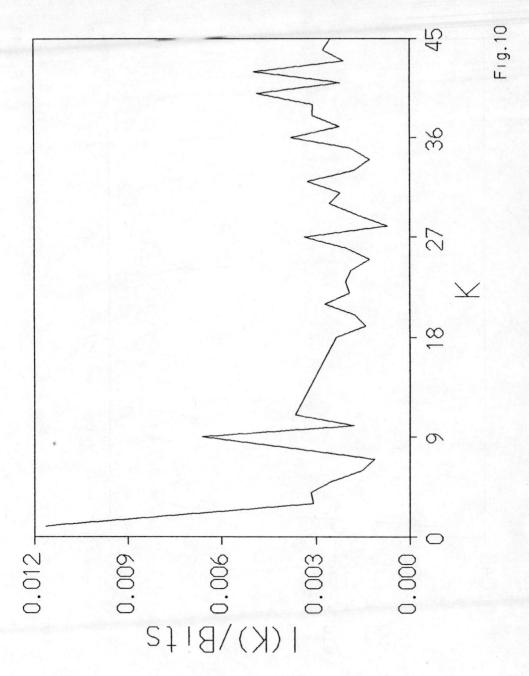

Fig.10

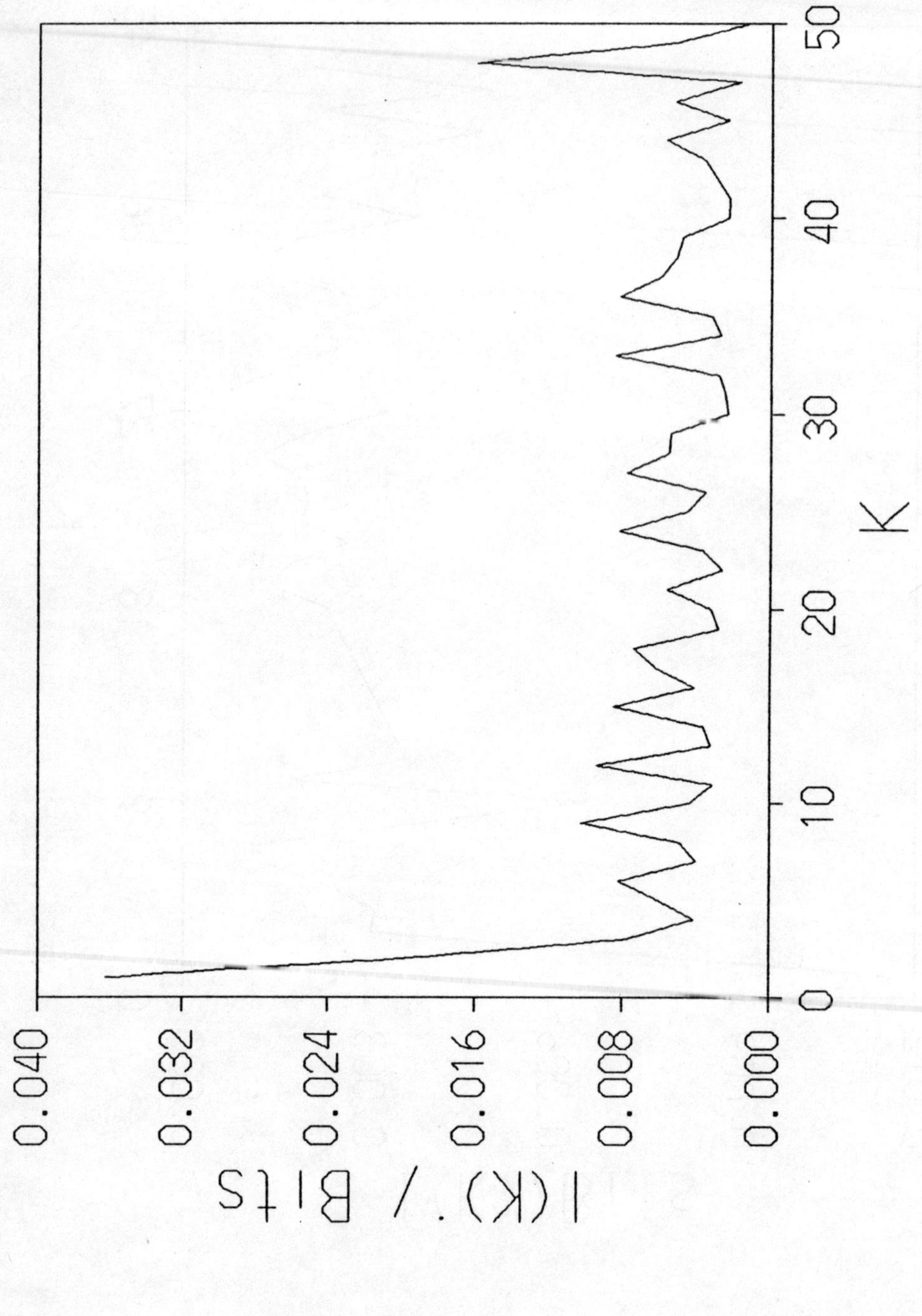

Fig.11

TEXT	1		TEXT	2	
WORD	Probability	Rank Order	WORD	Probability	Rank Order
KAI	0.0565	1	H	0.0633	1
NA	0.0310	2	KAI	0.0549	2
TO	0.0272	3	TO	0.0429	3
TOΥ	0.0199	4	O	0.0305	4
H	0.0191	5	NA	0.0220	5
ΠΟΥ	0.0185	6	ΠΟΥ	0.0189	6
THN	0.0165	7	EINAI	0.0174	7
EINAI	0.0161	8	ME	0.0121	8
ME	0.0134	9	AΠO	0.0112	9
THΣ	0.0128	10	TΩPA	0.0110	10
ΓH	0.0123	11	ΘA	0.0105	11

Table 1

CASE	1		CASE	2	
STRING	Probability	Rank Order	STRING	Probability	Rank Order
BD	0.26268	1	DC	0.24299	1
CD	0.23283	2	DAC	0.12461	2
ABD	0.13432	3	DBDC	0.07447	3
ACD	0.13283	4	DAAC	0.04672	4
AABD	0.06119	5	DBDBDC	0.03738	5
AACD	0.0447	6	DAAAC	0.03115	6
AAABD	0.04179	7	DAAAAC	0.02492	7
AAACD	0.03582	8	DBDAC	0.01557	8

Table II

Sequence	Probability deduced from counting	Probability inferred from a fifth order Markov chain
AAAAAAA	0.0049029417651	0.0048780563187
AAAAAAB	0.0025015009005	0.0025647512603
AAAAAAC	0.0023013808285	0.0022630158179
AAAAABD	0.0051030618371	0.005103061786
AAAAACD	0.004502701621	0.0045027015759
AAAABDA	0.004802801729	0.004716162984
AAAABDB	0.0024014408645	0.0023099573779
AAAABDC	0.0029017441046	0.0030799431732
AAAACDA	0.0056033620172	0.0054075923274
AAAACDB	0.0019011406844	0.0021056111717
AAAACDC	0.0024014408645	0.0023927399679
AAABDAA	0.004502701621	0.0050326491688
AAABDAB	0.0023011380825	0.0021790852071
AAABDAC	0.0030018010806	0.0025944149056l
AAABDBD	0.004802881729	0.004802881681
AAABDCD	0.0064038423054	0.0064038422413
AAACDAA	0.0056033620172	0.0055752344858
AAACDAB	0.0029017410446	0.0028657747357
AAACDAC	0.0028016810086	0.0028657747357
AAACDBD	0.004402641585	0.0044026415409
AAACDCD	0.0050030018011	0.005003001751
AABDAAA	0.0043025815489	0.00438006638919
AABDAAB	0.0024014408645	0.0023891257593
AABDAAC	0.0030018010806	0.0029366337457
AABDABD	0.0042025215129	0.0042025214708
AABDACD	0.0050030018011	0.005003001751
AABDBDA	0.0054032419452	0.0046527916284
AABDBDB	0.0021012607565	0.002651590928
AABDBDC	0.0024014408645	0.0026015609105
AABDCDA	0.0067040224134	0.006454339829
AABDCDB	0.0029017410446	0.0025491930417
AABDCDC	0.0020012007204	0.0026034311915
AACDAAA	0.0057034220532	0.0060354839751
AACDAAB	0.0025015009005	0.0023092286513

Sequence	Probability deduced from counting	Probability inferred from a fifth order Markov chain
AACDAAC	0.0025015009005	0.0023617111207
AACDABD	0.0055033019812	0.0055033019261
AACDACD	0.0055033019812	0.0055033019261
AACDBDA	0.0054032419452	0.0053067554287
AACDBDB	0.0022013207925	0.0027981074079
AACDBDC	0.0032019211527	0.0027016209455
AACDCDA	0.0052031218371	0.0054503855614
AACDCDB	0.0028016810086	0.0024278990228
AACDCDC	0.0023013808285	0.0024278990228
ABDAAAA	0.0040024014409	0.0040227181592
ABDAAAB	0.0024014408645	0.0024136308955
ABDAAAC	0.0024014408645	0.0023689340271
ABDAABD	0.004802881729	0.004802881681
ABDAACD	0.0059035421253	0.0059035420662
ABDABDA	0.0054032419452	0.0050040332978
ABDABDB	0.0020012007204	0.0025272895443
ABDABDC	0.0024014408645	0.0025272895443
ABDBDAA	0.0052031218731	0.0049867240419
ABDBDAB	0.0018010806484	0.0021371674465
ABDBDAC	0.0023013808285	0.0021816917684
ABDBDBD	0.0053031819091	0.0053031818S6
ABDBDCD	0.0052031218731	0.005203121821
ABDCDAA	0.0066039623774	0.005815116919
ABDCDAB	0.00200112007204	0.0028798674265
ABDCDAC	0.0033019811887	0.0032121598219
ABDCDBD	0.004702821693	0.0047028216459
ABDCDCD	0.004802881729	0.0048028811681
ACDAAAA	0.0052031218731	0.0057534520137
ACDAAAB	0.0033019811887	0.0028767260068
ACDAAAC	0.0030018010806	0.0028767260068
ACDAABD	0.004402641585	0.0044026415409
ACDAACD	0.004502701621	0.0045027015759
ACDABDA	0.004702821693	0.0053022289033
ACDABDB	0.0034020412247	0.0027254447634

Sequence	Probability deduced from counting	Probability inferred from a fifth order Markov chain
ACDABDC	0.0023013808285	0.0023785699753
ACDACDA	0.0040024014409	0.0041785080624
ACDACDB	0.0033019811887	0.0029297578254
ACDACDC	0.0023013808285	0.0024974984741
ACDBDAA	0.0058034820893	0.0056576001109
ACDBDAB	0.0022013207925	0.0024687709575
ACDBDAC	0.0030018010806	0.0028802327838
ACDBDBD	0.0058034820893	0.0058034820312
ACDBDCD	0.0056033620172	0.0056033619611
ACDCDAA	0.0055033019812	0.0056635922735
ACDCDAB	0.0028016810086	0.0028852262525
ACDCDAC	0.0027016209726	0.0024577853262
ACDCDBD	0.0049029417651	0.004902941716
ACDCDCD	0.0049029417651	0.004902941716
BDAAAAA	0.0038022813688	0.0043225935129
BDAAAAB	0.0030018010806	0.0022513507879
BDAAAAC	0.0022013207925	0.002431458851
BDAAABD	0.0054032419452	0.0054032418911
BDAAACD	0.0053031819091	0.005303181856
BDAABDA	0.0043025815489	0.0049282146114
BDAABDB	0.0030018010806	0.0027619664306
BDAABDC	0.0032019211527	0.0028161226351
BDAACDA	0.0054032419452	0.0054150136599
BDAACDB	0.0029017410446	0.0033323160984
BDAACDC	0.0032019211527	0.002759574269
BDABDAA	0.0052031218731	0.004689770337
BDABDAB	0.0026015060936	0.002679868764
BDABDAC	0.0021012607565	0.0025363043659
BDABDBD	0.004402641585	0.0033026415409
BDABDCD	0.0050030018011	0.0050030011751
BDACDAA	0:0049029417651	0.0051902798914
BDACDAB	0.0026015609366	0.0026218939658
BDACDAC	0.0025015009005	0.0021938296448
BDACDBD	0.0055033019812	0.0055033019261

Sequence	Probability deduced from counting	Probability inferred from a fifth order Markov chain
BDACDCD	0.0053031819091	0.005303181856
BDBDAAA	0.0055033019812	0.0055024004843
BDBDAAB	0.002877402088	0.0027016209726
BDBDAAC	0.0030018010806	0.0028269213498
BDBDABD	0.004802881729	0.004802881681
BDBDACD	0.0049029417651	0.004902941716
BDBDBDA	0.0058034820893	0.0057767378743
BDBDBDB	0.0029017410446	0.0027389705439
BDBDBDC	0.0021012607565	0.002290775364
BDBDCDA	0.0040024014409	0.0043581704142
BDBDCDB	0.0032019211527	0.0027238565089
BDBDCDC	0.0026015609366	0.0027238565089
BDCDAAA	0.004602761657	0.0052282553843
BDCDAAB	0.0021012607565	0.002240680879
BDCDAAC	0.0038022813688	0.0030373674138
BDCDABD	0.0052031218731	0.005203121821
BDCDACD	0.0058034820893	0.0058034820312
BDCDBDA	0.0058034820893	0.0051311892613
BDCDBDB	0.0022013207925	0.0025162562724
BDCDBDC	0.0027016209726	0.0030589782135
BDCDCDA	0.0043025815489	0.004780748587
BDCDCDB	0.0031018611167	0.0030377673305
BDCDCDC	0.0034020412247	0.002987967866
CDAAAAA	0.0058034820893	0.0052831698491
CDAAAAB	0.0020012007204	0.002751650963
CDAAAAC	0.0032019211727	0.0029717830401
CDAAABD	0.0055033019812	0.0055033019261
CDAAACD	0.0055033019812	0.0055033019261
CDAABDA	0.004802881729	0.00417722485754
CDAABDB	0.0021012606565	0.0023410953554
CDAABDC	0.0020012007204	0.0023410934445
CDAACDA	0.0050030018011	0.0049912299822
CDAACDB	0.0035021012608	0.0030715261429
CDAACDC	0.0021012607565	0.0025436075871

Sequence	Probability deduced from counting	Probability inferred from a fifth order Markov chain
CDABDAA	0.004602761657	0.050687416774
CDABDAB	0.0030018010806	0.0028964238157
CDABDAC	0.0031018611167	0.0027412582541
CDABDBD	0.0055033019812	0.00550330119261
CDABDCD	0.004802881729	0.004802881681
CDACDAA	0.004802881729	0.0045155435055
CDACDAB	0.0023013808285	0.0022810477502
CDACDAC	0.0016009605763	0.001908631791
CDACDBD	0.0061036621973	0.0061036621362
CDACDCD	0.0052031218731	0.005203121821
CDBDAAA	0.0054032419452	0.0054041433328
CDBDAAB	0.0030018010806	0.002826019908
CDBDAAC	0.0026015609366	0.0027764406114
CDBDABD	0.004802881729	0.004802881681
CDBDACD	0.0056033621172	0.0056033619611
CDBDBDA	0.005803482089	0.005830226188
CDBDBDB	0.0027643313822	0.0026015609366
CDBDBDC	0.0025015009005	0.002311986427
CDBDCDA	0.0056033620172	0.0052475929477
CDBDCDB	0.0028016810086	0.0032797455923
CDBDCDC	0.0034020412247	0.0032797455923
CDCDAAA	0.0059035421252	0.0052788048282
CDCDAAB	0.0024014408645	0.0022620206969
CDCDAAC	0.002301388285	0.0030662947225
CDCDABD	0.0054032419452	0.0054032418911
CDCDACD	0.004602761657	0.004602761609
CDCDBDA	0.004602761657	0.005250543808
CDCDBDB	0.002901740446	0.0025868055136
CDCDBDC	0.0035021012608	0.0031447439578
CDCDCDA	0.0053031819091	0.0048250147763
CDCDCDB	0.003001801806	0.0030658948058
CDCDCDC	0.0026015609366	0.0030156342352
DAAAAAA	0.004802881729	0.0048277670783
DAAAAAB	0.0026015609366	0.0025383105257

Sequence	Probability deduced from counting	Probability inferred from a fifth order Markov chain
DAAAAAC	0.0022013207925	0.002239685758
DAAAABD	0.0050030018011	0.005003001751
DAAAACD	0.0054032412945	0.0054032418911
DAAABDA	0.0050030018011	0.005089720448
DAAABDB	0.0024014408645	0.0024929243011
DAAABDC	0.0035021012608	0.0033238990681
DAAACDA	0.0057034220532	0.0058991916299
DAAACDB	0.0025015000905	0.0022970303691
DAAACDC	0.0026015609366	0.0026102617831
DAABDAA	0.0052031218731	0.004673142881
DAABDAB	0.0019011406844	0.0020234362637
DAABDAC	0.0020012007204	0.0024088526949
DAABDBD	0.0051030618371	0.005103061786
DAABDCD	0.0052031218731	0.005203121821
DAACDAA	0.0051030618371	0.0051311892613
DAACDAB	0.0026015609366	0.0026375271904
DAACDAC	0.0027016209726	0.0026375271904
DAACDBD	0.0064038423054	0.0064038422413
DAACDCD	0.0053031819091	0.005303181856
DABDAAA	0.004502701621	0.0044252191898
DABDAAB	0.0024014408645	0.0024137559217
DABDAAC	0.0029017410446	0.00296690832 04
DABDABD	0.0056033620172	0.0056033619611
DABDACD	0.0053031819091	0.005303181856
DABDBDA	0.0039023414048	0.0046527916284
DABDBDB	0.0032019211527	0.002651590928
DABDBDC	0.0028016810086	0.0026015609105
DABDCDA	0.0052031218731	0.0054528043383
DABDCDB	0.0018010806484	0.0021536286042
DABDCDC	0.0028016810086	0.0021994504894
DACDAAA	0.005803482093	0.0054714200522
DACDAAB	0.0019011406844	0.0020934128895
DACDAAC	0.0020012007204	0.0021409904552
DACDABD	0.0049029417651	0.004902941716

Sequence	Probability deduced from counting	Probability inferred from a fifth order Markov chain
DACDACD	0.0041024614769	0.0041024614358
DACDBDA	0.0056033620172	0.0056998484235
DACDBDB	0.0036021612968	0.0030053746233
DACDBDC	0.0024014408645	0.0029017410156
DACDCDA	0.0058034820893	0.005562182908
DACDCDB	0.0021012607565	0.0024750426932
DACDCDC	0.0026015609366	0.0024750426932
DBDAAAA	0.0050030018011	0.0049826849926
DBDAAAB	0.0030018010806	0.0029896109956
DBDAAAC	0.0029017410446	0.002934247829
DBDAABD	0.0057034220532	0.0057034219961
DBDAACD	0.00560336620172	0.0056033619611
DBDABDA	0.0045027016213	0.0049019101692
DBDABDB	0.0024014408645	0.0021786267419
DBDABDC	0.0026015609366	0.0024757122067
DBDACDA	0.004702821693	0.0050511075371
DBDACDB	0.0030018010806	0.0027781091454
DBDACDC	0.0028016810086	0.0026770869947
DBDBDAA	0.0060036021613	0.0062199998803
DBDBDAB	0.0030018010806	0.0026657142344
DBDBDAC	0.0026015609366	0.0027212499476
DBDBDBD	0.0055033019812	0.0055033019261
DBDBDCD	0.004602761657	0.0046027616109
DBDCDAA	0.0039023414048	0.0046911867582
DBDCDAB	0.0032019211527	0.0023232543945
DBDCDAC	0.0025015009005	0.0025913222093
DBDCDBD	0.0060036021613	0.0060036021012
DBDCDCD	0.0060036021613	0.0060036021012
DCDAAAA	0.0058034820893	0.0052531518386
DCDAAAB	0.0022013207925	0.0026265759193
DCDAAAC	0.0025015009005	0.0026265759193
DCDAABD	0.004502701621	0.0045027015759
DCDAACD	0.0061036621973	0.006103036621362
DCDABDA	0.0060036021613	0.0054041948438

Sequence	Probability deduced from counting	Probability inferred from a fifth order Markov chain
DCDABDB	0.0021012607565	0.0027778571627
DCDABDC	0.0025015009005	0.0024224311705
DCDACDA	0.004702821693	0.0045267159843
DCDACDB	0.0028016810086	0.0031739043108
DCDACDC	0.0029017410446	0.0027056233469
DCDBDAA	0.0052031218731	0.0053490037413
DCDBDAB	0.0026015609366	0.0023341107235
DCDBDAC	0.0026015609366	0.0027231291774
DCDBDBD	0.0051030618371	0.005103061786
DCDBDCD	0.0062037222333	0.0062037221713
DCDCDAA	0.0051030618371	0.005103061786
DCDCDAB	0.0026015609366	0.0025180156386
DCDCDAC	0.0019011406844	0.0021449762847
DCDCDBD	0.0061036621973	0.0061036621362
DCDCDCD	0.0060036021613	0.0060036021012

Table IV

Degrees of freedom	orders compared	x^2
1	1 - 0	9086
2	2 - 1	2793
4	3 - 2	503
8	4 - 3	123,1
16	5 - 4	36,5
32	6 - 5	0.629

Optimization and Complexity in Molecular Biology and Physics

Peter Schuster

Institut für Theoretische Chemie der Universität Wien, Währingerstraße 17
A-1090 Wien, Österreich

Abstract. Many optimization problems in physics and biology lead to highly *rugged* cost functions. The most important examples come from the theory of spin glasses and from biological evolution. During the last decade new algorithms were developed in order to deal with such complex systems. Two techniques – known as simulated annealing and genetic algorithms – are applied to study the optimization of polynucleotide replication. Ruggedness of value landscapes in evolutionary optimization has its origin in the complexity of genotype-phenotype-relations. Folding of RNA molecules into spatial structures represents an example of a complex relation which is particularly important in virology and cell metabolism. A model of polynucleotide folding and optimization of replication rates through mutation and selection is presented and discussed. It allows to extend the concepts of conventional population genetics into the realm of frequent mutations.

1. *Rugged* Cost Functions and Value Landscapes

During the last decade more and more problems from mathematics, physics, biology and other scientific disciplines were found to be especially hard and difficult to solve even with the assistence of powerful computers. Moreover, these problems have the naughty property that the numerical efforts to find a solution increase very fast, presumably exponentially with system size. According to this and other features which are used for classification [1,2] and which shall not be discussed here, these problems are characterized as *N(ondeterministic) P(olynomial time)-complete*. Problem solving techniques from physics and biology were generalized and applied to optimization tasks in other disciplines [3]. Three classes of methods became particularly important:

- *Simulated Annealing.* This method originated in spin glass research [4] and represents a modification of *Monte-Carlo* optimization methods [5].
- *Neural Networks.* They were originally conceived as models for the dynamics of central nervous systems. Recently, they were *rediscovered* and are used now in some optimization problems. The main field of application, however, is pattern recognition [6].
- *Genetic Algorithms.* These methods mimic biological evolution and apply its principles to optimization problems in other disciplines [7-9].

In this contribution we shall apply simulated annealing and a genetic algorithm which is derived from the evolution of simple organisms – so-called *prokaryonts*: viruses, bacteria and cyanobacteria – to problems of molecular evolution and compare the results of both techniques.

Most complex optimization problems share a common feature: one searches for the global maximum – or global minimum – of a highly structured objective function or cost function $V(x_1, x_2, \ldots, x_n)$ of many variables. The number n may be as large as several thousand and more. In some cases the qualitative behaviour in the limit $\lim n \to \infty$ is of interest. Using vector notation, $\mathbf{x} = (x_1, x_2, \ldots, x_n)$, we can define the problem to find the maximum of the cost function by

$$\mathbf{x}_m: \qquad V(\mathbf{x}_m) = \max\{V(\mathbf{x})\} \tag{1}$$

The main difficulty of this search is illustrated in figure 1. A typical cost function of a complex optimization problem has many maxima widely distributed with respect to height. Only a small percentage of these peaks is of interest. Most of them lie at small values of the cost function and are useless as approximate solutions of the optimization problem.

An example of complex optimization in physics is provided by the search for a minimum of the spin glass Hamiltonian [10,11]

$$\mathcal{H} = \sum_{i=1}^{n} h_i s_i + \sum_{i=1}^{n-1} \sum_{j>i}^{n} J_{ij} s_i s_j . \tag{2}$$

Variables of this cost function are the individual spins s_i which can take two values: $s_i \in \{+1, -1\}$; $i = 1, 2, \ldots, n$. The parameters h_i represent the local magnetic fields experienced by the individual spins s_i in a spin glass. The spin-spin constants J_{ij} are randomly distributed. A state of the spin system – a *configuration* – is described by a vector $\mathbf{S}_j = (s_1^{(j)}, s_2^{(j)}, \ldots, s_n^{(j)})$. For some applications it is useful to define a *configuration space* whose elements are the 2^n different vectors $\mathbf{S}_j$. This configuration space is identical to the *sequence space* of binary sequences of the same length (n) shown in figure 3. Configurations correspond to the corners of a hypercube of dimension n. The edges connect configurations which are transformed into each other by a single spin flip.

The problem is to find the configuration (S) which minimizes the Hamiltonian. The complexity, in essence, is caused by the coincidence of two conditions:
- exponentially increasing diversity of spin configurations as the number of spins goes up and
- lack of translational invariance of the Hamiltonian due to randomness in sign and value of the coupling constants J_{ij} in the spin glass.

In contrast to spin glasses magnetic *Ising* systems do not meet the second condition. All non-zero coupling constants are set equal. In the one dimensional nearest neighbour case these assumptions yield: $J_{ij} = J \cdot \delta_{j,i\pm 1}$. We make use here of Kronecker's symbol: $\delta_{ij} = 1$ if $i = j$ and $\delta_{ij} = 0$ if $i \neq j$ otherwise. Then, the cost function described by the Hamiltonian (2) has only a single minimum or two degenerate minima, and the optimization problem is fairly simple accordingly. Unlike in *Ising* systems it is impossible to achieve an optimum configuration by regular orientation of spins in spin glasses. Randomly distributed coupling constants that vary in sign cause *frustration* [12]: favourable orientation of one spin relative to a second creates inevitably an

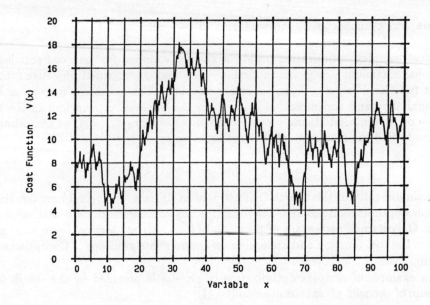

Figure 1: A cost function $V(\mathbf{x})$ characteristic of a complex optimization problem. An arbitrarily chosen degree of freedom is plotted on the x-axis. The cost function $V(\mathbf{x})$ commonly depends on many variables. It has a great number of maxima. Only the highest peaks, however, are useful as approximate solutions. Typical cost functions have bizarre structures and are often called *rugged* landscapes: close by lying vectors $\mathbf{x}$ and $\mathbf{x}+\delta\mathbf{x}$ may belong to very different values of the cost function.

unfavourable orientation relative to others. Frustration thus leads to the appearence of many minima of very different depth in the spin glass Hamiltonian.

Optimization by *simulated annealing* mimics cooling of a spin glass. At high temperatures all configurations have essentially the same probability to be realized. Slow cooling causes spin systems to approach configurations of low energy. The cooling process is used to modify the Monte-Carlo search for the global maximum of a cost function $V(\mathbf{x})$. The search is initiated by a start vector $\mathbf{x}_0$. New vectors are created by variation of $\mathbf{x}_0$ with a random element. Depending on the new value of the cost function and the selection principle applied the new vector is either accepted – $\mathbf{x}_1$ replaces $\mathbf{x}_0$ – or rejected. Repetition of this procedure leads to a sequence of temporarily accepted solutions

$$\mathbf{x}_0 \rightarrow \mathbf{x}_1 \rightarrow \ldots \rightarrow \mathbf{x}_k \rightarrow \mathbf{x}_{k+1} \rightarrow \ldots (\rightarrow \mathbf{x}_m) \ , \tag{3}$$

which converges eventually to the optimal solution. The probability of acceptance of one particular step, say $x_k \rightarrow x_{k+1}$, is determined by the relation

$$
\text{Prob}(x_k \rightarrow x_{k+1}) = \begin{cases} 1 & \text{if } V(x_k) \geq V(x_{k+1}) \\ \exp\left\{ \left(V(x_k) - V(x_{k+1}) \right) \Big/ T \right\} & \text{if } V(x_k) < V(x_{k+1}). \end{cases}
$$

Equation (3) implies, that a new solution is always accepted if it is better than, or as good as the current one. Worse solutions can be accepted too, but with a certain probability only which decreases exponentially with the quality of the solution. By this trick the optimization avoids being trapped in low lying local maxima. The probability of acceptance of a worse solution is given by a Boltzmann weight which uses a formal temperature T. The success of *simulated annealing* depends on the choice of an appropriate temperature program. A good program starts at sufficiently high temperature to reduce the sensitivity to starting conditions and then cooling is performed as close as possible to the formal thermodynamic equilibrium.

Evolutionary adaptation in biology can be understood as a complex optimization problem [13]. The object to be varied is a polynucleotide molecule – DNA or, in some classes of viruses, RNA. It is commonly called the *genotype*. The genotype is not evaluated as such, it is first mapped into a *phenotype* and then the selection principle operates on phenotypes. Charles Darwin's great merit was to find the universal mechanism of phenotype evaluation in nature: *natural selection* implies that those variants which have more – and/or more fertile – descendants in future generations increase in number whereas their less efficient competitors decrease. Efficiency in the natural evaluation process refers exclusively to reproductive success. The variant which produces the largest number of fertile descendants replaces ultimately all competitors and is selected. In population genetics the notion of *fitness* was introduced as a quantitative measure for the capability of being selected. The cost function of evolutionary optimization is therefore often characterized as *fitness landscape*. We shall show, however, that fitness in the sense of Darwin's *survival of the fittest* cannot be assigned to a single genotype. It is an ensemble property. Therefore we suggest to use the notion of *value landscape* rather than fitness landscape as an appropriate synonym for cost function in biology.

It is impossible to derive analytic expressions for realistic value landscapes in biology. This holds true even for the most simple conceivable cases like folding and replication of nucleic acid molecules. We shall study a model of this process here. In this model secondary structures are computed from known genotypes by means of a folding algorithm.

Genetic algorithms mimic evolutionary optimization. A *population* of test solutions is set up, individual solutions are *replicated* and subject to variations. The better a given solution, the more descendants it has. Two types of variations are commonly considered: mutations – variations within a solution – and recombinations – exchange of parts between different solutions. Selection is achieved by constant population size. Excess production is compensated by an appropriate dilution flux.

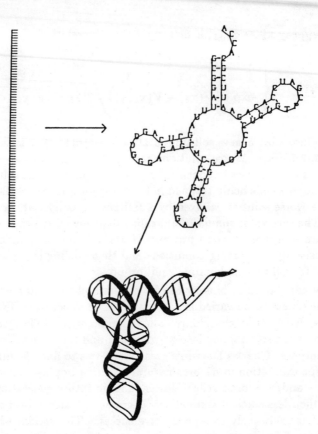

Figure 2: Folding of a polyribonucleotide chain into the twodimensional secondary and the threedimensional tertiary structure of the RNA molecule. *Phenylalanyl-transfer*-RNA, an RNA with a chain length of $\nu = 76$ bases, is chosen as an example. The symbols **D**, **Y**, **M**, **T** and **P** are used for modified bases which are unable to form base pairs. The major driving force of folding into secondary structures is the formation of thermodynamically favoured double helical substructures which contain stacks of complementary base pairs: **G≡C** and **A=U**. It is important to note that in double helices the two nucleotide strands run in opposite direction.

2. Genotype and Phenotype

In view of their chemical structure nucleic acids are visualized best as *strings* or sequences of symbols. There are four classes of symbols. In ribonucleic acid (RNA) they consist of **G**(uanine), **A**(denine), **C**(ytosine) and **U**(racil). Double helical structures (figure 2) are particularly stable in themodynamical terms. The symbols are grouped into two base pairs which constitute complementarity relations: **G≡C** and **A=U**. All other combinations of symbols opposing each other in the double helix cause low stability or instability.

In order to illustrate the source of ruggedness in the value landscapes of RNA folding and replication we distinguish the genotypic sequences or strings I_k and the phenotypic [14] spatial structures $\mathcal{G}_k(I_k)$. The phenotype is formed through folding to a threedimensional structure as shown in figure 2. The driving force for this spontaneous process in aqueous solution of proper ionic strength is minimization of Gibbs' free energy through base pairing, stacking and other types of molecular association. Polyribonucleotide folding can be partitioned into two steps:

- folding of the string into a quasiplanar – twodimensional – secondary structure through formation for as many basepairs as possible and
- formation of a spatial – threedimensional structure from the planar folding pattern.

The first of these two steps is modelled much more easily than the second one. Efficient algorithms are available for the computation of RNA secondary structures [15]. The prediction of tertiary structures of RNA molecules is much harder. At present there are no reliable theoretical models available for this purpose.

3. Evaluation of Phenotypes

The mean number of descendants is the evaluation criterion of Darwin's principle of selection in stationary populations: variants with more descendants than the mean increase in percentage of total population. Less successful competitors decrease in number until they die out. Elimination of less efficient variants causes the mean to increase. Then more and more variants fall under the failure criterion and ultimately only a single fittest type remains as expressed by the common – and often misused – catchphrase of the *survival of the fittest*. In order to make the model quantitative and to develop it further a measure of *fitness* is required. This measure has to be defined independently of the outcome of selection in order to avoid the long-known vicious circle of the *survival of the survivor*. In principle, chemical reaction kinetics is in a position to provide an independent access to the selection criterion if one would it apply to the biological multiplication process.

The main difficulty of the experimental investigation of Darwin's principle is caused by the enormous complexity of all organisms – even including bacteria – which is presently prohibitive for a study of multiplication by the methods of physics and chemistry. The simplest system in which selection could be observed consists of RNA molecules which are multiplied *in vitro* by a replication essay [16]. This essay contains an RNA replicase which replicates RNA of the bacteriophage $Q\beta$ in cells of the bacterium *Escherichia coli*, together with an excess of all molecular materials necessary for the synthesis of RNA. By means of the $Q\beta$ system the kinetics of RNA replication was explored in great detail [17]. These investigations can be understood as the basis of a molecular model of evolution [18,19] which we shall briefly discuss now.

Darwin's principle can be formulated on the molecular level by means of kinetic – ordinary – differential equations. The variables are the concentrations of genotypes: $[I_k] = c_k(t)$. The appropriate measure of genotype frequencies in populations are relative concentrations normalized to 1:

$$x_k(t) = c_k(t) \Big/ \sum_{i=1}^{n} c_i(t) \quad \text{with} \quad \sum_{i=1}^{n} x_i(t) = 1 \,. \tag{4}$$

Two classes of chemical reactions determine the time dependence of relative concentrations: replication with rate constant A_k

$$(\text{A}) + \text{I}_k \xrightarrow{A_k} 2\,\text{I}_k\,,$$

and degradation with rate constant D_k

$$\text{I}_k \xrightarrow{D_k} (\text{B})\,; \quad k = 1, 2, \ldots, n\,.$$

The symbols **A** and **B** denote low molecular weight materials which are either required for RNA synthesis or which are produced by degradation, respectively. The materials for the synthesis are assumed to be present in excess and their concentrations are essentially constant. Concentrations of degradation products do not enter the kinetic equations since the process is irreversible under the conditions of the experiment. Hence, neither the concentrations of **A** nor those of **B** represent variables in the kinetic equations – and both symbols were put in parantheses accordingly.

The kinetic differential equations depend only on differences in the rates of synthesis and degradation. It is useful therefore to introce an *excess production*, $E_k = A_k - D_k$, of the individual genotypes I_k:

$$\frac{dx_k}{dt} = (E_k - \bar{E})\,x_k \quad \text{with} \quad \bar{E}(t) = \sum_{i=1}^{n} x_i E_i\,; \quad k = 1, \vdots, n\,. \tag{5}$$

The mean excess production of the entire population is denoted here by $\bar{E}(t)$. It represents the molecular analogue of the mean number of descendants mentioned initially as the evaluation criterion for phenotypes.

The solutions of the differential equation (5) are given by

$$x_k(t) = x_k(0)\,\exp(E_k t)\Big/ \sum_{i=1}^{n} x_i(0)\,\exp(E_i t)\,; \quad k = 1, \ldots, n\,. \tag{6}$$

Provided that we waited for sufficiently long time all variables except one would have approached zero: $\lim_{t \to \infty} x_j(t) = 0$ for all $j \neq m$. The only non-vanishing variable, $\lim_{t \to \infty} x_m(t) = 1$, describes the concentration of that variant which has the maximum excess production, $E_m = \max\{E_i; i = 1, \ldots, n\}$, and hence reproduces itself most efficiently. Ultimately only the fittest genotype I_m remains.

The Darwinian scenario as well as conventional population genetics are built upon the assumption that mutations are rare events. Selection is fast – occuring on a time scale which is set by the rate constant of replication, or multiplication in general – and evolutionary adaptation is slow, since it is determined by the rate at which advantageous mutants are formed. Only then, we can expect the concept of *survival of the fittest* to reflect reality. Evolutionary optimization is characterized as a sequence of genotypes which were fittest at the instant of their selection.

In the light of the knowledge provided by present day molecular biology the simple evaluation scheme has to be modified in two points:

- The molecular biology of viruses – and to some extent also that of bacteria – has shown that mutations occur much more frequently than the classical scenario assumes. In case of frequent mutations no single fittest type survives. A distribution of genotypes generally consisting of a most frequent sequence, the *master sequence*, and its closely related mutants is selected instead. This distribution was called *quasispecies* [19] in order to point at some analogy to the notion of species with higher organisms.

- A high precentage of mutants is selectively neutral. Their proliferation in natural populations is the major source of information for phylogenetic studies on the molecular level. The dynamics of neutral selection is a stochastic phenomenon. It can be predicted by probability calculus only. Stochasticity is of general importance when particle numbers are small.

Proliferation of neutral mutants is generally observable in finite populations only. It can be modeled by means of stochastic processes but not by kinetic differential equations. Every model which uses kinetic differential equations is valid in the limit of large – in principle initely large – particle numbers. For most chemical reactions this is no real restriction, since all relevant results refer to 10^{10} or more molecules. In the domain of molecular evolution the limit of large numbers is not generally fulfilled. Every new genotype produced by mutation is at first present in a single copy only and the initial phase of growth is also dominated by stochastic phenomena.

The first problem has been extensively dealt with in the theory of the molecular quasispecies [20-22] . We restrict this contribution to a brief account on the most important results. The second question cannot be treated here in sufficient detail. A review of the theory of neutral mutants is found in [23] . The computer simulations introduced in section 5 comprise also the stochastic effects of evolution.

In the frequent mutation scenario mutant production must be accounted for explicitely. Using Q_{ki} for the frequency at which the genotype I_k is synthesized as an error copy of the sequence I_i the reactions leading to mutants can be written as

$$(A) + I_i \xrightarrow{A_i Q_{ki}} I_k + I_i .$$

Adding these reactions to error-free replication and degradation we end up at the kinetic differential equations of the replication-mutation system:

$$\frac{\mathrm{d}x_k}{\mathrm{d}t} = \sum_{i=1}^{n} A_i Q_{ki} x_i - \left(D_k + \bar{E}(t) \right) x_k ; \quad k = 1, \ldots, n . \tag{7}$$

We note that the mean excess production $\bar{E}(t)$ is defined precisely as in equation (5). Since every descendant is either a correct copy of the parental genotype or a mutant, a conservation law holds: $\sum_{k=1}^{n} Q_{ki} = 1$. The *mutation matrix* $Q \doteq \{Q_{ki}\}$ is a stochastic matrix. In case of symmetric mutation frequencies, $Q_{ik} = Q_{ki}$, it is a bistochastic matrix.

Equation (7) can be solved explicitly too [24,25]. As in (6) the solutions are superpositions of exponential functions, but now they are obtained as linear combinations of the eigenvectors $\vec{\ell}_j$ of a *value matrix* $W \doteq \{W_{ij} = A_j \cdot Q_{ij} - D_i \cdot \delta_{ij}\}$. The coefficients are time dependent: $u_j(t)$. The solution curves have the same general form as those shown in equation (6): the variables $x_k(t)$ have to be replaced by the coefficients $u_j(t)$ and and the eigenvectors of the value matrix, λ_j, appear in the exponential functions instead of the excess productions E_k.

The quasispecies Γ_0 can be defined now in precise mathematical terms. It is the combination of genotypes which is described by the dominant eigenvector of the value matrix, $\vec{\ell}_0 = (\ell_{10}, \ell_{20}, \dots, \ell_{n0})$ – this is the eigenvector which belongs to the largest eigenvalue: $\lambda_0 > \lambda_1 \geq \lambda_2 \geq \dots \geq \lambda_n$. The quasispecies can be written as

$$\Gamma_0 \;=\; \ell_{10}\,\mathbf{I}_1 \oplus \ell_{20}\,\mathbf{I}_2 \oplus \dots \oplus \ell_{n0}\,\mathbf{I}_n \;. \tag{8}$$

The components ℓ_{k0} measure the contributions of the individual genotypes $\mathbf{I}_k$ to the quasispecies. Their size is determined to a first approximation by the frequencies of mutation from the master sequence $\mathbf{I}_m$ to the genotype $\mathbf{I}_k$ and by the difference in selective values:

$$\ell_{k0} = \frac{Q_{km}}{W_{mm} - W_{kk}} \;; \quad k = 1, 2, \dots, n; \; k \neq m; \; W_{kk} < W_{mm} \;. \tag{9}$$

Here we characterize the diagonal elements of the value matrix, $W_{kk} = A_k Q_{kk} - D_k$, as selective values. The master sequence is usually the genotype with the largest selective value – for exceptions see [26] and section 4.

Making use of the structure of matrix W it can be shown that all components of the eigenvectors $\vec{\ell}_0$ are positive. The sum of relative concentrations is constant, $\sum_{k=1}^{n} x_k(t) = 1$, and the same holds true for the sum of the coefficients of the eigenvectors, $\sum_{j=1}^{n} u_j(t)$. Accordingly the contributions of all eigenvectors except $u_0(t)$, that of the dominant eigenvector $\vec{\ell}_0$ vanish in the limit $t \to \infty$. After sufficiently long time the mutant distribution converges towards the quasispecies.

The extension of the conventional selection model to the frequent mutation scenario has two important consequences:

- Goal of the selection process is not a fittest genotype but a clan of genotypes consisting of a master sequence – eventually of two or more master sequences – together with its frequent mutants. *Fitness* in the sense of Darwin's selection theory is a property of this clan, called the *quasispecies*. Fitness is not a property of the master sequence alone.

- Quasispecies in real – finite – populations become unstable at large mutation frequencies. Too many errors during replication accumulate and cause the loss of the master sequence and its mutants. Then, there exists no stable stationary mutant distribution.

The latter phenomenon is illustrated best in an abstract vector space of dimension ν – here ν is the length of the sequences. Those vectors which correspond to individual sequences are elements of the so-called *sequence space* [27]. Examples of low-dimensional sequence spaces of binary sequences are shown in figure 3. Sequence spaces are identical with the configuration spaces of spin systems. In the case of sufficiently accurate replication the support of the population – this is the set of all points whose sequences are present in the population – is stationary. The mutant distribution is localized [28] . When the error frequency exceeds some critical value the population starts to walk randomly through sequence space. The surprising result of the theory is not the phenomenon as such, it is the sharpness of the transition from a localized *quasispecies* to

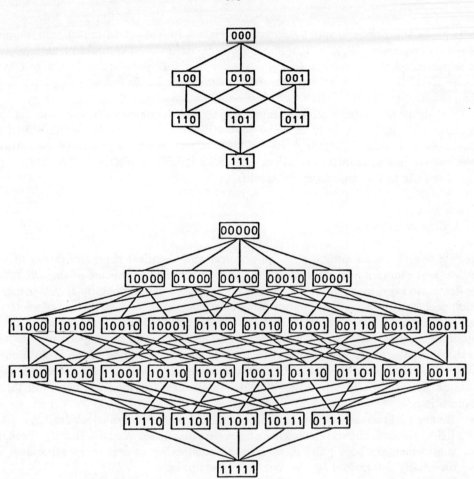

Figure 3: Sequence spaces of binary sequences of chain lengths $\nu = 3$ and $\nu = 5$. All pairs of sequences with Hamming distance $d = 1$ are connected by a straight line. The object drawn in that manner is a *hypercube* of dimension ν. The Hamming distance is the number of digits in which two sequences differ. Such a *vector-(-point-)-space* was used first by Richard Hamming [27] in information theory, The sequence space is identical with the configuration space of a spin system of the same dimension. The edges of the hypercube connect configurations which can be transformed into each other by a single spin flip.

a drifting mutant distribution which has much in common with phase transitions. It is well justified to speak of a critical error threshold at which the population becomes delocalized.

The relation between the error threshold of replication and phase transitions is much deeper than mere similarity. It was shown that statistical mechanics of equilibrium phase transitions in spin systems is equivalent to replication-mutation dynamics of binary sequences [29,30]. The localization-delocalization threshold of populations in sequence space corresponds to the order-disorder transition transition in spin systems. Examples are transitions from ferromagnetic or antiferromagnetic phases into the paramagnetic state.

Localization of quasispecies in sequence space is of fundamental importance for the evaluation of genotypes. Only localized populations conserve the information stored in sequences over sufficiently many generations. Only they can be evaluated by natural selection and are suitable for evolution. Delocalized, drifting populations cannot adapt evolutionarily to the environmental conditions.

4. A Simple Model

Despite the enormous progress in the physical and chemical characterization of selection and evolutionary adaptation – which was achieved by means of the $Q\beta$ RNA replication essay – many questions cannot be addressed experimentally as yet. Among other problems there are the experimental determination of the distribution of selective values in sequence space and the analysis of mutant distributions – during adaptation and in the case of stationarity – which exceed by far the present day capacities. Answer to these questions are, however, fundamental for an understanding of the evolutionary process. A simple model system was conceived which is suitable for numerical calculations and computer simulation [31,32]. In essence, the model is based on three assumptions:

- Binary $(0,1)$ sequences are considered instead of real polynucleotides. As with RNA, non-neighbouring elements of the polymer chain interact mainly through complementary base pairing, $0 \equiv 1$, and the stability of secondary structures is essentially determined by the number of base pairs.
- Variation of sequences is restricted to point mutations which exchange the two symbols at a given position in the polymer: $0 \to 1$ or $1 \to 0$. The chain length remains constant thereby.
- The accuracy of polymer synthesis – measured in terms of the frequency at which the correct symbol is incorporated into the growing polymer chain, $1 \geq q \geq 0$ – is assumed to be independent of the position in the chain.

As a consequence of these three assumptions we can express the frequencies of all mutations $I_i \to I_k$ – which are the previously mentioned elements of the mutation matrix Q – by the chain length ν, the Hamming distance of the two sequences, $d(I_k, I_i) = d_{ki}$ and a single parameter for the accuracy of replication, q:

$$Q_{ki} \;=\; q^\nu \cdot \left(\frac{1-q}{q}\right)^{d_{ki}}. \tag{10}$$

The accuracy parameter of replication, q, is a measure of the precision with which individual digits are incorporated into the polymer chain: $q = 1$ implies error-free copying – every symbol i the copy is identical to that in the template. In the case $q=0$ always the opposite symbol is incorporated – we are dealing with complementary copying which reminds one of the conventional photographic technique. Instead of positive and negative, *plus-* and *minus-*strands alternate here. Sequences occur in pairs, $(I_k{}^\oplus, I_k{}^\ominus)$, in complementary replication. Another extreme case is observed at $q = 0.5$: both symbols, 0 and 1, are incorporated with equal probabilty – and independently of the symbol which stands at the corresponding position in the template. The error frequency is at maximum and there is no correlation between the sequences of template and copy. We may characterize this case appropriately as *random replication*.

Most simple distributions of kinetic parameters in sequence space are particularly well suited to illustrate some fundamental properties of quasispecies. We restrict selection to the synthesis of macromolecules and set all degradation rate constants equal: $D_1 = D_2 = \ldots = D_n = D$. The solutions of equations (5) and (7) are invariant to additive constants and we choose $D = 0$ without losing generality. We consider here two examples of value landscapes for which the structures of quasispecies were studied as functions of the accuracy parameter q [26,31].

4.1. The Single-Peak-Landscape

On the *single-peak-landscape* a higher value of the replication parameter is assigned to the sequence I_0, all other replication parameters are assumed to be equal: $A_0 = \alpha$ and $A_1 = A_2 = \ldots = A_{2^\nu - 1} = 1$. Consequently I_0 is the master sequence ($m = 0$). The domain $1 \geq q \geq 0$ is partitioned into three ranges where direct, random and complementary replication domnate. The individual zones are separated by narrow transitions – even at chain lengths as short as $\nu = 10$. We may define two error thresholds, q_{min} for direct and q_{max} for complementary replication.

Direct replication dominates in the range $1 \geq q \geq q_{min}$. In the limit $q \to 1$ the quasispecies converges towards pure master sequence I_0.

In the range $0 \leq q \leq q_{max}$ complementary replication dominates. At $q = 0$ a master pair consisting of I_0 $(=I_0{}^\oplus)$ and its complementary sequence $I_0{}^\ominus$ is selected. For accuracy parameters $q > 0$ the master pair is accompanied by a distribution of its mutant pairs: $(I_1{}^\oplus, I_1{}^\ominus)$, $(I_2{}^\oplus, I_2{}^\ominus)$ etc. The goal of selection is a quasispecies of complementary pairs.

Chain elongation – increase in chain length ν – sharpens the two transitions remarkably. This can be understood as a strong hint that we are dealing with a co-operative process related to phase transitions. In figure 4 we show the stationary mutant distribution for binary sequences of chain length $\nu = 10$ in the entire domain $1 \geq q \geq 0$. The stationary concentrations of the individual genotypes are essentially constant betweem the two critical values, i.e. in the range $q_{min} > q > q_{max}$. In addition, all seqeunces are present at the same frequency. This implies, however, that stationary solutions cannot be approached by realistic populations when the chain lengths exceed $\nu = 80$. No population can be larger than one mole individuals, and $6 \cdot 10^{23}$ is roughly equal to 2^{80}. In case the population is too small, it cannot be localized in sequence space. Instead it drifts randomly and all genotypes have a finite life time only.

4.2. The Double-Peak-Landscape

Two genotypes are characterized by high replication efficiency on *double-peak-landscapes*. In case these two sequences are close relatives – having a Hamming distance

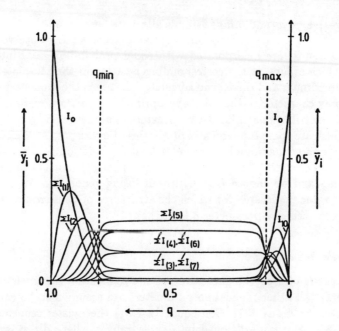

Figure 4: The stationary mutant distribution called *quasispecies* of binary sequences of chain length $\nu = 10$ on the *single-peak-landscape* as a function of the replication accuracy q. A replication parameter $\alpha = 10$ was used. We plot the relative stationary concentrations of error classes as functions of q. Error classes are the master sequence I_0, all one-error mutants $I_{(1)}$, all two-error mutants $I_{(2)}$, ... , and finally the ν-error mutant $I_{(\nu)}$ of the master sequence. Two transitions occuring at critical values of the replication accuracy, q_{min} and q_{max}, are easily recognized. In the domain of random replication which lies between the two transitions all individual genotypes are formed with essentially the same frequency – corresponding to a uniform distribution of sequences. Since we plot concentrations of error classes, the relative frequencies are proportional to the binomial coefficients in this range: $[I_{(k)}] = y_k = \binom{\nu}{k}$.

$d = 1, 2, \ldots$, up to some limit determined by population size, sequence length and structure of the value landscape – they are readily produced as mutants of each other and then a common quasispecies is formed at non-zero error rates, $q < 1$. It consists of the two highly efficient genotypes and a cloud of mutants surrounding both.

The case of more distant peaks appears more interesting. Then two distinct quasispecies – two distributions without common mutants – can exist. By an appropriate choice of the kinetic parameters both quasispecies can be stable. Each quasispecies has its own range of stability on the q-axis. These ranges don't overlap, instead they are separated by sharp transitions at critical replication accuracies q_{tr} which again resemble phase transitions.

A *double-peak-landscape* which is well suited for computation and illustration of transitions between quasispecies puts the two highest selective values at maximal Hamming distance $d = \nu$ – the two master sequences are complementary. Four different replication parameters are used for the two master sequences and their one-error mutants, in particular

(1) $A_0 = \alpha_0$, for the master seqeunce I_0,

(2) $A_{(1)} = \alpha_1$, for all one-error mutants $I_{(1)}$ – we consider all one-error mutants together as an error class and express this by the subscript "(1)" –

(3) $A_{(\nu)} = \alpha_\nu$ for the second master sequence $I_{(\nu)}$ and

(4) $A_{(\nu-1)} = \alpha_{\nu-1}$ for all one-error mutants $I_{(\nu-1)}$ of $I_{(\nu)}$ – these are the $(\nu - 1)$-error mutants of I_0.

The replication constant $A_k = 1$ is assigned to all other sequences.

An illustrative example of quasispecies rearrangement on a *double-peak-landscape* is shown in figure 5. The two genotypes with high replication rates were assumed to be almost selectively neutral – the difference in replication rates is very small. The less efficient genotype has more efficient mutants than its competitor. Then, the more efficient master sequence dominates at low, the less efficient one at high error rates.

The mechanism of the rearrangement of stationary mutant distributions at some critical replication accuracy, q_{tr}, is easily interpreted. The genotype with the somewhat higher replication parameter and the less efficient neighbours in sequence space dominates at small error rates because the mutation backflow from mutants to master sequence is negligible. With increasing error rates – decreasing q-values – mutation backflow becomes more and more important, and at the critical replication accuracy, q_{tr}, the difference in replication rates of the two master sequences is just compensated by the difference in mutation backflow. Further increase in error rates makes the less efficient master sequence superior to that with the higher replication rate constant. The conventional selection principle – as visualized by the rare mutation scenario – has to be modified in one aspect of primary importance:

- Fitness is a collective property and it cannot be attributed to a single genotype. Fitness is determined by all sequences of the quasispecies and precisely to the extent in which they contribute to the stationary distribution.

Accordingly fitness is also an implicit function of the replication accuracy q.

Studies on simple model landscapes are well suited for illustration of phenomena like error thresholds or rearrangements of quasispecies. They suffer, of course, from the fact that they have little in common with real value landscapes. In order to be able to proceed towards an understanding of selection and adaptation on the molecular level it is inevitable to investigate more realistic models.

5. A Model of a Realistic Value Landscape

Relations between genotypes and phenotypes are highly complex even in the most simple systems which are realized in test-tube RNA replication and evolution experiments. The phenotype is formed in a many step process through decoding of the genotype. The outcome of this process is usually unequivocal in constant environment, but many genotypes may yield the same phenotype.

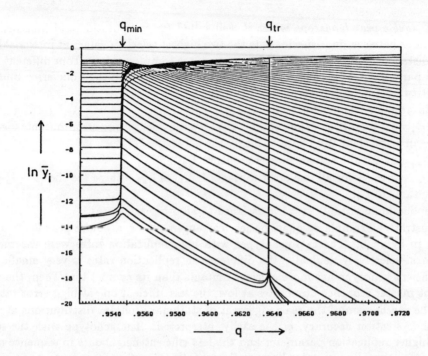

Figure 5: The stationary mutant distribution called *quasispecies* of binary sequences of chain length $\nu = 50$ on the *double-peak-landscape* as a function of the replication accuracy q. Relative concentrations of the individual error-classes are plotted on a logarithmic scale in order to illustrate the sharpness of the two transitions. The two almost neutral master sequences are positioned at maximal Hamming distance, $d = 50$. The master sequence with higher replication efficiency, I_0 with $\alpha_0 = 10$, is surrounded by less efficient one-error mutants, $\alpha_1 = 1$, whereas the less efficient master sequence, $I_{(50)}$ with $\alpha_{50} = 9.9$, has better one-error mutants, $\alpha_{49} = 2$. Two sharp transitions are observed: a rearrangement of quasispecies at $q_{tr} = 0.9638$, where the two master sequences I_0 and $I_{(50)}$ exchange there roles, and the error threshold at $q_{min} = 0.9546$. I_0 is the most frequent genotype in the range $1 \geq q > q_{tr}$ and $I_{(50)}$ dominates in the neighbouring range $q_{tr} > q > q_{min}$.

The genotype-phenotype relation in the model [32,33] considered here is reduced to the folding of a binary string into a secondary structure following a minimum free energy criterion as described in section 3. In the actual computations we applied base pairing and base pair stacking parameters of pure **G,C**-sequences [34,35]. The folding process is restricted to the formation of unknotted planar structures. This restriction is common in all presently used folding algorithms, and is also well justified on an empirical basis of the presently known structural data.

An important common feature of all correlations between genotypes and phenotypic properties is the *bizarre* structure of value landscapes. This means that closely related genotypes may – but need not – lead to phenotypes of very different properties. Distant genotypes, on the other hand, may yield almost identical phenotypic properties. This peculiar feature of genotype-phenotype relations is illustrated by means of secondary structures of binary sequences in figure 6.

Rate constants of replication (A_k) and degradation (D_k) are particularly important properties in molecular evolution. The present knowledge on the dependence of these constants on RNA structures is not sufficiently detailed in order to allow reliable predictions. Nevertheless, it is possible to conceive models which reproduce qualitative dependences correctly. Such an ansatz was used in the computation of a value landscape [32] which describes the distribution of excess productions ($E_k = A_k - D_k$) in sequence space. In order to approach an optimal excess production the replication rate should be as large as possible ($A_k \to$ max) and, at the same time, the degradation rate should be at minimum ($D_k \to$ min). The two constants depend on secondary structures in such a way that both conditions cannot be fulfilled: the two trends contradict eachother. An increase of A_k is commonly accompanied by an increase rather than a decrease in D_k. We thus observe conflicting structural dependences which remind of the frustration phenomenon in spin glasses [12] and we expect a value landscape of high complexity.

In order to explore the details of the value landscape two kinds of *computer experiments* were carried out:

- A population of 3000 binary sequences of chain length $\nu = 70$ was put on the value landscape in order to study the optimization by means of a genetic algorithm. The accuracy parameter was kept constant: $q = 0.999$. The stochastic dynamics of a chemical reaction network consisting of replication, mutation and degradation processes was simulated by means of an efficient computer algorithm [36]. Sequences produced in excess are removed by a dilution flux.
- The distribution of optimal and near optimal regions of excess production (E_k) in sequence space was investigated by simulated annealing [4]. In this search one- and two-error mutants were admitted as variations of the current strings.

The mean excess production $\bar{E}(t)$ increases steadily – apart from small random fluctuations – during an optimization run with the genetic algorithm and finally approaches a plateau value. The secondary structures obtained in three independent optimization runs which were performed with the genetic algorithm are shown in figure 7. Despite identical initial conditions – 3000 all-0-sequences of chain length $\nu = 70$ – the three runs reach different regions in sequence space. These three regions of high excess production are far away from each other. The Hamming distances between these finally selected genotypes are roughly equal, and lie around $d = 30$. Although the kinetic properties of the three optimization products are very similar, they differ largely in sequence and secondary structure.

Depending on population size, replication accuracy and structure of the value landscape two different scenarios of optimization were found:

- The population resides for rather long time in some part of sequence space and moves quickly and in jumplike manner into another area. The optimal solution is approached by stepwise improvements.
- The population approaches the optimum gradually through steady improvement.

0110001101001001101000001011010011000011101100101010001110010101000100

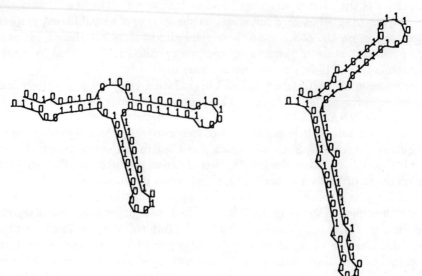

001011010001101010011011101100110000001010000000001010000101100001101 1

Figure 6: Examples of secondary structures of binary sequences. One main reason for the *ruggedness* of value landscapes is to be seen in the complexity of the folding process which presumably cannot be casted into any simpler algorithm. In the upper part of the figure we show two sequences of Hamming distance $d = 1$ which have very different secondary structures and below that two rather distant sequences with $d = 9$ which fold into indentical secondary structures. Mutated positions are indicated by *.

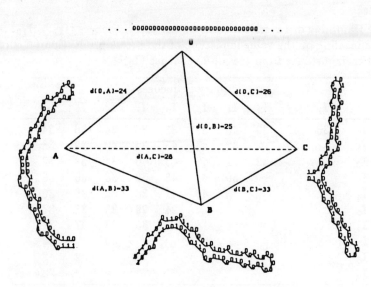

Figure 7: Secondary structures obtained in three independent optimization runs performed with the genetic algorithm. All three runs started with identical initial conditions, 3000 all-0-sequences of chain length $\nu = 70$, but used different seeds for the random number generator. We show the most frequent genotypes at the end of the optimization processes. Interestingly, these three sequences and the all-0-sequence at the start have pairwise almost the same Hamming distances. The four sequences span an approximate tetrahedron in a threedimensional subspace of the sequence space.

The first scenario is characteristic for small populations, low error rates and distant local maxima in sequence space whereas optimization dynamics of the second kind dominates in large populations, at high error rates and with close by lying local maxima. There is a steady transition from one scenario into the other: the gradual approach changes, for example, into a sequence of small jumps when the error rate is reduced. Larger jumps appear on further decrease of the mutation frequency and finally the population is *caught* in some local maximum.

Optimization runs with the genetic algorithm were also carried out at replication accuracies which lie below the critical accuracy for a population of 3000 molecules – the error threshold was found to depend on population size [37] and occurs at higher replication accuracies in smaller populations. No optimization on the value landscape was observed. The populations drift randomly in sequence space. This general behaviour was observed even when we started with *near optimal* genotypes which had been optimized in previous runs. The initial master sequence is first surrounded by a growing cloud of mutants and then it is displaced by its own error copies. All other sequences in the population are lost likewise after sufficiently long time. In other words,

Table 1: The Hamming distances between the ten *best* sequences ($I_1, \ldots, I_{10}$) obtained by simulated annealing on the value landscape of the excess production (E_k) of binary sequences and their distance from the all-0-sequence (I_0)[t).

	Binary Sequences									
	I_1	I_2	I_3	I_4	I_5	I_6	I_7	I_8	I_9	I_{10}
I_0	30	40	30	40	37	33	36	34	36	34
I_1		30	2	32	43	45	34	40	32	38
I_2			32	2	45	43	40	34	38	32
I_3				30	41	43	32	38	30	36
I_4					43	41	38	32	36	30
I_5						24	29	35	31	37
I_6							35	28	37	31
I_7								30	2	32
I_8									32	2
I_9										30

[t) The binary sequences are arranged in pairs of complementary and inverted sequeneces: I_2 is the inverted complementary sequence of I_1 ($I_2 = \neg I_1 ^\Theta$), I_4 is inverse complementary to I_3 ($I_4 = \neg I_3 ^\Theta$), etc.

every genotype has a finite lifetime only. Therefore, the predictions of the molecular theory of evolution as introduced in section 3 are valid also in populations as small as a few thousand molecules.

In order to complement optimization by genetic algorithms the value landscape was investigated by means of simulated annealing [4]. The main goal of this study was to explore the distribution of high maxima of the excess production (E_k) and to compare the model value landscape of polynucleotide replication with a spin glass landscape. The highest possible value of the excess production of binary sequences with chain length $\nu = 70$ can be computed from the criteria used in the evaluation of replication and degradation rate constants: it amounts to $E_{max} = 2245\,[t^{-1}]$ in arbitrary reciprocal time units. Such a *best* structure, however, does not occur as a stable folding pattern. All sequences which, in principle, could yield the *best* structure fold into other secondary structures because of the minimun free energy criterion – which of course need not meet a maximum excess production criterion. The highest value of excess production obtained by simulated annealing was $E_{opt} = 2045\,[t^{-1}]$. This optimal value is degenerate: it is found with ten distinct sequences. The Hamming distances between the ten sequences and the all-0-sequence are given in table 1. The ten genotypes are related in pairs by symmetry: every sequence is transformed into one with identical structure and properties by complementation and inversion – inserting complementary symbols at all positions and swapping both ends (figure 8). The remaining five sequences (I_1, I_3, I_5, I_7 and I_9) form two pairs, (I_1, I_3) and (I_7, I_9), which are close relatives at Hamming distance $d = 2$ and one *solitary* sequence (I_5). These potential centers for quasispecies are well separated in sequence space – all Hamming distances are larger than $d \geq 30$.

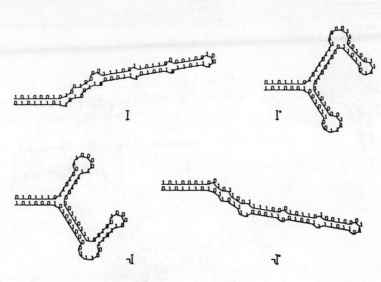

Figure 8: *Symmetry of sequences with respect to secondary sturcture.* Every sequence I at distance d from the all-0-sequence origin has a structure which is identical to that of the complementary and inverted sequence $(\neg I^{\ominus})$ at distance $\nu - d$ from the origin.

In total 879 sequences with exces productions $E_k \geq 2011\,[t^{-1}]$ were identified. Their distribution in sequence space is shown in figure 9 by means of a *minimum spanning tree* [38]. The clusters of peaks with high excess production have rich internal structures. High peaks are more probable in the neighbourhood of other high peaks. Like in mountainous areas on the surface of the earth we find ridges as well as saddles and valleys separating zones of high excess production.

The distribution of near optimal configurations of the spin glass Hamiltonian (2) shows chracteristic features of *ultrametricity*. Arbitrarily chosen triangles of three near optimal configurations are either equilateral or isosceles with small basis. The distribution in sequence space of the high clusters of excess production, on the other hand, shows no detectable bias towards ultrametricity. This distribution of clusters does not deviate significantly from a random distribution. Two causes may be responsible for the difference between spin glass Hamiltonians and value landscapes of polynucleotides. Either binary sequences with a chain length $\nu = 70$ are too short to reveal higher order structures in the value landscape, or polynucleotide folding and evaluation of folded structures have some fundamental internal structure which is different from that intrinsic to the spin glass Hamiltonian.

Finally, it is worth mentioning that optimization of excess production (E_k) by simulated annealing was not only faster – measured in CPU computer time units – but led also to significantly higher values than the genetic algorithm. This may be – at least in part – a result of the constant environmental conditions applied here. Constant enviroment favours algorithms with no or little memory. Storing of previously *best*

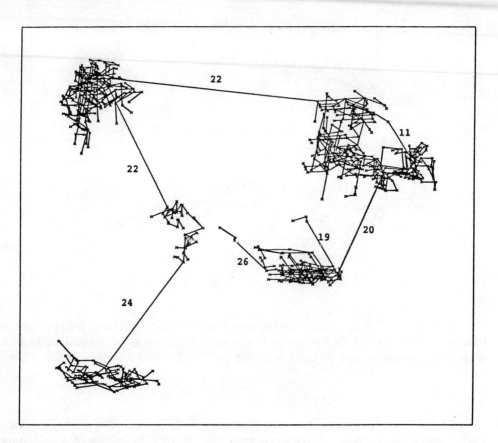

Figure 9: Clustering of near optimal values of the excess production (E_k). The projection shows the *minimum spanning tree* of the frozen out optima of seven annealing runs operating under identical conditions (except random number seeds). The optimal regions reached by each run consist of an ensemble of selectively neutral mutants and qualify as clusters in the minimum spanning tree. The shortest Hamming distances between clusters are indicated.

solutions – as it is naturally done by populations – is advantageous only in variable environments.

6. Concluding Remarks

Rugged value landscapes derived from RNA secondary structures – modeled here by means of binary sequences – have much in common with energy distributions in spin glasses. It seems that (some) complex optimization problems from different scientific disciplines have a common formal origin. Eventually future developments will lead to a notion of *complexity* which is universal in nature. Apart from the apparent similarities there are, of course, substantial differences in the microscopic origins of complexity. In spin glasses spin-spin coupling constants are randomly distributed – or they are at least much less regular than in solids – and this leads to *frustration*. In biological evolution complextiy can be traced down to the properties of polynucleotides. In the simplest cases it is the folding of strings into graphs which follows the rules of molecular biophysics – base pairing and stacking of base pairs such that they yield a minimal free energy – and which creates the complex structure of value landscapes. It is straight to use the notion of frustration here too: structural restrictions have to be fulfilled and therefore it is usually not possible to form the maximal number of base pairs.

Frustration appears again when folded structures are evaluated according to a maximum excess production criterion ($E_k = A_k - D_k$). Here the origin of frustration is twofold. Firstly, according to model assumptions – certainly met by reality – the genotypes which replicate fastest are also readily degraded. Genotypes which are particularly stable do not replicate very well. Optimization of excess production implies searching for a non-trivial compromise between these two trends.

The second source of frustration is typical for biological problems and was found – as a surprise – already in our simple model. The *best* structure having maximal excess production cannot be realized because of principal restrictions. All secondary structures have to meet a minimum free energy criterion, and no genotype was found which can fulfil both criteria – maximum excess production and minimum free energy – at the same time. Boundary conditions restrict the wealth of possible solutions.

An important question concerns the significance of results derived from binary sequences for realistic RNA molecules representing (**G,A,C,U**)-sequences. Depending on the thermodynamic parameters used, binary sequences correspond either to pure (**G,C**)-sequences or pure (**A,U**)-sequences. In addition RNA molecules can be described by twice as long binary sequences. Each base is then encoded by two binary digits, for example by **G=11, A=10, U=01** and **C=00**. Complementarity relations hold – with the restriction that odd numbered positions have to be opposite to odd numbered positions in double helical structures, and the same holds true for the even numbered positions – but the evaluation of Hamming distances has to be modified and is not as straight as with binary sequences. The most important difference, however, concerns the richness in structures which is certainly larger with binary sequences. Point mutations in true RNA molecules will therefore somewhat less likely lead to drastic changes in secondary structures than they do with binary sequences. Nevertheless we can expect that all qualitative findings reported here remain valid in four base sequences.

The occurence of rugged value landscapes was found to be a fundamental property of biological macromolecules which can be traced down to the process of folding strings of digits into threedimensional molecular structures. This fact is the ultimate cause, why transformation of genotypes into phenotypes is an exceedingly complex process – even in the most simple cases. Unfolding of genotypes cannot be reduced to simple algorithms therefore. Genotype-phenotype relations are an important – if not the most important – source of complexity in biology.

Acknowledgements

The work reported here was supported financially by the *Fonds zur Förderung der wissenschaftlichen Forschung in Österreich* (Projects No.5286 and No.6864), the *Stiftung Volkswagenwerk* (B.R.D.) and the *Hochschuljubiläumsstiftung Wien*. Numerical computations were performed on the IBM 3090 mainframe of the *EDV-Zentrum der Universität Wien* as part of the IBM Supercomputing Program for Europe. Assistence in preparation of figures and reading of the manuscript by Dr.Walter Fontana and Mag.Wolfgang Schnabl is gratefully acknowledged.

References

1. J.L.Gersting. *Mathematical Structures for Computer Science*. 2nd Ed., pp.501-504. W.H.Freeman & Co., New York 1987.

2. M.R.Garey, D.S.Johnson. *Computers and Intractability: A Guide to the Theory of NP-Completeness*. W.H.Freeman & Co., San Francisco 1979.

3. D.G.Bounds. *Nature* 329(1987), 215-219.

4. S.Kirkpatrick, D.C.Gelatt, M.P.Vecchi. *Science* 220(1983), 671-680.

5. N.Metropolis, A.Rosenbluth, M.Rosenbluth, A.Teller, E.Teller. *J.Chem.Phys.* 21(1952), 1087-1092

6. J.J.Hopfield, D.W.Tank. *Biol.Cybern.* 52(1985), 141-152.

7. I.Rechenberg. *Evolutionsstrategie*. Friedrich Frommann Verlag, Stuttgart und Bad Cannstatt 1973.

8. J.H.Holland. *Adaptation in Natural and Artificial Systems*. The University of Michigan Press, Ann Arbor 1975.

9. D.E.Goldberg. *Genetic Algorithms in Search, Optimization and Machine Learning*. Addison-Wesley Publ.Co., Reading (Mass.), 1989.

10. S.Edwards, P.W.Anderson. *J.Phys.F* 5(1975), 965-974.

11. K.Binder, A.P.Young. *Rev.Mod.Phys.* 58(1986), 801-976.

12. P.W.Anderson. *J.Less-Common Metals* 62(1978), 291-294.

13. S.Kauffman, S.Levin. *J.theor.Biol.* 128(1987), 11-45.

14. Here we generalize the notion of phenotype commonly used for organisms to molecules.

15. M.Zuker, P.Stiegler. *Nucleic Acids Res.* 9(1981), 133-148.

16. S.Spiegelman. *Quart.Rev.Biol.* 4(1971), 213-253.

17. C.K.Biebricher, M.Eigen. *Kinetics of RNA Replication by Qβ Replicase*. In: E.Domingo, J.J.Holland, P.Ahlquist, eds. RNA Genetics. Vol.I: RNA-directed Virus Replication, pp.1-21. CRC Press. Boca Raton (Florida), 1988.

18. M.Eigen. *Naturwissenschaften* 59(1971), 465-523.

19. M.Eigen, P.Schuster. *Naturwissenschaften* 64(1977), 541-565.

20. M.Eigen, J.McCaskill, P.Schuster. *J.Phys.Chem.* 92(1988), 6881-6891

21. M.Eigen, C.K.Biebricher. *Sequence Space and Quasispecies Distribution.* In: E.Domingo, J.J.Holland, P.Ahlquist, eds. RNA Genetics. Vol.III: Variability of Virus Genomes, pp.212-245. CRC Press. Boca Raton (Florida), 1988.

22. M.Eigen, J.McCaskill, P.Schuster. *Adv.Chem.Phys.* 75(1989), 149-263.

23. M.Kimura. *The Neutral Theory of Molecular Evolution.* Cambridge University Press. Cambridge (U.K.), 1983.

24. C.J.Thompson, J.L.McBride. *Math.Biosc.* 21(1974), 127-142.

25. B.L.Jones, R.H.Enns, S.S.Ragnekar. *Bull.Math.Biol.* 38(1976), 12-28.

26. P.Schuster, J.Swetina. *Bull.Math.Biol.* 50(1988), 635-660.

27. R.W.Hamming. *Coding and Information Theory.* 2nd Ed., pp.44-47. Prentice Hall. Englewood Cliffs (N.J.), 1986.

28. J.S.McCaskill. *J.Chem.Phys.* 80(1984), 5194-5202.

29. L.Demetrius. *J.Chem.Phys.* 87(1987), 6939-6946.

30. I.Leuthäusser. *J.Statist.Phys.* 48(1987), 343-360.

31. J.Swetina, P.Schuster. *Biophys.Chem.* 16(1982), 329-353.

32. W.Fontana, P.Schuster. *Biophys.Chem.* 26(1987), 123-147.

33. W.Fontana, W.Schnabl, P.Schuster. *Phys.Rev.A,* Sept. 1989

34. M.Zuker, D.Sankoff. *Bull.Math.Biol.* 46(1984), 591-621.

35. S.M.Freier, R.Kierzek, J.A.Jaeger, N.Sugimoto, M.H.Caruthers, T.Neilson, D.H.Turner. *Proc.Natl.Acad.Sci.USA* 83(1986), 9373-9377.

36. D.T.Gillespie. *J.Comp.Phys.* 22(1976), 403-434.

37. M.Nowak, P.Schuster. *J.Theor.Biol.,* 137 (1989), 375–395

38. J.B.Kruskal. *Proc.Amer.Math.Soc.* 7(1956), 48-50.

Understanding Evolution as a Collective Strategy
for Groping in the Dark

by H. P. Schwefel, University of Dortmund, Dept. of Computer Science,
P. O. Box 50 05 00, D-4600 Dortmund 50, F. R. Germany

Abstract

On the one hand side many people admire the often strikingly efficient results of organic evolution. On the other hand side, however, they presuppose mutation and selection to be a rather prodigal and unefficient trial-and-error strategy. Taking into account the parallel processing of a heterogeneous population and sexual propagation with recombination as well as the endogenous adaptation of strategy characteristics, simulated evolution reveals a couple of interesting, sometimes surprising, properties of nature's learning-by-doing algorithm. 'Survival of the fittest', often taken as Darwin's view, turns out to be a bad advice. Individual death, forgetting, and even regression show up to be necessary ingredients of the life game. Whether the process should be named gradualistic or punctualistic, is a matter of the observer's point of view. He even may observe 'long waves'.

Introduction

Evolution can be looked at from a large variety of positions. Beginning with the closest physico-analytic viewpoint, one might focus attention to the molecular and cellular processes. A more distant point of view centers on the behaviour of populations and species. Another difference emerges from whether one emphasizes the homeostatic aspect of the adaptation to a given environment, which is more relevant in the short term, or the euphemistic view of development to the more complex, higher, or even better in the long term.

The instruments used here, will be a macroscope and a time accelerator. Moreover, for methodological reasons, an optimistic point of view will be shared by comparing macroevolution with iterative optimization, or, even more adequately, with permanent meliorization techniques, i.e. hill-climbing or ridge-following procedures. By means of a simple algorithmic formulation of the main evolutionary principles, it is possible to reveal some properties of the process which in some cases are striking at the first glance. These findings may not only be helpful for better understanding 'nature's intelligence' but also be beneficial for global long-term planning and other groping-in-the-dark situations.

Modelling evolution

Ashby's homeostat /1/ was a device which should find back to a feasible state by a sequence of random trials, uniformly distributed over a given parameter space. Many people have made the mistake of thinking of mutations as 'pure' random trials. A couple of them was malignant. They wanted to show that evolution theory never will be able to explain how 'nature' found a way to complex living beings within about 10^{17} seconds - the age of our globe. Montroll's random walk /2/ paradigm, on the other hand, neglects the selection principle of evolution. Both mutation and selection (chance and necessity) are the first principles, which, of course, have to be programmed properly.

Broadly accepted hereditary evidence has led to the proverb 'The apple does not fall far off of the tree'. A better model of mutations therefore is a normal distribution for parameter changements between generations, its maximum being centered at the respective ancestor's position. The rôle of chance in such a model is not explicative, however, but only descriptive. An important question now is the suitable size of the standard deviation(s) of the changes, which may be addressed as mean step size(s) from one generation to the next. This question arises with all optimization or meliorization schemes.

Modelling the selection principle is far more easy, as it seems first. 'Survival of the fittest' is the maxime which Spencer derived from Darwin's observations. Some evolution programmers have taken it for granted: According to a given selection criterion, a descendant is rejected if its vitality is less than that of its ancestor, the ancestor otherwise. This scheme may be called a (1+1)- or two membered evolution strategy (E.S. in the following), resembling the 'struggle for life' between one ancestor <u>and</u> one descendant. Rechenberg /3/ has derived theoretical results for the convergence velocity of that process in an n-dimensional parameter space. Most important was his finding that for an endless ridge following situation as well as for a minimum (or maximum) approaching situation the convergence rate is inversely proportional to the number of parameters. Distances growing with the square root of n, the number of iterations or generations needed to proceed from one to the other arbitrary point in space, increases with $O(n^{1.5})$ only, and not exponentially as in the case of simple Monte-Carlo strategies.

This type of creeping random search strategy (see e.g. Brooks /4/, Schumer and Steiglitz /5/, or Rastrigin /6/) was first devised for experimental optimization, where measurement inaccuracies drop out one-variable-at-a-time and gradient-following procedures due to their inability of non-local operation.

Bremermann's 'simulated evolution' /7/ does not differ so much from Rechenberg's as e.g. G.E.P. Box's 'Evolutionary Operation' EVOP /8/

does, an experimental design technique, and the so-called Simplex and Complex strategies of Nelder and Mead /9/ and M.J. Box /10/ for numerical optimization. Whereas random trials are vividly rejected by G.E.P. Box, he centers several experiments (at the same time, on principle) in a deterministic way around the position of the current best point in a low-dimensional parameter space. The best of all then is taken as the center of the next trial series. Nelder, Mead, and M.J. Box, however, using a polyhedron for placing the trials, reject the worst position and find a new one by reflecting the worst with respect to the center of the remaining points of the simplex or complex.

The first concept may be called a (1+lambda)-, the latter a (mu+1)-evolutionary scheme, mu denoting the number of parents, lambda the number of children within one generation. More general, therefore, is a (mu+lambda)- cheme with mu ancestors which have lambda descendants, the mu best of all become parents of the next generation. The fact that individual life times are limited is reflected by the (mu,lambda)- version, first introduced by Schwefel /11/. Now the mu parents are no longer included into the selection, thus lambda must be greater than mu. Theoretical results so long are available for the (1+lambda)- and (1,lambda)- evolution strategies only. All further observations in the following, therefore, were found by computer simulation only.

Self-adaptation of strategy parameters

As for all optimization techniques, the appropriate step size adjustment is of crucial importance. Rechenberg found that there is a 'window' of one decade only within which the (1+1)- evolution strategy has a reasonable convergence velocity. He devised a simple rule for exogenously adjusting a near optimum performance of the process, i.e. to control the success probability which should be in the vicinity of one success among five trials. This advice is good for many but not all situations. Moreover, it does not give any hints to adapt the standard deviations of the parameter changements individually. Some may be too large, others too small, at the same time. Only within the multimembered strategy, one can include the step size or even different step sizes (mutation rates) into the set of the individual's genes and adapt them endogenously. There is some evidence that by means of repair enzymes the effective mutation rates are controlled, the rate of premutations due to environmental conditions being constant over long periods.

Let us think at first, however, of one common step size for all object parameters. Within a (1,lambda)- E.S. the correct step size turns out to be even more important than within a (1+lambda)- version /11/. In the first case regression takes place instead of progress when the step size is too large, whereas stagnation is the worst case in the latter.

At a first glance, therefore, 'survival' of an ancestor might be a good advioc. Simulation results, however, show that the opposite is true. This is the first surprise. Figure 1 demonstrates the difference between a (1+10)- and a (1,10)- E.S. when

$$F_1 = \sum_{i=1}^{n} x_i^2 \quad \text{with } n=30.$$

minimizing the function

The number of variables, n, was taken as large as 30 in order to avoid improper conclusions. In lower dimensional cases nearly every strategy may achieve good results. One common step size sigma (or standard deviation, more precisely) for all x_i is changed by mutation, i.e. by multiplying the ancestor's value with a random number, drawn

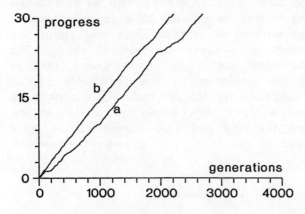

Figure 1:
Self-learning of the mutation rate (gross mean step size here) for function F_1, n= 30
a) (1+10)- evolution strategy
b) (1,10)- E.S.

The 'progress' is measured in terms of log $((F^o/F^g)^{1/2})$, where F^o denotes the start value, F^g the current value of the objective function.

from a logarithmic normal distribution in order to avoid exogenous drift. The (1,10)- strategy turns out to be superior. An explanation for this surprising fact is the following: if an ancestor happens to arrive at a superior position, this might be - by chance - in spite of a non-optimum step size, or a step size which is not suitable for further generations. The (1+lambda)- scheme preserves the unsuitable step size as long as with it a further success is placed. This leads to periods of stagnation. Within a (1,lambda)- E.S. the good position, occasionally won with an unsuitable step size, is lost, together with the latter, during the next generation. This short term regression, however, enhances the long term velocity of the whole process by a stronger selection with respect to the suitable step size (strategy parameter). Simply speaking: Forgetting is as important as learning, the first must be seen as a necessary ingredient of the latter. One might interpret the fact of an inherent finite life time (preprogrammed maximum number of cell divisions) of living beings as an appropriate measure of nature to overcome the difficulties of undeserved success - or, in a changing environment, of forgetting obsolete 'knowledge'.

Collective learning of proper scalings

In most cases it is not sufficient to adapt one common step size for all object parameters. For an objective function

$$\text{like } F_2 = \sum_{i=1}^{30} (i \, x_i^2),$$ for example, individual standard deviations sigma_i,

appropriately scaled, could speed up the progress rate considerably. To achieve this kind of flexibility within the multimembered evolution strategy, each individual is characterized by a set of n step sizes in addition to the n object parameters. They are mutated by multiplication with two random factors, one being common for all step sizes as before, the other acting individually, however. Thus general and specific scaling can be learned at the same time. Operating with an (1,lambda)-strategy – however large lambda may be – leads to a second surprise: this kind of process does not work at all, it gets stuck prematurely by approaching a relative optimum in a lower dimensional space. The reason is rather simple: As said above, the convergence rate is inversely proportional to n, the dimension of the parameter space. Descendants operating in a subspace by sharply reducing one or even some of the step sizes have a short term advantage. Selecting the fittest descendant to become the one and only parent of the next generation, is counterproductive in the long term, as was the survival of the ancestor.

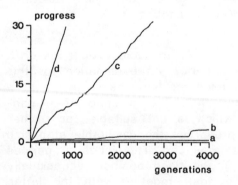

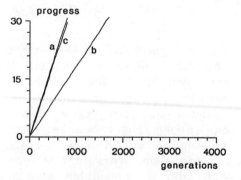

Figure 2:
Learning of the scaling I
a)　(1,10)- E.S.
b)　(3,10)- E.S. with recombination
c)　(6,30)- E.S. with recombination
d)　(15,100)- E.S. with recombination
(curve 2d equals 3c)

Figure 3:
Learning of the scaling II
a) (1,100)- E.S.
b) (15,100)- E.S. without recomb.
c) (15,100)- E.S. with recomb.
a) & b) with prefixed optimum
scalings, c) with scaling learned

Figure 2 demonstrates how to overcome the difficulty. If more than one, i. e. not only the best of the descendants, become parents of the next generation <u>and</u> recombination by sexual propagation takes place, i.e. mixing of the information gathered by different individuals during the course of evolution, then stagnation can be overcome. Now the convergence rate steeply goes up with the population size. On a conventional one-processor computer the parallelism of that scheme cannot be realized, but multi-processor machines are coming. That is why all figures show the progress over the number of generations and not over computing time. The total number of individuals must not increase with the number of object parameters, however, if the problem complexity remains constant like with objective functions F_1 and F_2 above.

The overwhelming success of recombination demonstrated here, may explain the early appearance of sexual propagation on earth. But it is unprobable that only the additional variability provokes the success. A better explanation might be the following: The typical situation during the meliorization process is ridge following. Within a population some individuals have a position on one side, others on the other side of the ridge. Mixing genetic information is a means of riding the ridge more efficiently. A similar argument holds for mixing the step sizes: Individuals on one side of the ridge have 'internal models' (made up of the set of step size relations) of the response surface which are different from those on the other side. Even if both models are wrong, at least in the long term, since both may be locally adapted only (if the 'model' learned is not a permanent natural law), some 'mean' model (or better: hypothesis) may turn out to be more useful for the future. Simply speaking, one may say that 'natural intelligence' is distributed.

Now the question of the appropriate selection pressure prevails: How many of the descendants should be selected as new parents. The answer - at least for the objective function chosen - is given by

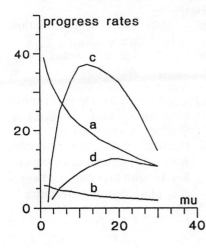

Figure 4:
Comparing progress rates per 1000 generations of different (mu,100)-E.S.'s over mu, the number of parents
a) with prefixed optimum scaling ($sigma_i^0 = c/(i)^{1/2}$)
b) with prefixed arbitrary scaling ($sigma_i^0 = c$)
c) with adaptive scaling by means of individual learning plus recombination
d) with adaptive scaling by recombination only

figure 4. All other conditions being held constant, including the number of descendants lambda=100 within one generation, only mu, the number of parents, was changed. Three cases were investigated for function F_2:

Whereas in both cases a) and b) mu=1 is the best choice, in a learning situation (cases c and d), it is better, even necessary, to increase mu far beyond 1. The diagram moreover demonstrates the effectivity of the collective learning process. Under proper conditions nearly the same convergence rate as with total knowledge of the optimum scaling can be achieved - even with lower selection pressure. This is the third surprise. Recombination alone (case d) is not as successful, however, as together with individual mutations (case c) of the genetically transmitted information about the different step sizes (which represent a simple internal model of the current/local environment).

Learning of a metric, and the epigenetic apparatus

Topologies of vitality response surfaces normally are not as simple as assumed above. The next possible complication is to incline the main axes of the hyper-ellipsoids which form the subspaces F = const. Now scaling alone does not help in achieving optimum performance. What can nature do, what has it done, to overcome the difficulty? In many cases one has found that a single gene influences several phenotypic characteristics (pleiotropy) and vice versa (polygeny). These are the two sides of the same coin, which is correlation, the perhaps best known example of it being allometric growth. The transmission mechanism between genotype and phenotype, called epigenetic apparatus, in a first order may be approximated by linear correlation. In addition to individual step sizes, now correlation coefficientes or

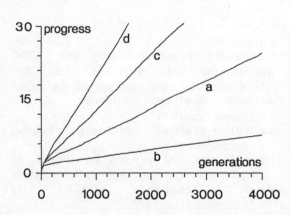

Figure 5: Learning of a metric
a) (1,100)- E.S. with constant but arbitrary scaling
b) (15,100)- E.S. under the same conditions, i. e. without recombination
c) (15,100)- E.S. with recombination and adaptive individual step sizes
d) (15,100)- E.S. as before, with additional learning of linear correlations between the phenotypic mutations

angles of inclination of the ellipsoid, forming tho surface of constant probability density of a mutation, have to be learned. Figure 5 shows first results for the objective function

$$F_3 = \sum_{i=1}^{n} \left(\sum_{j=1}^{i} x_j \right)^2 .$$ Four cases were simulated, three of them corresponding to those of figure 3.

In both cases c) and d), recombination, intermediate for both the object parameters and the step sizes, was used. It is obvious that these sampling conditions bear a variety of possibilities with respect to the mutabilities of step sizes and correlation angles so that simulations c) and d) might not yet respresent the best choices. Nevertheless the results show how much may be gained in terms of progress velocity by allowing to learn a simple 'internal model' of the topology of the environment, the 'real world'. Nevertheless, such a model in most cases will not be a correct 'theory', but simply a useful local or temporal hypothesis. Especially and again, no single individual has the best long-term knowledge about the 'correct' world model. This knowledge is partial, temporal only, and distributed.

Gradualism or punctualism, and so-called long waves

Up until now all figures showing the evolutionary progress over time or generations were drawn for objective function values only every 50th generation and for the mean of the (parents) population moreover. If we take a closer look by picking out one of the decision variables, e.g. x_{15} for function F_2, and look at it at every generation (figure 6, curve a) then the picture reveals more details.

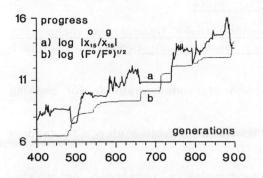

Figure 6: Time cut of one of the variables within a (3,100)-E.S. for F_2.

It depends on the density of the historical record whether we may speak of a gradual or a punctualistic process (see e. g. Stanley /12/). Due to the fact that the objective function (curve b) depends on many parameters, a single one of them, like x_{15}, must not resemble the progress as a whole. Great success in one direction forgives regression in others. And, for sure, looking at some arbitrary cut of the time record, one might get the impression of stochastically disturbed 'long waves' (normally three) with a more or less constant period. Even more aggregated sub-objectives, like the GNP of a nation, could exhibit such behaviour, if the underlying process operates left of the maximum of curve c) in diagram 4, as was the case here with mu=3 and lambda=100.

Conclusions

Many people today, when speaking about long term planning, environmental forecasting, technology assessment etc., are embarrassed by the degree of our ignorance. Very often then they speak of the uncertainties involved. But looking more precisely, isn't it a fact that there are, at the same time and with access to the same data, different certainties, i.e. different interpretations of the past and different aspirations for the future, or, in other words, different 'internal models' of the world? In the light of the simulation results above, one should appreciate, not regret, that. Due to the findings of a rather new field of science, i.e. nonlinear dynamics, we must admit that knowledge about the long-term future is principally unavailable for an open dissipative system operating far off of equilibria. We really are groping in the dark /13/. Therefore we should not try to establish one best model of the world, but make the best of the different individual ones of the ridge we are trying to follow without clearvoyance. Even if all of them were wrong they might as well be worthwhile to be recombined with each other. Instead of relying upon too strong competition, which leads to stagnation, as we have seen, we better should agree upon further experiments in cooperation. Solidarity should help those who have exploited the situation by means of a-posteriori failures, however.

References

1.   Ashby, W.R. (1960) Design for a brain. New York: Wiley, 2nd ed.

2.   Montroll, E.W. and Shuler, K.E. (1979) Dynamics of technological evolution: random walk model for the research enterprise. Proc. Natl. Acad. Sci. USA, 76, 6030- 6034.

3.   Rechenberg, I. (1973) Evolutionsstrategie: Optimierung technischer Systeme nach Prinzipien der biologischen Evolution. Stuttgart: Frommann-Holzboog.

4.   Brooks, S.H. (1958) A discussion of random methods for seeking maxima. Oper. Res., 6, 244-251.

5.   Schumer, M.A. and Steiglitz, K. (1968) Adaptive step size random search. IEEE Trans., AC-13, 270-276.

6.   Rastrigin, L.A. (1965) Sluchainyi poisk v zadachakh optimisatsii mnogoparametricheskikh sistem. Riga: Zinatne.

7.   Bremermann, H.J. (1968) Numerical optimization procedures derived from biological evolution processes. In: Cybernetic problems in bionics. (H.L. Oestreicher and D.R. Moore, eds.), New York: Gordon and Breach.

8.   Box, G.E.P. and Draper, N.R. (1969) Evolutionary operation: A statistical method for process improvement. New York: Wiley.

9.   Nelder, J.A. and Mead, R. (1965) A simplex method for function minimization. Computer Journal, 7, 308-313.

10. Box, M.J. (1965) A new method of constrained optimization and a comparison with other methods. Computer Journal, 8, 42-52.

11. Schwefel, H.P. (1981) Numerical optimization of computer models. Chichester: Wiley.

12. Stanley, S.M. (1981) The new evolutionary timetable. New York: Basic Books.

13. Meadows, D., Richardson, J. and Bruckmann, G. (1982) Groping in the dark-The first decade of Global Modelling. Chichester: Wiley.

Parallel Genetic Algorithms, Population Genetics and Combinatorial Optimization

H. Mühlenbein

GMD

Postfach 1240, D-5205 St. Augustin 1

Abstract

In this paper we introduce our asynchronous parallel genetic algorithm ASPARAGOS. The two major extensions compared to genetic algorithms are the following. First, individuals live on a 2-D grid and selection is done locally in the neighborhood. Second, each individual does local hill climbing. The rationale for these extensions is discussed within the framework of population genetics. We have applied ASPARAGOS to an important combinatorial optimization problem, the quadratic assignment problem. ASPARAGOS found a new optimum for the largest published problem. It is able to solve much larger problems. The algorithm uses a polysexual voting recombination operator.

1 Introduction

Genetic algorithms have been applied successfully to many optimization problems. Parallel genetic algorithms are an extension of genetic algorithms. They are designed for parallel computers. Whereas genetic algorithms are based on an idealized population model, our parallel asynchronous algorithm simulates the evolution of a finite population more realistically. The algorithm has been introduced in [MGK88].

The rationale of our algorithm can be explained by the system theory of evolution which has been described elsewhere [MK89]. One of the important aspects of the system theory is the following. Each individual is part of the environment of other individuals, selection is totally distributed. The evolution process is driven by the individuals themselves.

In addition to the optimization of artificial problems, parallel genetic algorithms can be used also to investigate population genetics problems. In fact, genetic algorithm and population genetics research face the same problem: to understand why evolution with powerful genetic operators and selection works so effectively. In this paper we therefor discuss our parallel genetic algorithm from an algorithmic and a population genetics point of view.

The outline of the paper is as follows. In chapter 2 we describe our parallel genetic algorithm ASPARAGOS. The next two chapters look at selection and recombination in ASPARAGOS. In chapter 5 an important combinatorial problem is introduced, the quadratic assignment problem QAP. We then describe the application of ASPARAGOS to the QAP. In the algorithm we use a polysexual voting scheme for recombination. In chapter 7 numerical results are given. ASPARAGOS found a new optimum for the largest published QAP. It is able to solve much larger problems.

2 Parallel Genetic Algorithms

The basic algorithm can be described as follows:

Asynchronous Parallel Genetic Algorithm ASPARAGOS

GE0: Define a genetic representation of the optimization problem

GE1: Create N individuals

GE2: Each individual does local hill climbing (increases its fitness)

GE3: Each individual chooses a partner for mating in the 'neighborhood' — local selection

GE4: Create a new offspring with genetic operators (e.g. crossover, mutation)

GE5: Replace the individual

GE6: If not finished, go to **GE2**

The two extensions to the classical genetic algorithms are [Gol89]

- Selection is done locally in a neighborhood

- each individual does local hill climbing

The neighborhood is defined as follows: Individuals live in an environment e.g. a two dimensional grid. Each individual occupies a grid point. The neighborhood is then given by some distance. Mating and selection of each individual is done in its neighborhood. Note that the neighborhoods of the individuals overlap!

Local hill climbing has the following effect: Recombination is now done in the space of all genotypes representing some local fitness maxima.

We will give the rationale for the extensions in the next two chapters. Both extensions have been proposed also by Goldberg [Gol89]). He calls the neighborhood model the pollination model and local hill climbing a hybrid genetic algorithms.

Other parallel genetic algorithms have been described in [C*87,P*87,Tan87]. In these algorithms, a standard genetic algorithm is run in a subpopulation. Subpopulations are evaluated in parallel. Our algorithm is fine grained. Each individual is evaluated asynchronously in parallel.

3 Selection revisited

Most of the genetic algorithms developed so far are based on an idealized population model. This model describes the development of an infinite population with random mating and selection according to

$$p_i(t+1) = p_i(t)\frac{f_i}{\bar{f}}$$

These assumptions have been useful for mathematical analysis, but should they also be used in computer simulations? The basic problem of random mating (intercrossing) was already pointed out by Darwin [Dar59] "Intercrossing plays a very important part in nature in keeping the individuals uniform in character. In animals which unite for each birth, but which wander a little, a new and improved variety might be quickly formed on any spot ... . A local variety when once thus formed might subsequently slowly spread to other districts". In genetic algorithms we want a fast evolution and not keeping the individuals uniform!

Darwin's conjecture that local populations with some isolation evolve faster than a large population was confirmed later by experiments and qualitative analysis [Wri32]. Wright summarizes: "There must be selection, but too severe a process destroys the field of variability, and thus the basis for further advance; prevalence of local inbreeding within a species has extremely important evolutionary consequences, but too close inbreeding leads merely to extinction. A certain amount of crossbreeding is favorable but not too much".

A recent discussion of the importance of the population structure can be found in [Cro86]. Crow describes in addition to the idealized population model two other models - the island model and the neighborhood model. In the island model subpopulations are isolated. Sometimes immigrants from other subpopulations arrive.

In the neighborhood model we have an 'isolation by distance'. The neighborhoods of different individuals overlap. Selection is done only in the neighborhood.

This model has been used extensively in ASPARAGOS. In accordance to Darwin, in such a population we find more variability than in a random mating population. Furthermore, the evolution is driven totally by the system itself. There is no need for artificial external control parameters. Especially there is no need for a sharing function [Gol89] to maintain variability. The neighborhood model is totally asynchronous and therefor very efficient on parallel computers.

Different selection schemes within the neighborhood model have been evaluated with large combinatorial problems. To obtain good solutions, a "soft" selection scheme is necessary. A hard selection scheme leads to rapid progress in the beginning, but very soon it converges to suboptimal solutions [Nap89,Gor89]. Similar successful experiments have been reported for the island model [Spi88].

4 Recombination revisited

The recombination or crossover operator is less well understood than selection. This statement is true as well for theoretical biology as for genetic algorithms. The basic problem still to be solved is the following: Why should a complex crossover operator lead to a faster evolution than mutation? Or in more algorithmic terms: Should we use a large population which evolves by small mutations or a small population evolving by sexual reproduction and crossover? Which algorithm is faster (in number of computer instructions)?

There is lots of evidence that for many applications the crossover operator is the key to the success of the genetic algorithm [Hol75,DJ80] but there are only some qualitative arguments explaining the above observation.

Recombination is also a big problem in biomathematics. Maynard Smith sees major problems for a population with sexual reproduction [MS87]. "Second, a large asexual population can evolve the adaptation, whereas a sexual one cannot or does so excessively slow. This illustrates the general fact that, when there are epistatic fitness interactions, sexual reproduction can actually slow down evolutionary progress by breaking up co-adapted groups of genes as soon as they arise". (In epistatic interactions genes do contribute nonlinearly to the fitness function).

In genetic algorithms for combinatorial optimization we always have very complex epistatic interactions. These interactions arise because of the constraints of the problem. This will be shown later with the QAP. ASPARAGOS is especially efficient with sexual reproduction and "intelligent" crossover. I believe that the opposite to Maynard Smith's statement may be true. His conjecture is based on a (too) simple population genetics model. Furthermore, crossover is analysed in isolation and not in connection with the other forces driving the evolution.

Let me give an example which shows the problem. We consider the fitness function in figure 1.

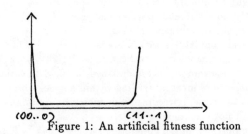

Figure 1: An artificial fitness function

Let a population be given, consisting half of (00..0) and half of (11..1) genotypes. Sexual reproduction with random mating and crossover will create offsprings where half of the individuals have a fitness of almost zero! The average fitness of the population therefor decreases rapidly. The population is not able to remain at the fitness maxima.

ASPARAGOS circumvents the above problem by two substrategies. First, the neighborhood confines mating to similar individuals. Second, crossover with local hill climbing prevents the population from falling into fitness minima. Third, in the argument of Maynard Smith [MS87] too much emphasis is put on a static equilibrium. Indeed, all the classical population genetic models deal with equilibrium in the space of all possible genotypes. But this makes no sense with a large number of genes. 100 genes with 100 alleles give 100^{100} different genotypes! Most of the genotypes will never be tested as phenotypes!

Evolution should therefor be seen as an open ended process, where a small number of genotypes is navigating in an almost infinite space. Natural evolution and genetic algorithms face the same problem: How can a very small population search such a huge space in an efficient way?

I believe that the interaction between genetic algorithm research and biological research will be fruitful for both sides. In genetic algorithms we can test evolutionary strategies, even strategies not yet found in biology. Our crossover with local hill climbing for instance is against the classical doctrine: All mutations have to arise random before there can be any intimitation of their consequences. But in a recent discussion, Cairn [Nat88] writes the following: "It is easy to imagine mechanism that might test the utility of mutations before they come irrevocably fixed into the genome ... We now know that, in the processing of biological information, almost anything is possible. Sequences are spliced, rearranged, cast aside, resurrected, and so it should not be too difficult for an organism to devise a way of testing phenotype before adopting any new genotype". Local hill climbing or directed mutations are considered as possible by Cairn!

This example shows that genetic algorithm research can guide biological research to look into areas where it seems worthwhile.

5 The quadratic assignment problem

Parallel genetic algorithms have been applied to a variety of combinatorial optimization problems e.g. the famous traveling salesman problem (TSP) and the graph partitioning problem [MGK88,Gor89]. In this chapter we will describe a parallel genetic algorithm for a general combinatorial problem - the quadratic assignment problem QAP. The QAP arises e.g. in VLSI module placement, design of factories and process - processor mapping in parallel processing. It can be formulated as follows:

Definition: Let two $n \times n$ matrices A, B be given. Let $x_{i,j}$ denote

$$x_{i,j} = \begin{cases} 1 & \text{if process } i \text{ is on processor } j \\ 0 & \text{otherwise} \end{cases}$$

Then the QAP is defined as:

$$\min \sum_{i=1}^{n}\sum_{j=1}^{n}\sum_{k=1}^{n}\sum_{l=1}^{n} a_{i,k}b_{j,l}x_{ij}x_{kl}$$

$$\sum_i x_{i,j} = 1$$

$$\sum_j x_{i,j} = 1$$

The QAP is NP-complete. A QAP of size n has n! different placements. A recent survey of the QAP can be found in [Bur84]. Several combinatorial problems can be formulated as a QAP, e.g. the TSP and the graph partitioning problem. We take the TSP as an example. Let $x_{i,j}$ denote that city i is on position j of the tour. Then the matrix $A = (a_{i,k})$ is given by the distances. The matrix B defines a valid tour e.g.

$$B = \begin{array}{ccccc} 0 & 1 & 0 & \dots & 0 \\ 0 & 0 & 1 & \dots & 0 \\ 0 & 0 & 0 & \dots & 1 \\ \vdots & \vdots & \vdots & & \vdots \\ 1 & 0 & 0 & \dots & 0 \end{array}$$

It is easily seen that the QAP is now equivalent to the TSP.

The largest QAP where the exact solution has been computed has a problem size of $n = 15$, the largest published problem is Steinberg's problem with n = 36 [BR84,SR86].

6 A parallel genetic algorithm for the QAP

The genetic representation of the problem is straightforward. The processors are arranged on a chromosome where the genes specify which process is placed on the processor. Note, that only a fraction of all possible genotypes (n^n) gives a valid phenotype. In a valid phenotype each process is placed only once. This constraint gives a nonlinear epistatic interaction. Local hill climbing of the individuals is done with a simple 2-opt exchange [Lin65]. The placement of two processes is exchanged. The exchange is accepted if it decrements the function to be minimized. This is repeated until no improvement occurs. 2-opt is a very simple heuristic and can be applied to almost any combinatorial problem. Its complexity is $0(n^2)$.

In order to test an evolutionary strategy not found in biology, we designed a recombination operator which can be called p-sexual voting recombination. In Figure 2 the case $p = 4$ is shown. For each processor, the process placements of the parents are counted. If the same process is placed on the processor more often than a threshold, it is also placed in the offspring.

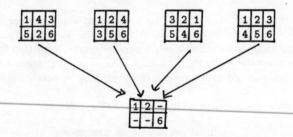

Figure 2: Voting recombination with threshold = 3

The remaining places will be filled in with mutations. What is the rationale for voting recombination? We analysed the space of all 2-opt solutions. We found that there is a strong similarity between the solutions. Only a small fraction of all possible permutations is realized. Moreover, components of good 2-opt solutions are represented very often in other solutions. In such spaces, recombination of solutions should be successful (see the same argument for the TSP [MGK88]. It has to be mentioned that the number of different 2-opt solutions is still exponential in the problem size.

A genetic algorithm with voting has been also proposed by Ackley [Ack87].

7 Implementation and numerical result

The individuals are placed on two rings. Each individual has four neighbors, two on each ring (Figure 3).

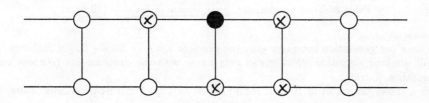

Figure 3: Neighborhood for a parallel genetic algorithm

This neighborhood was especially easy to implement on our 64 processor system with distributed memory. The interconnection of the processor can be configured to fit the application. The raw performance of the system is about 10 Mips and 1 Mflops per processor [MKPR89].

Voting recombination is done with the four neighbors. In addition the global best is included in the neighborhood with a weight of two. In total we have seven virtual individuals used for recombination. The voting threshold was set to five.

The global best individual was included in the neighborhood because this led to a fast convergence. The global best individual can easily be distributed asynchronously on a parallel processor. Therefore the efficency of the parallel algorithms does not suffer. For bigger QAP's we have to use a larger local neighborhood with local selection only. This topic is discussed in detail in [Gor89].

The following table shows the quality of the solutions for selected problems. NUG30 and STEIN36 are the two largest QAP's which have been published [BR84,SR86]. TSP40 and TSP50 are traveling salesman problems where the solutions are known.

BEST	NUG30	STEIN36	TSP40	TSP50
KNOWN	6124	9548	1071	1403
2-opt	6130	9642	1182	1548
ASP*	6124	9526	1071	1403

Table 1: Quality of QAP solutions

Note that our simple parallel genetic algorithm found a new optimum for Steinberg's problem! The new solution is given by

—	5	6	12	11	27	26	25	24
9	4	1	13	20	14	23	21	22
2	8	10	7	28	19	32	34	33
17	18	3	15	16	29	30	31	—

Table 2: New optimal solution for STEIN36

Table 3 shows the time to compute the optimum:

BEST	NUG30	STEIN36	TSP40	TSP50
Fastest	363	279	9.9	731
Average	528	579	37.9	3185
Gener.	4.7	6.0	9.8	22.2

Table 3: Time to compute the optimum in seconds (10 runs)

The time per generation increases with the problem size. To handle larger problems we will use a hill climbing algorithm which grows only linear with the problem size (see also our TSP implementation [Gor89]).

Table 4 shows the result of 10 runs for STEIN36 on 16, 32 and 64 processors. Note, that in our implementation the population size is equal to the number of processors.

Processors	64	32	16
Optimum	10	8	6
Fastest	279s	502s	580s
Slowest	1151s	1711s	3256s
Average	579s	993s	1163s
AverageW	579s	1794s	2465s

Table 4: Time to compute the optimum for STEIN36 (10 runs)

In table 4 we terminated a run after 5000 seconds. With a population of 16, ASPARAGOS found the optimum only six times within 5000 seconds. AverageW is computed by counting each run which has not been succesful with 5000 seconds. 64 processors give by far the best results. But note the difference between fastest and slowest solution time.

Table 5 gives statistical data about 1000 2-opt solutions and ASPARAGOS with a population of 64 individuals after 80 generations.

Stein36	2 OPT	ASPARAGOS
Best	9682	9526
Worst	12006	9972
Average	10745	9726
S.D.	406	60

Table 5: Statistics of ASPARAGOS

ASPARAGOS shifts the total population to good solutions. The solutions are much better and the variation of the solutions much smaller.

8   Conclusion

Genetic algorithms will have an impact in computer science as well as in biology. They provide a different intellectual perspective on evolution. Instead of having to rely only on observations of biological systems or standard mathematical models, we will be able to approach genetics and evolution by large scale simulations. We support Axelrod's [Axe87] view: "We can begin investigating alternative ways genetics might have evolved and see just which properties are arbitrary and which are not." Whereas the importance of the selection scheme and the population structure is now well understood in evolution, the basic problem still is sexual recombination. Our simple voting recombination operator worked surprisingly well. We will now compare this scheme with the usual bisexual recombination with crossover. Which recombination operator will lead to faster evolution?

9 Acknowledgement

ASPARAGOS has been designed together with Martina Gorges-Schleuter. I thank Rainer Becker for making the experiments with the QAP.

References

[Ack87]   Ackley D. H. *An Empirical Study of Bit Vector Function Optimization.* in Dav87, 87.

[Axe87]   Axelrod R. *The Evolution of strategies in the Iterated Prisoner's Dilemma.* in Dav87, 87.

[BR84]   Burkard R. E., Rendl F. *A thermodynamically motivated simulation procedure for combinatorial optimization problems.* Europ. Journ. of Operat. Research, 17:169–174, 84.

[Bur84]   Burkard R. *Quadratic Assignment Problems.* Europ. Journal of Operations Research, 15:283–289, 84.

[C*87]   Cohoon D. P. et al. *Punctuated Equilibria: A Parallel Genetic Algorithm.* in Gre87, 87.

[Cro86]   Crow J. E. *Basic Concepts in Population, Quantitative, and Evolutionary Genetics.* Freeman, New York, 86.

[Dar59]   Darwin C. *The Origin of Species by Means of Natural.* Penguin Books, London, 1859.

[Dav87]   Davis L. *Genetic Algorithms and Simulated Annealing.* Morgan Kaufmann, Los Altos, 87.

[DJ80]   De Jong K. *Adaptive system design: A genetic approach.* IEEE Trans. Syst., Man and Cybern., 10:566–574, 80.

[Gol89]   Goldberg D. E. *Genetic Algorithms in Search, Optimization, and Machine Learning.* Adison-Wesley, 89.

[Gor89]   Gorges-Schleuter M. *ASPARAGOS: Simulation of the TSP.* to be published, 89.

[Gre87]   Grefenstette J. J., editor. *Genetic Algorithms and their Applications,* Hillsdale Lawrence Erlbaum Ass., Proc. 2nd Conf. on Genetic Algorithms, 87.

[Hol75]   Holland J. H. *Adaptation in natural and artificial systems.* Ann Arbor, University of Michigan Press, 75.

[Lin65]   Lin S. *Computer solution of the traveling salesman problem.* Bell. Sys, Tech. Journ., 44:2245–2269, 65.

[MGK88]   Mühlenbein H., Gorges-Schleuter M., Krämer O. *Evolution Algorithms in Combinatorial Optimization.* Parallel Computing, 7(1):65–88, 88.

[MK89]   Mühlenbein H., Kindermann J. *The Dynamics of Evolution and Learning – Towards Genetic Neural Networks.* in: Connectionism in Perspectives, J. Pfeiffer ed., 89

[MKPR89]   Mühlenbein H., Krämer O., Peise G., Rinn R. *The MEGAFRAME HYPERCLUSTER — A Reconfigurable Architecture for Massively Parallel Computers.* 89.
to be published.

[MS87]   Maynard Smith J. *When learning guides evolution.* Nature, 329:761–762, 87.

[Nap89]   Napierala G. *Ein paralleler Ansatz zur Lösung des TSP.* Diplomarbeit, University Bonn, Jan 89.

[Nat88]   Nature. *Scientific correspondence.* Nature 336, 527-528, 88.

[P*87]   Pettey C. B. et al. *A parallel genetic algorithm.* in Gre87, 87.

[Spi88]   Spiessens P. *Genetic Algorithms.* AI-MEMO 88-19, VUB Brussels, 88.

[SR86]   Sheraldi H. D., Rajgopal P. *A Flexible, Polynomial-time Construction and Improvement Heuristic for the Quadratic Assignment Problem.* Operations Research, 13:587–600, 86.

[Tan87]   Tanese R. *Parallel Genetic Algorithms for the Hypercube.* in Gre87, 87.

[Wri32]   Wright S. *The Roles of Mutation, Inbreeding, Crossbreeding and Selection in Evolution.* In Proc. 6th Int. Congr. Genetics, pages 356–366, 32.

ASPARAGOS
A Parallel Genetic Algorithm and Population Genetics

Martina Gorges-Schleuter
GMD, Postfach 1240, D–5205 St. Augustin 1

Abstract

ASPARAGOS is an implementation of an asynchronous parallel genetic algorithm. It simulates a continous population structure. Instead of modeling one large population AS-PARAGOS introduces a continuous population structure. Individuals behave independent and due to the limited distance an individual may move, there are only local interactions. We get an algorithm which is very robust in parameter setting and which may deal with much smaller population sizes than needed with only one large population to overcome the problem of preconvergence. These results have been derived with the traveling salesman problem as testbed. The high quality solutions yielded show the effectiveness of ASPARA-GOS especially with large problem sizes and give hope that it may serve as a general purpose optimization algorithm suitable for a wide range of applications.

1 Introduction

Genetic Algorithms are inherently parallel. To get a maximum of parallelity it is not possible to use a supervisor having global knowledge of all individuals in the population. This fact leaded us to the view of a genetic algorithm as a population of individuals each having only information about individuals living nearby.

Our implementation focuses on massively parallel systems. To achieve an efficient and reliable program one should fulfill the following requirements :

- no central control
- local interactions
- fault tolerance

With only one large population these expectations are hard to fulfill. Therefore we introduce a degree of inhomogeneity which we refer to as a *population structure*. Dealing with several sub-populations at the same time also has advantages in overcoming the problem of preconvergence. The genepool may evaluate different regions of the searchspace at the same time and maintain variability for a longer period.

According to Wright [16] the most favorable population structure is one "subdividing the species into local populations (demes) sufficiently small and isolated that accidents of sampling may overwhelm many weak selection pressures. There must, however, be enough diffusion that a deme that happens to acquire a favorable interaction system may transform its neighbors to the point of autonomous establishment of the same peak, and thus ultimately transform the whole species or at least that portion of it in which the new system actually is favorable."

When looking at different population structures there is one principal distinction; either the models are discontinuous or continuous. For example in the island model demes are completely separated and migration is described by an immigration rate specifying the number of individuals moving between demes per generation. The advantage of discontinuous models is their relative easy theoretical analysis.

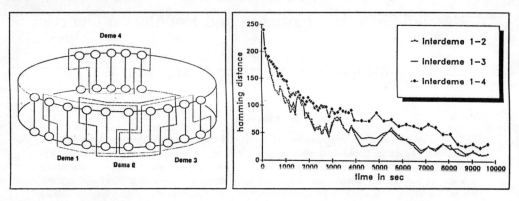

Figure 1:
The ladder-population with deme size 8

Figure 2:
Mean hamming distance between demes

In most actual cases, immigrants come largely from neighboring groups and there is more or less "isolation by distance". At the extreme end the population is viewed as a continuous habited area. This model is called the "neighborhood model". We define the *neighborhood* (deme, "tribe") of an individual to be the subpopulation in the immediate locality, as determined by the topological structure of the manifold (i.e. environment) within which we imagine the whole population to be distributed. One can view the neighborhood of an individual as its set of potential partners.

The dispersion of the individuals in the environment may be either linear, planar (surface of the earth) or spatial. The neighborhood model may be classified according to the dimension of the environmental manifold. We decided to simulate a system inspired by the wellknown ring-species. Because the architecture looks like a ladder we called that specific neighborhood model the ladder-population (Figure 1). In ASPARAGOS the population number of the neighborhood is determined by an individual's mobility. For example, with the ladder-population a moving radius of two yields a neighborhood of size eight. Hence, the population is separated in demes by the limited distance an individual can move.

Consider now the demes. By definition a deme is the reproduction community of an individual. As demonstrated in Figure 1 demes are overlapping in space. Therefore a better view of migration in the continuous model is to think of it as a diffusion process. Individuals with better fitness have a better chance of being chosen for mating, thus reproducing their genes more often and finally propagating their genes through the whole population.

In the continous neighborhood model demes are likely to change their genetic characteristics "smoothly" as we move across the manifold. This means that mates differ less than they may in the discontinous model. Consequently one might expect a more continous exploration and exploitation in the neighborhood model.

The mean hamming distance between demes give us an inside view of the exploitation of different niches. Figure 2 shows the averaged hamming distance between demes. Demes 1 and 2 are very close and have 4 individuals in common. Demes 1 and 3 are close together but do not overlap. Deme 4 is as far as possible from deme 1 (Figure 1).

2   ASPARAGOS

ASPARAGOS uses extensively the inherent advantages of genetic algorithms extended by population structure:

- natural parallelism
- no synchronisation, neither global nor local
- relatively few communications (cpu cycles vs. send/receive requests)

The basic element of ASPARAGOS is the individual and its local behavior. All individuals together form the population, this is a selforganizing process. There is no global knowledge of the entire system. Therefore we describe ASPARAGOS from the individuals local point of view.

First we describe the representation of an individual used in the testbed traveling salesman problem (TSP).

Let $C = \{c_1, c_2, \ldots, c_N\}$ be the cities to be visited and

$$\forall c_i, c_j \in C : d(c_i, c_j) \text{ denotes the distance.}$$

Then the genotype is a permutation of C

$$g = (c_{\pi(1)}, c_{\pi(2)}, \ldots, c_{\pi(N)}),$$

and the phenotype is

$$p = f(g) = \sum_{i=1}^{N-1} d(c_{\pi(i)}, c_{\pi(i+1)}) + d(c_{\pi(n)}, c_{\pi(1)})$$

We considered only a symmetrical TSP, that is

$$d(c_i, c_j) = d(c_j, c_i).$$

With the neighborhood model, the fitness of an individual is determined by its own phenotype and the phenotypes of all the other individuals in its neighborhood, which together form the selective environment where the individual resides. This environment is steadily changing, so the fitness of a phenotype is both spatially and time dependent.

Now we can describe the overall layout of ASPARAGOS. Each individual executes the following cycle until the simulation is terminated:

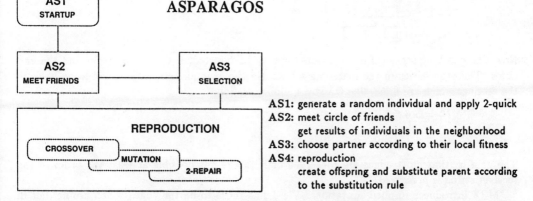

AS1: generate a random individual and apply 2-quick
AS2: meet circle of friends
     get results of individuals in the neighborhood
AS3: choose partner according to their local fitness
AS4: reproduction
     create offspring and substitute parent according
     to the substitution rule

After the creation of an individual it is decided if the individual may reproduce. An individual dies immediately if its fitness is worst in its neighborhood. If it survives (is accepted) it replaces the parent on its place, otherwise the parent chooses a new mate and reproduces again. Typically at the start of the simulation the first offspring produced was accepted. After a dozen or so generations one needs to generate about two offspring, i.e. the second is usually accepted. Only after 1500 or more generations does it become necessary to generate at least three offspring before one is accepted.

To obtain very good solutions it seems important to enforce the early death of weak individuals. Accepting all offspring leads to the disadvantage that individuals with "high" fitness can't profit. In Table 3 the column *accept all* shows the lower quality of the solution and in Figure 3 the curve *accept all* shows that the variability of the genepool stagnates at an early point in the simulation. Individuals can't settle down in local optima.

There is no synchronous update of the population. That means different individuals could be in different generations and usually are. *Therefore the generation given in all figures is the average number of generations in the population at time t.*

From the viewpoint of the neighborhood model an individual's fitness is environmentally dependent; an individual may be very attractive in one neighborhood and only average in another. In general the better an individual is the better its chance to be chosen as mate. A globally best individual would be the most attractive in every neighborhood to which it may belong, and therfore will be chosen statistically most often. In this way the diffusion process inherent in the neighborhood model gives schemata belonging to better individuals a higher chance to propagate than schemata belonging to weaker individuals.

3 The genetic operators

The crossover operator is a kind of order crossover (OX) [5], with the difference that it keeps the receiver's relative city order as far as possible. Let us see how the maximal preservative crossover (MPX) works. For a description we use the example from Goldberg p. 174 [5].

$$A = 9\ 8\ 4\ \boxed{5\ 6\ 7}\ 1\ 3\ 2\ 0$$

$$B = 8\ 7\ 1\ 2\ 3\ 0\ 9\ 5\ 4\ 6$$

A is the donator and B the receiver. Before the crossover can take place a swap makes the begin and the end of the donator's matching section (enclosed in a frame) lay together in the receiver. Swapping can either bring the 5 nearby the 7 or the 7 nearby the 5.

$$B_1 = 5\ \boxed{7\ 1\ 2\ 3\ 0\ 9}\ 8\ 6\ 4$$

$$B_2 = 6\ 4\ \boxed{1\ 2\ 3\ 0\ 9\ 5}\ 7\ 8$$

Now the matching section from the donator may be inserted, between the begin and the end cities. The other cities in the matching section (in the example only city 6) will leave holes in the receiver which are filled, like OX, by a sliding motion.

$$B_1' = 5\ 6\ 7\ 1\ 2\ 3\ 0\ 9\ 8\ 4$$

$$B_2' = 4\ 1\ 2\ 3\ 0\ 9\ 5\ 6\ 7\ 8$$

The gain is, that most edges are preserved or if this is not possible at least the order of the receiver string is preserved.

MPX introduces implicit mutations. We define these to be the edges which are neither in the donator nor in the receiver. For example B_1' has no implicit mutations and B_2' has two. To get now a mutation rate as specified by the mutation parameter the implicit mutations count as mutations already done and only as many additional mutations as necessary are done.

The problem specific algorithms 2–quick and 2–repair both work in linear time. 2 quick is a Lin 2–opt [8] without checkout. It gives much weaker solutions than 2–opt, but the goal was to get start points which are well distributed in the search space and carry already some information. Table 1 gives a comparison with Lin's 2–opt. The cpu-time is the mean value for 100 runs on a T800 transputer.

```
PROC two.quick ([] INT path, VAL INT nodes)
   -- path  is the random initilization
   -- nodes is the problem size
 ... declarations
 SEQ
   i := 0
   WHILE i < (nodes - 2)
     SEQ
       j := i + 2
       WHILE j <= (nodes - 1)
         SEQ
           IF
             d(path[i],path[i+1]) + d(path[j],path[j+1]) <
             d(path[i],path[j]) + d(path[i+1],path[j+1])
             SEQ
               ... exchange edges
                 j := i + 1
             TRUE
               SKIP
           j := j + 1
       i := i + 1
```

2–repair in contrast has only the basic move, swapping of a subtour, in common with Lin's algorithm. The design is based on the fundamental principle of the continuity of evolution. 2–repair starts from the hypothesis that the ancestors are well adapted to their environment. Offspring should therefore be not too much different from their parents. So changes in the genotype at positions which carry new information, introduced by crossover and mutation, are checked for consistency. Without computing the phenotype, these new edges are checked and if it is possible to exchange them by better edges this step is performed. These better edges are again marked as new. This is done until all new edges are checked. Repairing of genotypes without knowledge of the actual phenotype takes also place in nature. See Mühlenbein [11] for a discussion.

	442		532	
	2–opt	2–quick	2–opt	2–quick
best value	5394	5661	29423	30260
worst value	5986	6267	32013	33869
average value	5712	5936	30729	31940
cpu-time[sec]	51	8	100	12

Table 1: Comparison 2–opt versus 2–quick

```
PROC two.repair ( [] INT path, mark, VAL INT nodes)
    -- path  is the offspring after crossover and mutation
    -- mark  indicates for every edge if it belongs to a partner
    -- nodes is the problem size
  ... declarations
  SEQ
    ... initialization
    start := path[0]
    i := 0
    WHILE (path[(i+1) REM nodes] <> start)
      SEQ
        IF
          ... edge is marked
            SEQ
              j := (i + 2) REM nodes
              WHILE ((j + 1) REM nodes <> i)
                SEQ
                  IF
                    d(path[i],path[j]) + d(path[i+1],path[j+1]) <
                    d(path[i],path[i+1]) + d(path[j],path[j+1])
                      SEQ
                        ... exchange edges
                        ... mark j and j+1
                        j := (i + 1) REM nodes
                        start := path[i]
                    TRUE
                      SKIP
                  j := (j + 1) REM nodes
              ... remove mark for i and i+1
          TRUE
            SKIP
        i := (i + 1) REM nodes
```

4 ASPARAGOS Parameters

We define ASPARAGOS as follows:

$$\mathbf{AS} = (M, D, C, P_M, W, S) \text{ where}$$

- M is the population size
- D is the neighborhood (deme) size
- $C = [c_1..c_2]$ is the size of the crossover interval
- P_M is the mutation rate
- W is the window size
- S is the selection strategy

This definition is a slightly changed version from Grefenstette [6]. The standard parameter setting is

$$\mathbf{AS_S} = (64, 8, [N/3..N/2], 0.01, 10, ES)$$

where N is the problem size.

We experimented with population sizes of 16, 32 and 64 and population number of neigh borhoods of 8 and 5.

The size of the substring to be crossed over could be of any fixed length or of varying size equally distributed between the lower and the upper bound.

The mutation rate was set to 0.02, 0.01 or 0.005 respectively.

A window size $W = n$ indicates that the base value from which local fitness is computed is determined by the locally least fit individual from $n - 1$ generations in the past. We use window sizes W in the range $1 - 10$.

The selection strategy may be an adaptive or a ranking strategy [1]. Both may be either a pure strategy (PS) or elitist (ES) [3]. The elitist strategy preserves the best individual of each neighborhood. To get results for a model which has some global knowledge of the whole population the distribution of the global best individual may be chosen (+gb). This super-individual is then potential partner in all neighborhoods.

5 Performance Measures

All simulations results are shown for Padberg's 532 cities of USA [13], if not explicitly specified. There is a misprint in [13]; the coordinates of the 265th city should be (8583, 8619) instead of (8483, 8619). We used the published coordinates. The optimal tourlength with the "wrong" coordinate is 27694. The results for Grötschel's 442-problem [14] are added in some Tables and Figures to show results for a problem which has a regular structure. The optimal tour length is 5069.

Figure 3 shows the best results for the 442-problem (left) and for the 532-problem (right) obtained with APARAGOS.

To give an impression of the running time Table 2 gives the cpu-time needed for AS_S (with different mutation rates) running on a system with 64 T800 transputers.

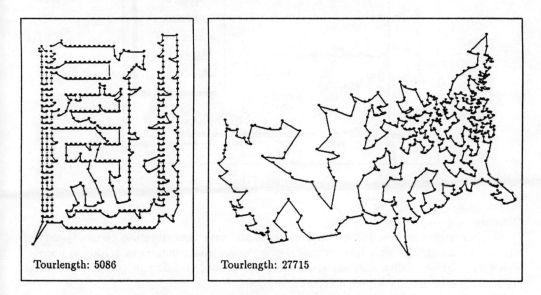

Tourlength: 5086

Tourlength: 27715

Figure 3: Best solutions obtained by ASPARAGOS

Gener-ation	442		532		
	$P_M=0.005$	0.01	0.005	0.01	0.02
100	110	110	175	175	225
200	190	200	300	310	430
400	340	400	530	580	930
800	580	750	950	1080	1880
1600	1100	1500	1630	2750	4350
2400	1630	2200	3000	4000	6350

Table 2: cpu-time [sec] needed for ASPARAGOS

Robustness

We considered the robustness of the algorithm with different parameter settings. Robust means how certain we can be of achieving a high quality solution. The quality depends not only on the parameter settings but also on the number of generations computed. The stopping criterion therefore plays an important role. The algorithm may be either stopped if time is up, a certain number of generations is computed or if the variation of the gene pool drops. To see that this criterion makes sense Figure 4 shows the variation of the gene pool. Due to the rapid elimination of weak individuals there is always a rapid progression towards a minimally diverse gene pool. We decided to stop after 1600 generations, because it takes a long time with AS_S until there is no variation in the gene pool; and experiments with long runs showed a very small probability that better solutions are found thereafter. Thus one achieves better quality solutions with several short runs than with one extremely long run.

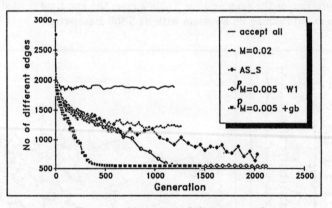

Figure 4: Variation of the gene pool

Convergence

Another important criterion on performance is the time needed until a certain quality of solution is reached. The main interest may then be to get good solutions as quickly as possible even if the highest quality solutions are missed. We decided therefore to take the best-so-far performance measure to compare different strategies.

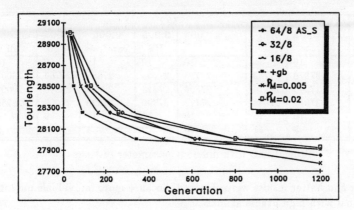

Figure 5: Convergence of ASPARAGOS

6   Simulation Results

The effects of mutation and migration are nearly the same. Both introduce variation into the gene pool. However, migration is *less destructive* than mutation, since substrings of co-adapted alleles flow through the population introducing useful variation, rather than producing variation by random alteration.

It is a poor strategy to attempt to use a high mutation rate ($P_M = 0.02$) to prevent loss of diversity; the price is too high. In fact a high mutation rate slows down the convergence and ends up yielding significantly worse solutions (Figure 5). In comparing the effects of mutation and migration we conclude that chance is less important than cooperation.

Table 3 gives the best, worst and average results and the number of runs which have final results below 28000 (1% above optimum) and below 27875 (0.5% above optimum) for various parameter settings. Figure 5 shows the convergence of AS_S with different mutation rates and for the strategy with global best knowledge. It shows the averaged quality over all runs performed versus the generations performed.

The algorithm is very robust with respect to variation of the crossover parameter. A fixed string length with about 40% of N gave the best convergence. A string length less than $N/4$ leads to slower convergence and lower quality results. A long string length of $N/2$ was not very successful because too much information from one parent got lost. This is due to the fact that the implicit mutations destroy only information in the receiver string whereas no information in the matching string will be destroyed. A varying string length between $N/3$ and $N/2$ ([160..260] for the 532-problem) gave the best result of all and only a little less convergence. This is clear, because at the beginning of the simulation longer matching strings of better partners give a quicker convergence, but later on it is advantageous to make individuals not too equal and thus giving a varying length matching string the opportunity to perform better.

The most important parameters are the selection strategy and the population size together with the chosen environment. First we consider the results with different population and deme size (Table 4). The quality of the solution gained depends heavily on the population size, due to the higher variation in the gene pool. Hence, a larger population gives better results with a higher probability. A deme size of 8 always gave better results, than a deme size of 5.

The population size needed to achieve good solutions with high probability is problem size dependent. For example with a 100-city problem a population size of only 16 is sufficient to always converge to the global optimum [7,10].

With regard to the convergence, a deme size of 8 was superior to the smaller deme size of 5

	532							442
	$P_M=0.02$	AS_S	$P_M=0.005$	0.005 PS	0.005 random	0.005 +gb	accept all	AS_S
no of runs	10	15	10	10	12	10	10	10
best value	27870	27753	27715	27755	27852	27783	28009	50.86
worst value	27948	27974	27804	27913	28006	28048	29009	50.90
average value	27919	27845	27770	27790	27926	27903	28780	50.87
runs < 28000	10	15	10	10	11	9	0	
runs < 27875	3	11	10	9	1	3	0	

Table 3: AS with different parameter settings

(Figure 5). Faster and better results were obtained, because more individuals may immediately choose an individual with high fitness as mate.

We have tested three different selection strategies: random mating, ranking and an adaptive strategy with different window sizes. What showed up is that nature's strategy, better individuals should have better chances to reproduce, is the best. An adaptive strategy with window size 1 has the problem that the weakest individual in each deme has no chance to get chosen as mate, resulting in an overestimation of the local best individuals. With window size 10 this problem didn't arise, the weakest had some chance to mate. Experiments with an infinite window size showed that the relative fitness of an individual was not distinguishable, leading quickly to a random mate selection within the neighborhood. If the values for the ranking strategy are set up appropriately, ranking is as good as the adaptive strategy with window size 10. For the 532-problem with deme size 8 (7 neighbors) this is a selective setting of 25%, 20%, 15%, 15%, 10%, 10%, 10% and 5% ranked from best to weakest. An extensive study of ASPARAGOS with a ranking strategy can be found in Napierala [12].

	532/8			532/5		
	64	32	16	64	32	16
no of runs	15	10	10	5	5	5
best value	27753	27839	27898	27818	27851	27938
worst value	27974	28031	28082	27890	27995	28139
average value	27845	27904	28000	27858	27921	28009
std deviation	68.67	67.49	64.31	27.04	55.93	81.74
runs < 28000	15	8	6	5	5	3

Table 4: AS_S with different population and deme sizes

With the elitist strategy it depends on the mutation rate. With high mutation rates ($P_M = 0.01$) the elitist strategy is more robust than a strategy which can forget about the locally best individual developed. But with a mutation rate of $P_M = 0.005$ the elitist strategy is neither more robust nor does it have a higher convergence rate than the pure strategy.

To get high quality solutions it is important that most demes are separated, and only locally near demes overlap. This means local information should only propagate to other demes through a diffusion process. Indeed the knowledge of some global best information leads quickly to good results, but later the very best local optima are not explored. This is because genes get fixed in early generations and mutation offers little hope of a shift from one local optimum to another (Figure 4). Thus a strategy using global-best information is an interesting possibility due to its fast convergence to a 'fairly good' solution.

7   Conclusions

Selection and acceptance of (only) locally better individuals eventually leads to a fixation of all genes. The fixation of genes shows us how fast the algorithm settles down. In our testbed problem the variation of the gene pool may be described by the total number of different edges. Figure 6 shows at the abscissa the different edges sorted by frequency for the 532-problem with default parameter setting AS_S. At the beginning of a simulation run with ASPARAGOS most individual's genes are different. The variation is very high. As more and more generations are computed the individuals get more equal. With no variation only 532 edges with a frequency of 64 would be present. This final state wasn't reached after 1200 generations.

What Figure 6 does not show is most remarkable: how the population evolves to better and better solutions, ending up with an astonishing good result. This is shown in Figure 7.

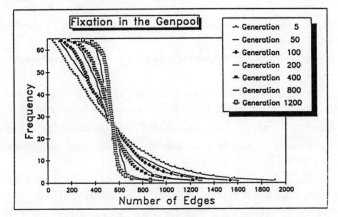

Figure 6: Fixation of edges for an AS_S run

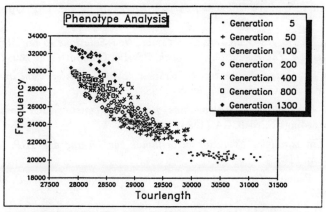

Figure 7: Evolution of a high quality solution

This gives us an inside view of the power of evolution. Demes may develop and exploit a region of the search space without underlying strong competition in a global sense. The diffusion process leads on the one hand side to soft competition between demes and on the other side to a cooperation. Demes which got fixed in some local optimum get the chance to shift from one local optimum to another, thereby exploring new regions of the search space.

The approach of introducing a population structure in addition to a usual genetic algorithm made, for the first time, a genetic algorithm for the TSP comparable to other heuristics. With large problem sizes ASPARAGOS got better solutions in very reasonable times than simulated annealing [4]. Compared to k-opt [8] the heuristic with the best solutions, ASPARAGOS reached the same quality but much faster.

Acknowlegements

I would like to thank Dr. Antonia J. Jones, Imperial College London, Dr. H. Mühlenbein, GMD St.Augustin, and Prof. Dr. Hans-Paul Schwefel, University of Dortmund for their cooperative discussions.

References

[1] J.E. Baker, *Adaptive Selection Methods for Genetic Algorithms* , Proceedings first ICGA, Pittsburgh (1985) 101–111

[2] J. Davis, *Genetic Algorithms and Simulated Annealing* , Morgan Kaufman Publ., Los Altos (1987)

[3] K. A. De Jong, *An analysis of the behavior of a class of genetic adaptive systems*, Doctoral dissertation, University of Michigan, Dissertation Abstracts International 36(10), 5140B

[4] G. Dueck, T. Scheuer, *Threshold Accepting*, IBM HDSC Technical Report 88.10.011 (1988)

[5] D.E. Goldberg, *Genetic Algorithms in Search, Optimization and Machine Learning*, Addison–Wesley (1989)

[6] J.J. Grefenstette, *Optimization of Control Parameters for Genetic Algorithms*, IEEE Transactions on Systems, Man and Cybernetics 16(1), 122 – 128 (1986)

[7] M. Gorges-Schleuter, H. Mühlenbein, *The Traveling Salesman Problem – An Evolutionary Approach*, Annual Report of the GMD 1988 (in german)

[8] B. W. Kernighan, S. Lin, *An Efficient Heuristic Procedure for the TSP*, Operation Res. 21, 498–516 (1973)

[9] H. Mühlenbein, M. Gorges–Schleuter, O. Krämer, *New Solutions to the Mapping Problem of Parallel Systems – The Evolution Approach*, Parallel Computing 6, 269–279 (1987)

[10] H. Mühlenbein, M. Gorges–Schleuter, O. Krämer, *Evolution Algorithm in Combinatorial Optimization* , Parallel Computing 7, 65–88 (1988)

[11] H. Mühlenbein, *Parallel Genetic Algorithms, Population genetics, and Combinatorial Optimization*, Proceedings third ICGA (1989)

[12] G. Napierala, *Ein paralleler Evolutionsalgorithmus zur Lösung des TSP*, Master Thesis, University of Bonn (1989)

[13] W. Padberg, G. Rinaldi, *Optimization of a 532 – City Symmetric TSP*, Operation Res. Lett. 6, 1–7 (1987)

[14] Y. Rossier, M. Troyon, T. M. Liebling, *Probabilistic Exchange Algorithms and Euclidean Traveling Salesman Problems*, OR Spektrum 8, 151–164 (1986)

[15] H.P. Schwefel, *Collective Phenomena in Evolutionary Systems*, in Problems of Constancy and Change, vol.2, Int. Society for General System Res., Budapest (1987)

[16] S. Wright, *Character Change, Speciation, and the Higher Taxa*, Evolution 36(3), 427–443 (1982)

Evolution: Travelling in an Imaginary Landscape

by P.M.ALLEN and M.J.LESSER

International Ecotechnology Research Centre.

Cranfield Institute of Technology,
Bedford, MK43 OAL, U.K.

```
SATOR
AREPO
TENET
OPERA
ROTAS
```

The Ploughman's furrow winds the Heavens' clock

ABSTRACT

The evolutionary process is shown both to result from, and be driven by, a dialogue between average dynamics and the non-average detail and diversity that necessarily underlies, but is necessarily ignored by, any reduced description of the system. The consequences of this view are presented and discussed. The notion that evolution can be represented as hill climbing in a landscape where elevation represents increased opportunity is explored and rejected. We show that the surface of any adaptive landscape is itself dynamically variable, being formed by the particular exploration processes that happen to occur. Rather than viewing evolution as a process of discovering pre-existing niches and improvements we adopt the view that evolution is fundamentally a creative process, resolving the problems and taking advantage of the opportunities which mutual coexistence presents over time.

INTRODUCTION

What is the consequence of evolution? Is it the steady improvement in performance of the individual parts of a system? Is it the improvement in functioning of the system as a whole? Does one imply the other, or is improvement merely whatever happens to happen?

What does improvement mean? Is it a rationalization of subjective values or is evolutionary change characterized by some fundamental, perhaps thermodynamic, property or measure? If the latter is so, then can individuals oppose such an imperative ?

Mechanical images have traditionally been used to convey scientific understanding (1,2). In this view, the behaviour of a system can be understood, and anticipated, by classifying and identifying its components and the causal links, or mechanisms, that act between them. In physical systems the fundamental laws of nature govern these mechanisms and determine what must happen.

In models of isolated or closed physical systems these restrictions apparently limit behaviour so that we can predict the properties of the final state, thermodynamic equilibrium, for almost any system however complex. This was the triumphal achievement of classical science.

The analogy was falsely applied to the biological and human sciences. It was argued that the processes of evolution and interaction present in such systems would inevitably lead to equilibria, which were not only predictable, but also, in some sense, optimal. The assertion that natural evolution must lead to optimal outcomes was used as a justification for the unregulated working of free markets - the "invisible hand" of

Adam Smith - (3,4). Such ideas were useless, being based on superstition rather than on scientific investigation.

Biological and human systems are always complex. Understanding such systems requires the reduction of this complexity to a simplified representation by words and other symbols. We try to model complex biological or human system in terms of a few variables and parameters. The particular set of variables chosen to represent the state of the system depends on the questions being considered (e.g. flows of money, goods, services, carbon, nitrogen, energy etc). But none of these can provide a model or description of the system which will predict its future.

Reality, whatever else it is, is all detail. Each part is unique. In our minds we carry, as a reduced description of this, elements typical of the system, according to the classification scheme that we have decided to apply. Models will always lack the particularity and diversity of reality.

If it is assumed that the elements making up the variables i.e. individuals within a species, firms in a sector,etc., are all identical to the average then the model reduces to a mechanical device which represents the system in terms of a set of differential (perhaps non-linear) equations which govern the values of its variables. The future of the system can now apparently but wrongly be predicted by the simple expedient of considering the behaviour of the model.

This Newtonian vision of the world as a vast and complicated clockwork mechanism allows predictions to be made by simply running the model forward in time. General statements about the future, under given conditions, can be made by studying the types of solutions that are possible for the equations in the long term. Explanation of the world is then obtained only in terms of the internal functioning of the system, and this functioning itself is viewed as a long term solution of the dynamical equations expressing the maximum or minimum of some a priori potential.

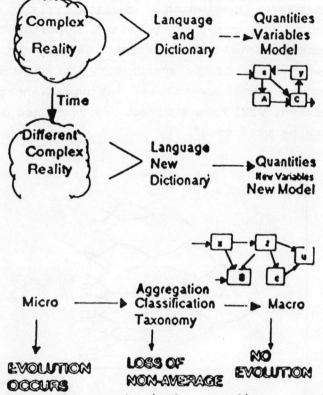

Figure (1). The mechanical paradigm consists in identifying the components of a system, and the causal links which act between them. But evolution concerns the change in this structure.

This view suggested that evolutionary superiority must result from just letting the system run. This was expressed as the "survival of the fittest" and "laisser faire" economics, which supposed that observed behaviour in a mature system must manifest some measure of optimality. While offering the convenience of avoiding the need to take difficult decisions, this view is defective.

When non-linear mechanisms are present, the system may continue to change indefinitely - executing either a cyclic path in state space or possibly a chaotic movement caused by a strange attractor. Its evolution may involve structural changes of spatial and hierarchical organization, in which qualitatively different characteristics appear. New problems, opportunities and issues emerge invalidating former values and formerly useful behaviours. If the system survives the transition, new behaviours and new values, taking advantage of new opportunities, will have evolved. How can these behaviours and values be predicted?

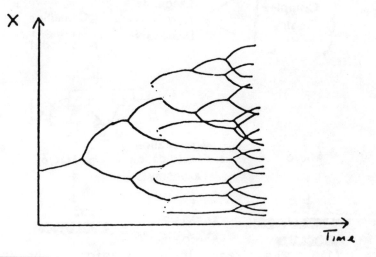

Figure (3). Multiple solutions characterize systems of non-linear equations, which differ in their symmetries and dimensions.

What actually occurs will depend on local, non-average details - the accidents of history - and the evolution that results is really about creativity. These details are ignored in the dynamical equations which appear on the right hand side of figure (1). Such models are capable of functioning but not of evolving.

Evolutionary change must therefore be the consequence of

what has been lost in the reduction to the deterministic
description. If this is so then the future of any system
will be subject to terms of two kinds: changes brought
about by the deterministic action of the typical behaviour
of its average components, and structural qualitative
changes brought about by the presence of non-average
components and conditions within the system.

The process that results in what we call selection emerges
from a dialogue between the describable average dynamics
and the unpredictable, exploratory non-average
perturbations around this average which result from the
inevitable occurrence of non- average events and
components. These perturbations search or explore a
stochastic process that generates information.

2. The Evolutionary Paradigm

In recent papers (6,7) the consequences of viewing
evolution as a dialogue between average processes and
non-average detail has led to the concept of evolutionary
drive, which shows that evolution selects for populations
with the ability to learn, rather than for populations
exhibiting allegedly optimal behaviours. This corresponds
to the appearance of diversity creating mechanisms in the
behaviour of populations, initially involving genetics
and, later, cognitive processes.

The model whose output is shown in figs. 4 and 5 was based
on the most simple population equation possible:

$$dx/dt = bx(1-x/N)-mx$$

Where: x is the size of a population
         b is the birth rate of the population
         m is the mortality rate of the population
         N is the maximum population allowed by a scarce
            resource (niche).

Populations were modeled in a two dimensional matrix or adaptive landscape in which distance from the origin in one dimension represented behaviours or strategies which resulted in an increasing birth rate (more offspring at each unit of time) and in the other in a decreasing mortality (more units of time in each life), giving the formula below.

$$dx(i,j)/dt =$$

$$[\{b(i)(fx(i,j)+3/8(1-f)x(i,j+1)+1/8(1-f)x(i,j-1)\}$$

$$+b(i+1)3/8(1-f)x(i+1,j)+b(i-1)1/8(1-f)x(i-1,j)].$$
$$(1-totpop/N)-m(j)x(i,j)$$

> Where: f is the fidelity of reproduction.
> i and j are axes of birth (b) and death (m) space.
> totpop is the sum of all the populations in b/m space.

Evolutionary progress corresponds to diagonal movement towards the lower right hand corner of figure 4(a), as both the birth rate and the longevity of the population x increase in this direction. But this begs the question "Do populations improve ?" and, if so, how ?

In the model we have introduced the possibility of error making in reproduction into the dynamics of the logistic equation. That is to say, the parental character that corresponds to particular values of b and m is not reproduced faithfully. Offspring therefore appear at different points in the b/m landscape from their parents. In accordance with common sense we have further supposed that random errors in reproduction of a complex organism more frequently give rise to a lower performance than to a higher one.

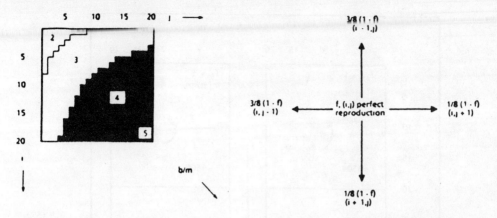

Figure 4(a) The adaptive landscape of the logistic
equation. 4(b) Errors in reproduction lead to off-spring
defusing into neighbouring cells of character space.

Following the occurrence of errors over successive
generations, an initially homogeneous population, which at
first occupies a single cell in b/m space, gradual spreads
into a cloud of populations, mostly located lower on the
hill than the initial position. However, populations which
are lower on the hill reproduce more slowly and have
higher mortality rates than those higher up, and so,
because of the differential performance of the populations
present, the cloud gradually slides up the hill. We may
interpret this as resulting from the forces of selection,
but it is important to realize that there has been no
intervention from outside. The model itself has determined
which types should be retained and which rejected. The
system structures itself both by diffusing populations
throughout b/m space due to error making and by
differential selection resulting from the dynamical
equations.

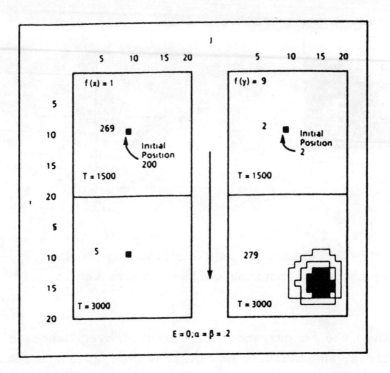

Figure (5). Evolution selects for imperfect reproduction even though at any moment it would always be better not to make errors.

The model was used to examine the most effective amount of error making necessary to climb a hill of a given slope and also to determine how easy or difficult it was for a new population to invade a system. However, the main point illuminated was that any particular population would always be within an ecosystem, with prey, predators, parasites and competitors all using a similar mechanism to search behaviour space and all using differential reinforcement to further their own success. In short, for any population, position in b/m space depends on the performance of that population relative to the performances of the other populations in the ecosystem, which will also be evolving.

Instead of a hill which can be climbed we have a down escalator up which it is necessary to climb just to stay in the same place. This leads to a view very similar to the Red Queen Hypothesis of Van Valen (8), except that we have quantitative models which can be used to examine the question.

The manner in which evolution can adjust the degree of error making which leads to evolutionary change was also considered. It was supposed that fidelity of reproduction was an hereditary property which could be passed on imperfectly. It was shown that in such systems a population increased its error making quite violently, climbed an adaptive slope and then, after it had crested a local maximum, increased its fidelity of reproduction.

In another experiment the effects of error making on a positive feedback term, a different dimension of evolution which might enhance the reproduction of a particular character or type, was explored. In other words, we examined what would happen if there existed a character or strategy which would increase the faithful reproduction of that particular character or strategy. In such a case the birth rate b is enhanced :

Where: All values same as previous equation
         pf is intensity of positive feedback
         totpop is the sum of all the populations in b/pf
         space.

If we consider a new space b/pf, then, depending on the relative slopes of the dimensions b and pf, a population can move to a certain value of pf and then be unable to move any further. It would not explore the state space, being neither able to increase its own positive feedback nor to climb the slope of improved reproduction. The mechanism of self-reinforcement expressed by pf becomes so dominant that it outweighs the advantages in improved birthrate that mutated populations exploring the state space might discover.

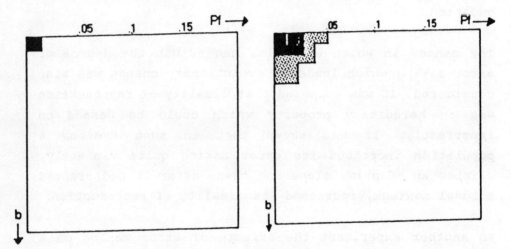

Figure (6). The positive feedback can trap a population, neither allowing the improvement of the birth rate, nor increased positive feedback.

We have called this phenomenon the Positive Feedback Trap (PFT).

The PFT suggests that, although in many dimensions evolution will lead to error making and adaptability, enabling populations to adjust to their changing circumstances, some characteristic or strategy could exist which would result in positive feedback enhancing this characteristic and trapping it. Such a strategy could involve, for example, mutual recognition and cooperation.

An example of this from biology is the peacock's tail. A gene produces the beautiful tail in the male and makes such a tail attractive to the female. In sexual reproduction, anything which enhances the probability of mating produces a positive feedback and fixes itself. However, this occurs at the expense of functionality with respect to the environment. Peacocks' tails are not an aid to finding food or escaping predators but merely the marker of a positive feedback trap.

Human systems abound with positive feedback. Much of culture may well be behaviours which are fixed in this way. In most situations imitative strategies are not eliminated by the evolutionary process and so fashions, styles and indeed entire cultures rise and decline without necessarily enjoying any clear functional advantages. Indeed, culture can be viewed as being an expression of ignorance of other ways of doing things.

Similarly, academic disciplines in the human sciences offer almost perfect examples of such autocatalytic, self confirming systems, where theorems and proofs are rewarded with prestige and prizes, even though it is quite clear that they produce no functional advantage. Much human activity is role playing in groups where values are generated internally and resources and evolutionary opportunities are largely ignored.

In some further experiments our models were extended to show that adaptive landscapes are generated by the mutual interaction of populations. In character space we suppose that similar populations are most in competition with each other, since they feed off the same resources and are fed on by the same predators. We also suppose that there is some amount of separation in character space, some degree of dissimilarity, at which two populations do not compete with each other.

Such an idea can be expressed in a 1-D character space by modifying the competition term in the logistic bracket so that it expresses the decay of niche overlap with increasing separation in character space.

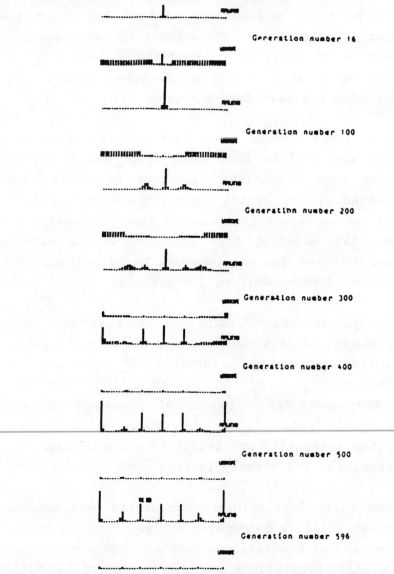

Figure (7). In a 1 dimensional character space, imperfect reproduction creates a simple ecology of populations. We see successive population distributions, and self-induced adaptive landscapes.

Our equations now express the dynamics of imperfectly reproducing populations diffusing to their nearest neighbours within this 1-D character space. We start the model with a population at a single position.

Initially, this population grows until it reaches the limits set by competition for resources. At this point there is an opportunity for error makers because of the negative feedback caused by competition. We could say that, although initially there was no hill to climb, the population digs a valley creating a hill on either side of itself. However, over some distance in this space error making populations cannot multiply, because they are still in the resource shadow of the original population. They diffuse in small numbers up the slope. After a certain distance in character space populations are sufficiently differentiated to grow on some other resource.

In time, such a diffusing population will increase until it too is limited by scarcity of resource. Once again there is an opportunity for error makers, particularly for those on the outside of the distribution of the population in character space, to climb another self-made hill towards unpopulated regions. In this way concentrations of population appear successively as a result of the diffusion of error makers from other populations which have reached their growth limit. Character space gradually fills with concentrations of populations separated by distances which are characteristic of resource diversity.

From a single population our model generates a dynamic ecology. The identity of each population is maintained by the balance between a diffusion of error makers outwards into character space and the differential birth and mortality rates caused by the presence of other populations. In addition, local conditions encountered

during the occupation of character space will affect
these events. It is not true that evolution represents the
discovery of pre-existing niches. The niches are
themselves the creation of the evolutionary process.

3. Evolutionary Human Systems

Complex systems dynamics resulting from the dynamic
interplay of average processes and non-average detail have
been used to explore how we can model and understand the
evolution of human systems. In early models (9,10) a
system of non-linear dynamical equations expressing the
evolution of the supply and demand of different products
was driven by the occurrence of entrepreneurs at random
points and times in the system. Supply was characterized
by a non-convex production function for different economic
activities. Consumer demand was assumed to reflect
relative prices. The random parachuting of entrepreneurs
onto the plain of potential demand resulted in the gradual
emergence of a market structure and a pattern of
settlement.

Even in this very preliminary form the models showed
important principles: many final states are possible,
precise prediction in the early stages is impossible,
approximate rules appear (separations of concentrations of
population, etc.) but with considerable deviation and
local individuality. The results are affected by the
details as well as by the data flows informing the
consumers. The evolution of structure as a result of
changing patterns of demand and supply, technology,
transportation, resource availability, etc. was explored.

The bases of the models are the decisions of the different
types of participants, reflecting either their cognitive
maps, or the particular heuristic rules that they have
developed. The spatial dynamics give rise to complex
patterns of flow and structure and to the emergence of

concentrations of population and other collective entities. In such systems the microscopic and macroscopic levels are not related in a simple fashion. It is not true that a big structure is a little one writ large. The emergence of macroscopic structure affects the circumstances of the microscopic parts. They find themselves in a larger, collective entity, each co-evolving with the others.

Models have evolved from these early simulations which provide tools for understanding regional and urban evolution, in which the patterns of flow and structure are the result of a process of self-organization (11,12). In addition, this approach has been used to develop dynamical models for the management of natural resources. The work connected with the simulation of fishing activities proved to be particularly fruitful (13). The problem contained all the essential aspects of the evolutionary situation in a strikingly simple way. This work has been discussed elsewhere in detail but it is worth summarizing some of the main conclusions.

The essential point that emerged was that success in fishing requires two apparently contradictory facets of behaviour. First, the ability to organize one's behaviour so as to exploit the information available concerning net benefits, a rational approach, which we have called "Cartesian" behaviour. Second, the ability to ignore present information and to explore beyond present knowledge, an apparently irrational approach, which we have called "Stochastic" behaviour. Fishermen who have the first ability, the Cartesians, make good use of information, but fishermen with the second, the Stochasts, generate it !

Stochastic behaviour is the root of creativity.

In the short term the more rational must outperform the less rational. To take steps to maximise present profits must, rationally, be better than not doing so. Nevertheless, over a longer period the best performance will not come from the most rational but instead from behaviour which is a complex compromise. For example, a fleet of Cartesians which goes where available information indicates highest profits will in fact lock into zones of the fishery for much too long, remaining in ignorance of the existence of other, more profitable, zones simply because there is no information available about them - "You don't know what it is you don't know".

New information can only come from fishermen who have chosen not to fish in the, rationally, best zones, or who do not subscribe to consensus values, technology or behaviour, and who thus generate information. They behave as risk takers but may not see themselves as such. They may act as they do through ignorance or knowledge. Whatever the reason or lack of it they are vital to the success of the fishing endeavour. They explore the value of the pattern of the existing fishing effort and lay the foundations for a new one (14,15).

The model can be used as a simulator providing information either for the management of the entire system or for the benefit of any particular fleet wishing to improve its performance. The parameter values which appear in the mechanisms governing the fishermen's decisions are calibrated to give realistic behaviour (16).

This approach can be inverted. Instead of supposing that we know, from observation, the parameters which characterize the fishing strategies of different fleets, we can use the model to discover robust fishing strategies for us. The model can simulate up to eight fleets. The fleets, calibrated with otherwise identical initial conditions, differ in the values of the parameters governing their fishing strategy and in whether or not they spy on or copy other fleets. Our model reinforces the more effective strategies, gradually eliminating the others. By including a stochastic change in strategies for losing fleets, our model gradually evolves sets of compatible behaviours. Each of these will be effective only in the context of the others, none are objectively optimal. In addition, such a system can adapt to changed external circumstances. It could be used to show us robust heuristics for the exploitation of renewable natural resources. This is clearly similar in aim to the work on learning algorithms (17,18).

Discussion

Evolution has often been viewed as an optimising process, whereby mutations perform a random search in some parameter space, and through the differential selection of survivors, the performance of a population is steadily improved. Indeed, this perspective stimulated the work of Schwefel (19) for example, who suggested an 'evoutionary approach' to finding optimal solutions in engineering problems. In a recent paper (20) he returned to the problem of representing the evolutionary process, but still maintained a fixed fitness function, corresponding to the existence of a hill up which the populations were climbing.

The view emerging from this paper however, is that instead of viewing evolutionary dynamics as the ascent by a population up a predetermined landscape of opportunity, the landscape itself is produced by the populations in interaction and that details of the exploration process may significantly affect the outcome. The evolutionary dynamic is characterized by the emergence of selfconsistant sets of populations posing and solving the problems and opportunities of their mutual existence.

Innovation occurs because of the initiatives of non-average individuals. When this leads to exploration of an area where positive feedback outweighs negative feedback then population growth will occur. Stochastic processes, ignorance of the future, and differing opinions lead to exploration and discovery which change the system and result in new ignorance of the future and new differing opinions. In human systems ignorance permits learning, but learning creates new ignorance.

More generally non-average behaviour generates adaptation which in its turn generates new non-average behaviour. Because biological and human systems are dominated by internal interactions, it is the endogenously created, dynamical hills and valleys which are most significant.

Within each new situation we retrospectively assign values which we assert are improvements of objective parameters of the system. It seems problematic, given the possibility of backsliding, collapse or emergence of greater levels of complexity caused by structural instability, even to assert that, by the invasion and animation of an increasing fraction of inorganic matter, evolution exhibits any net progress.

When we rationalize our behaviour we induce, from the false premise of the pre-existence of a hill, an exogenous potential which drove events. Hills and hill-climbing are a rationalization of what has occurred. The landscape of opportunity is a dynamic variable which itself depends on the unfolding of events. The alteration of apparently insignificant details may produce a quite different future.

We can learn from the past. We can, in so far as the
future resembles the past, learn about the future. However
there are elements of the evolutionary process,
perceptible in the present as discoveries or creations,
which cause radical and unpredictable differences between
the past and the future. In this situation nothing can be
certainly identified as being of future advantage. There
is a hill only in so far as there are eager climbers.

References

1). I.Prigogine and I.Stengers, 1987, "Order out of Chaos", Bantam Books, New York

2). P.M.Allen, 1988 "Evolution: Why the Whole is greater than the sum of its parts", in Ecodynamics, Springer Verlag, Berlin.

3). K.Arrow and G.Debreu,1954, "Existence of an equilibrium for a competitive economy", Econometrica.

4). G.Debreu, 1959, "Theory of Value", John Wiley and sons.

5) G.Nicolis and I.Prigogine, 1977, "Self-Organization in Non-Equilibrium Systems", Wiley Interscience, New York.

6) P.M.Allen and J.M.McGlade, 1987, "Evolutionary Drive: The Effect of Microscopic Diversity, Error Making and Noise", Foundations of Physics, Vol.17, No.7, July.

7) P.M.Allen and J.M.McGlade, 1989, "Optimality, Adequacy and the Evolution of Complexity", in "Structure, Coherence and Chaos in Dynamical Systems", Eds. Christiansen and Parmentier, Manchester University Press, Manchester.

8) L. Van Valen, 1973, "A New Evolutionary Law", Evolutionary theory, Vol 1. p1-30

9) P.M.Allen and M.Sanglier,1979, "Dynamic Model of growth in a Central place System", Geographical Analysis, Vol 11, No 3.

10) P.M.Allen and M.Sanglier, 1981, "Urban Evolution, Self-Organization and Decision Making", Environment and Planning A.

11) P.M.Allen, 1985, "Towards a new Science of Complex Systems", in "the Science and praxis of Complexity", United Nations University Press, Tokyo.

12). M.Sanglier and P.M.Allen, 1989, "Evolutionary Models of Urban Systems: an Application to the Belgian Provinces", Environment and Planning, A, Vol.21. p477-498.

13) J.M.McGlade and P.M.Allen, 1985, "The Fishing Industry

as a Complex System", Can.Tech.Fish and Aquat. Sci. No 1347, Fisheries and Oceans, Ottawa.

14) P.M.Allen and J.M.McGlade, 1986, "Dynamics of Discovery and Exploitation: the Scotian Shelf Fisheries", Can.J. of Fish. and Aquat. Sci. Vol.43, NO.6.

15) P.M.Allen and J.M.McGlade, 1987, "Modelling Complex Human Systems: a Fisheries Example", European Journal of Operations Research, vol.30. p147-167

16) P.M.Allen and J.M.McGlade, 1987, "Managing Complexity a Fisheries Example", Report to the United Nations University, Tokyo.

17) J.Holland, 1986, "Escaping Brittleness: The possibilities of General Purpose Machine Learning Algorithms Applied to Parallel Rule Based Systems", in Michalski et al, "Machine Learning: An Artificial Intelligence Approach", Vol 2. Los Altos, California, Kauffmann.

18) J.H.Miller, 1988, "The Evolution of Automata in the Repeated Prisoner's Dilemma",

19) H.P.Schwefel, 1988, "Optimum Seeking by Imitating Natural Intelligence", Proceedings of IMACS 12th World Congress on Scientific Computation, July 18-22, Paris.

20) H.P.Schwefel, 1988, "Collective Intelligence in Evolving Systems", in Ecodynamics, Springer Verlag, Berlin.

FINDING THE GLOBAL MINIMUM OF A LOW-DIMENSIONAL
SPIN-GLASS MODEL

W. Banzhaf
Institut für Theoretische Physik und Synergetik
Universität Stuttgart
Pfaffenwaldring 57/IV D-7000 Stuttgart 80

ABSTRACT

We present a search method based on a variation of natural molecular evolution to find the global minimum of low-dimensional spin-glasses. A group of different machines operates on a population of suitably defined data strings. Consecutive selection of "better" strings allows to find the global minimum of the energy landscape.

1. Some words on Spin-glasses

A spin-glass [1] is a system of spins,

$$\vec{s} = \{s_1, s_2, ..., s_N\}$$

i.e. variables with only two states

$$s_i \in \{+1, -1\}$$

It possesses an energy function or Hamiltonian $H(\vec{s})$ determining its dynamics:

$$H(\vec{s}) = -\sum_{ij} s_i J_{ij} s_j \qquad (1)$$

with couplings J_{ij} between components s_i, s_j. For a system to be a spin-glass, these couplings should become disordered by choosing them from random numbers (by random physical interactions):

$$J_{ij} = \{ \text{ randomly distributed} \in [-1, +1], \text{ e.g. gaussian } \}.$$

This leads to a "frustration" of states, i.e. to a situation where not all constraints of the system may be fulfilled simultaneously. An immediate consequence of frustration is the existence of a huge number of local minima in the energy landscape formed by equ. (1). Locality is defined here in terms of the "Hamming" topology valid in spin systems. A neighborhood of a state is determined by those states which differ exactly in one component's spin from the original state.

Among other features, the ultrametric organization of these states is typical for spin-glasses. The notion of ultrametricity [2] often used in this context means that the local minima of the system are organized in a hierarchical fashion.

Many problems of combinatorial optimization have been related to the physics of spin-glasses [1]. The high-dimensional energy landscape of spin-glasses corresponds to a presumably complicated cost function landscape in optimization. Frustration is encountered, too, in combinatorial problems such as the travelling salesman problem. Specifically, the fact that the number of local minima in spin-glasses explodes exponentially with the number of spins involved is similar to combinatorial explosion in optimization problems. In general it is assumed that finding the ground-state (the true or global minimum) of a spin-glass may provide us with better understanding of the nature of other optimization problems.

In physics there exists an enormous bulk of literature on methods of thermo-dynamics and statistical mechanics to find the ground-state of a physical system. Many scientists have thus tried to map combinatorial optimization problems to spin-glasses in order to obtain good solutions for them [3] - [9].

Mainly, this goes back to the Metropolis algorithm [10] and its application to the search for ground-states of systems. Based on the introduction of a parameter analogue to the temperature of a physical system, a stochastic relaxation process is created to reach Boltzmann distribution. By starting at a high value of the temperature parameter and slowly cooling it down, it is possible to find nearly minimal-energy-states in the so-called method of simulated annealing (SA) [11] which was derived from the physical annealing process.

At this workshop, we want to concentrate on evolutionary strategies and we shall consider therefore methods derived from natural evolution to solve the ground-state problem of a particular spin-glass model.

2. The philosophy of the Ansatz

The process of evolution is a rather successful optimization method used by nature. We are just beginning to realize its enormous potentiality. Though the concepts of "evolutionary strategies" (ES) [12,13,14] or "genetic algorithms" (GA) [15] have been developed more than decades ago, interest in application to combinatorial optimization problems has risen just recently [16] - [22] .

A common characteristics of many proposals in this context is to use natural evolution as a guideline, but not to copy its features in any detail. Essentially, three issues are needed to cover at least roughly the natural evolution process:

+ Randomness to generate variations
+ Selection of better variants
+ Populations of varying subjects to search in parallel

Let us discuss these three points in greater detail:

Randomness is used by nature in local and non-local environments, as may be derived from the fact of point mutations at gene-strings and of chromosome mutations at large accumulations of gene-strings [23]. Randomness is used to escape from local minima as well as to proceed to better regions in state space.

Local randomness is diffusion-like [24] and resembles in some aspects the tunneling through barriers which is encountered in quantum physics. Just recently, optimization procedures based on tunneling were proposed by Rujan [25].

Non-local randomness is Monte-Carlo-like and allows the system to search the state space without respect to neighborhood relations. It is used, too, to escape from local minima. And indeed, it is by far more effective in escaping from local minima than tunneling, since the operations involved in some sense try to cover the whole space.

Whereas with the element of randomness a general mechanism for generating variability is at hand, *selection* of the better states enables the search process to proceed rather systematical. Though there may be different strategies (like hard and soft selection), the common effect is to drive the system continuously towards the more promising regions of state space. There is a close analogy between selection of better exemplars and the method of steepest descent known since many years. Steepest descent methods are search strategies which always converge by an exponential decay of progress during they proceed. A fast convergence at the beginning is followed by a slower and slower convergence in the later stages of the process. The same is true in selectionist strategies where not always the best though at least better states are chosen.

The last issue to be mentioned is the fact that usually a *population* does execute the search in evolution. In nature this term is often related to redundancy or even wastefulness. It would, however, be a crude tendency of nature if no reason for searching by a population would exist. Evidently using a population of differing individuals is a parallel search. It makes sense under the condition that some means of communications between the individuals are present to enable them to profit from information which other individuals were able to accumulate in separated regions of state space (see Figure 1). Thus we conclude that the intrinsic

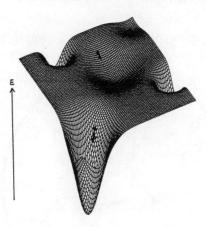

Figure 1: 3 individuals of a population search for a minimum in a 2-dimensional state space.

parallelity of natural evolution is closely related to the usage of populations during the search.

3. Application of an evolutionary search strategy to a spin-glass model: The Algorithm

We now want to turn to the concrete problem. As energy or cost function to be minimized we take the Hamiltonian of the N-dimensional SK-model [26]

$$E(\vec{s}) = -\frac{1}{N} \sum_{j \leq i=1}^{N} s_i J_{ij} s_j \qquad (2)$$

where $s_i \in \{+1, -1\}, i = 1, ...N$.

The coupling coefficients J_{ij} are chosen with equal probability from three possible values

$$J_{ij} \in \{+1, 0, -1\}, \quad i, j = 1, ..., N.$$

We use a population of data strings

$$S \equiv \{\vec{s}^\mu(t)\}, \quad s_i^\mu(t) \in \{+1, -1\}, \quad i = 1, ..., N, \quad \mu = 1, ..., M \qquad (3)$$

searching in state space for the lowest energy value.

Since $\vec{s}^\mu(t)$ will be subject to changes it is parametrized by a generation counter t defined below. The μth data string $\vec{s}^\mu$ is enlarged by one component s_0^μ which carries the actual value $E(\vec{s}^\mu(t))$ according to equ. (2) (cf. Figure 2).

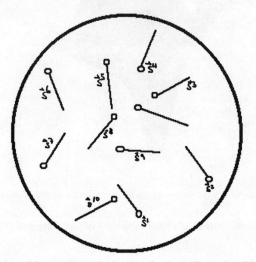

Figure 2: A population of data strings. Each string is $N+1$-dimensional having an 1-dimensional extension to carry its quality signal

The population as a whole is characterized by an extra string $\vec{s}^{pop}$ defined as

$$s_i^{pop} = sign(\sum_{\mu=1}^{M} s_i^\mu), \quad s_0^{pop} = E(\vec{s}^{pop}) \qquad (4)$$

Changes in strings are brought in using an aggregation of p specialized machines or processors. These machines have their own internal time-scales $t_l, l = 1, ..., p$

and perform operations on strings they have chosen randomly out of the string
population, see Figure 3.

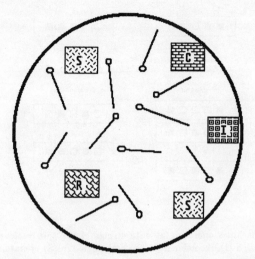

Figure 3: Different machines are operating on data strings they choose by means of a random
generator.

To maintain the parallel search, there are fewer machines than data strings
$p_{eff} \leq M$, where we had to define an effective number p_{eff} of machines:

$$p_{eff} = \sum_{l=1}^{p} t_l \times m_l \tag{5}$$

m_l : *number of machines of sort l.*

A generation is defined as that number c of operation cycles which suffices to give
every string a chance to be changed:

$t \rightarrow t + 1$ when $c = \lfloor \frac{M}{p_{eff}} \rfloor$ *operations are performed.*

Our four sorts of machines are called the C-, I-, R- and S-machines performing
complement, inversion, recombination and single-flip operations (see Figure 4).
Depending on the size of string population, machines of the same sort could act
in a parallel manner.

An operation cycle consists of:

1. picking up the number n_{op} of strings necessary to perform the prescribed
   operation (C, I, S: $n_{op} = 1$; R: $n_{op} = 2$),

2. performing the operation controlled by a random generator,

3. evaluating the newly generated string(s) according to equ. (1),

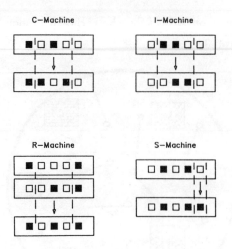

Figure 4: The four machines operating at data strings of the spin vectors $\vec{s}^{\mu}$. They perform complement (C), inversion (I), recombination (R) and single-flip (S) operations.

4. comparing them to the original string(s),

5. releasing the n_{op} best one(s) into the population.

Hence it is a totally local activity of machines without any reference to the global state of the system that leads to the desired progress.

Before coming to results let us briefly discuss the effects of the different machines. What is the rationale behind implementing just these functions? Though the general concept of the algorithm is rather universal, data strings and particular operations on them should be adapted to the problem at hand.

In our case , the symmetry of the energy function

$$E(\vec{s}) = E(-\vec{s}) \tag{6}$$

suggests quite naturally to use the complement operation. Valuable information originally accumulated in wrong context may thus be activated by complementing parts of the data strings. The inversion operation may be seen as an implicit large-scale mutation. It decorrelates spins and considerably pushes strings around in the state space. The recombination operation provides for continuous data exchange between strings and shows clearly the parallel character of the search procedure. In order to use the advantage of parallel search recombination was introduced. It is recombination that accelerates the convergence velocity significantly profitting hy information of different strings. If, on the other hand, strings are recombined too often, the parallel character of search gets lost and a variance collapse may be the consequence leading to an early trap in one of the numerous local minima. Lastly, random single-spin flips are used which support search in an ultrametric landscape as is that of spin-glasses.

Though all operations introduced make sense in the context of spin-glasses, it is their particular combination that allows us to find global minima in highly rugged energy landscapes.

4. Application of an evolutionary search strategy to a spin-glass model: Simulation Results

Our results are grouped around a standard configuration of $p = 4$ different machines applying operations onto $M = 9$ data strings of lengths between $N = 10$ and $N = 20$. The time-scale for recombination is $t_R = \frac{1}{30}$, for single spin-flip $t_S = \frac{1}{10}$. C- and I-operations are performed as often as possible, i.e. $t_C = t_I = 1$.

We first demonstrate a few characteristic features which become evident already with the standard configuration. In Figure 5 three stages of the search process are shown. In every stage, the strings of the population have changed due to machine operations.

Figure 5: Three stages of the search in standard configuration: Left: Starting state, Middle: after 200 generations, Right: after 1000 generations.

Figure 6 shows the decrease in energy of one individual as well as the average and the population string defined in (4). Whereas individuals proceed in leaps, the average energy of the population decreases more smoothly on an exponentially

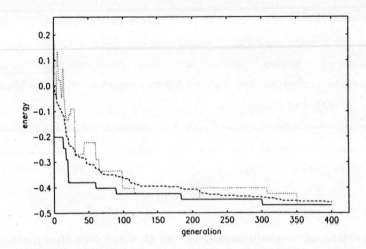

Figure 6: Decreasing of energy during a search process. **Solid line:** Sample individual, **Dashed line:** Average energy, **Dotted line:** Population string $\bar{s}^{pop}$

decaying curve. This behaviour reflects the similarity of selection strategies with steepest descent strategies only allowing downhill evolution.

The population string, however, derived from the population by nonlinear operations also goes uphill in search space. The population as a whole ignores local hills in the region over which its different individuals are distributed. In other words, the population variance smoothes out the energy landscape and allows the population string to jump over those local hills. In this role, we could compare the population variance with the temperature parameter of simulated annealing. It does, however, not require operation near equilibrium and therefore will be considerably faster. A better model would be a ball of diameter of the population variance moving downhill in search space [27] (cf. Figure 7).

In Figure 8 the different contribution of our machines are measured by summing up their respective effects. One can see that the inversion works mainly at the beginning (spreading the states to all regions) whereas the complement operation continuously contributes to the progress. At different times R- and S-machines intervene to accelerate te process. This depends mainly on the random number generator applied. Note that the figure stems from a single run.

Figure 9 qualitatively exemplifies the fact that short after beginning of search the population concentrates on better regions.

It shows the state density over different energy levels during successive generations. After approximately 20 generations, the energetically disfavourable half of

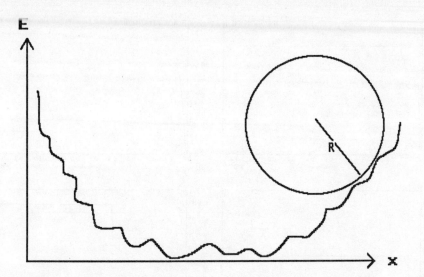

Figure 7: Smoothing out of a landscape by the population of search subjects

the states has become extinct. The population approaches the global minimum by a slow density increase of the corresponding state from its first appearance near generation 50.

To see the scaling behaviour we now vary the dimension N of data strings where we have constrained ourselves to provable cases, $N \leq 20$. Table 1 shows the average computing time to find the global minimum by our standard configuration on a particular serial computer (MicroVax). Statistics is based on 10 different landscapes $E(J_{ij})$ for every N.

$N =$	10	12	14	16	18	20
$\langle \bar{T}^{pop} \rangle =$	6.2	9.7	14.2	24.5	36.1	55.7
$T_{min}^{pop} =$	0.5	1.9	0.7	6.0	9.9	5.5
$T_{max}^{pop} =$	29.3	39.6	53.2	38.0	159.6	155.0
$\bar{t}^{exh} =$	1.7	8.3	41.5	202.6	973.3	4598.5

Table 1: Comparison of computing time between $N = 10$ and $N = 20$. $\langle \rangle$ means averaged over 10 runs with different starting state. $\bar{T}$ means additional averaging over 10 different landscapes J_{ij}. t^{exh} gives computing time for exhaustive search.

The computing time T^{pop} is defined as the physical cpu-time needed to fulfill the following condition:

$$T^{pop}: \quad \vec{s}^{pop} = \pm \vec{s}^{Min} \tag{7}$$

where $\vec{s}^{Min}$ is the global minimum state determined by an exhaustive search on $E(J_{ij})$. Thus, when the string that represents the total population state is iden-

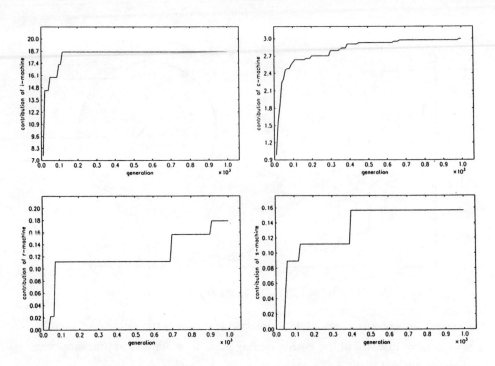

Figure 8: Different contribution of machines during search process. Standard configuration

tical to the minimum string, computation is stopped. Note that our simulations were executed on a serial computer. Parallel computing times are derivable by dividing through a factor of at least $c \times p_{eff}$.

In the considered range, the scaling behaviour may be described by a rough exponential law

$$T^{pop}(N) = T_0^{pop} \times exp \ [\alpha(\frac{N}{N_0} - 1)] \qquad (8)$$

with $\alpha = 2.18$ and $T_0^{pop} = T(N = 10)$ as basis time. Of course, due to the fact that we constrained ourselves to cases where $N \leq 20$, the exponential scaling law is considerably restricted in its value.

It is interesting to observe the spreading of T^{pop} values. The cpu-time depends critically on starting conditions, thus a factor of 20 between slow and fast runs for the same landscape is not seldom. Consequently, it is generally better to limit computation time and to restart with newly initialized strings, rather than to continue the run in a disadvantageous search region.

The effect of an increase in population size M is twofold. Firstly, - as it is to be expected on a serial computer - , physical cpu-time goes up (cf. Table 2). A tripling of M, however, does not result in a tripling of cpu-time T^{pop}, since a larger search region is covered by the total population which is activated during operation by the R-machine. Therefore, the simultaneously tripling of the number

453

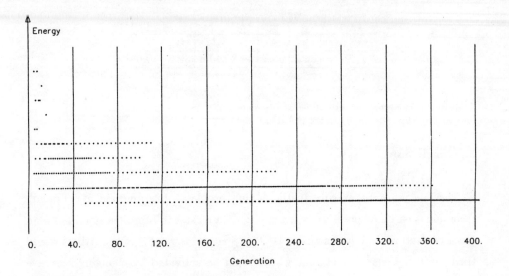

Figure 9: State density in different energy levels during search. Every circle symbols a data string.

$M =$	9	27	81	243	729
$\langle \bar{T}^{pop} \rangle =$	55.7	67.1	107.8	165.2	253.6
$T^{pop}_{min} =$	5.5	8.0	16.4	40.1	71.9
$T^{pop}_{max} =$	155.0	98.3	153.5	214.7	223.4
$\sigma_{rel} =$	0.45	0.31	0.25	0.28	0.02

Table 2: Effects of choosing different population sizes M. σ_{rel} is relative variance of results.

p_{eff} of installed machines will produce a decrease in computation time to one third of the corresponding value T^{pop} in Table 2. The second effect is less obvious, although not less important: The variance in T^{pop} between different runs with the same landscape decreases, again as a consequence of the larger region in search space covered by participating individuals.

It is tempting to accelerate convergence by increasing the recombination rate of the system. Nevertheless, as is shown in Table 3, this disequilibrates operations and leads more often to traps in local minima. The reason is disappearance of population variance due to the high recombination frequency. Theoretically, a tuning of recombination rate to an optimal value would be possible. Quantitative statements would, however, require more detailed studies on a fairly enlarged data base.

454

$t_R =$	$\frac{1}{60}$	$\frac{1}{30}$	$\frac{1}{10}$	$\frac{1}{5}$	$\frac{1}{2}$	1
$n_0 =$	2	4	12	17	19	25
$\langle \bar{T}^{pop} \rangle =$	64.2	55.7	44.8	40.8	39.8	30.6

Table 3: Searching process with different recombination rate t_R. n_0 means the number of runs (out of 100) which did not reach the global minimum.

5. Conclusion

There is yet another cure to the tuning problem: Introduction of "diploid" recombination made possible by extending data strings to represent dominant and recessive features in any dimension [28,29]. The recessive part of the strings would then work as a variance storage which could be activated by changing recessive and dominant features of strings. Though it is beyond the scope of this letter, the advantage of this approach is its relative stability against parameter detuning.

The search method discussed here has shown its ability to find the ground-state of a low-dimensional spin-glass in reasonable time. Next is an extension to cases which are not provable by exhaustive search, as e.g. $N = 50$ or $N = 100$. In such situations, a comparison with SA or other methods [25] are the only practical way to get a measure of the quality of the proposed algorithm. In addition, the criterion for stopping the search process has to be redefined. Besides other possible measures, the variance breakdown always occurring at some point in the evolution is a natural signal [28].

The concept introduced here is ideally suited to be implemented in parallel computing structures. Moreover, the completely local character of the search procedure opens the way to use such concepts as In/Out-processes introduced by parallel programming languages like LINDA [30]. It is suspected that ample search problems in different branches of science may be accessible by algorithms of the kind described here.

References

[1] M. Mezard, G. Parisi, M.A. Virasoro, *Spin glass theory and beyond*, World Scientific, Singapore, 1987

[2] R. Rammal. G. Toulouse, M.A. Virasoro, Rev. Mod. Phys. **58** (1986) 765

[3] J.J. Hopfield, Proc. Natl. Acad. Sci. USA **79** 1982 2554

[4] M.P. Vecchi, S. Kirkpatrick, IEEE TA **CAD-2** (1983) 215

[5] P.W. Anderson, Natl. Acad. Sci. USA **80** 1983 3386

[6] Y. Fu, P.W. Anderson, J. Phys. A **19** (1986) 1605

[7] M.Mezard, G.Parisi, J. Physique **47** (1986) 1285

[8] Z. Li, H.A. Scheraga, Proc. Natl. Acad. Sci. (USA) **84** (1987) 6611

[9] J.D. Bryngelson, P.G. Wolynes, Proc. Natl. Acad. Sci. (USA) **84** (1987) 7524

[10] N. Metropolis, A.W. Rosenbluth, M.N. Rosenbluth, A.H. Teller, E.J. Teller, J. Chem. Phys. **21** (1953) 1087

[11] S. Kirkpatrick, C.D. Gelatt, M.P. Vecchi, Science **220** (1983) 671

[12] H.J. Bremermann, M. Rogson, S. Salaff in: M. Maxfield, A. Callahan, L.J. Fogel (Eds.), it Biophysics and Cybernetic Systems, Spartan Books, Washington, 1965

[13] I. Rechenberg, *Evolutionsstrategien*, Frommann-Holzboog Verlag, Stuttgart, 1973

[14] H.P. Schwefel, *Numerical Optimization of Computer Models*, Wiley, Chichester, 1981

[15] J.H. Holland, *Adaption in natural and artificial Systems*, University of Michigan Press, Ann Arbor, 1975

[16] D.H. Ackley, *A Connectionist Machine for Genetic Hillclimbing*, Kluwer Academic Publishers, Boston, 1987

[17] T. Boseniuk, W. Ebeling, A. Engel, Phys. Lett. **125A** (1987) 307

[18] R.M. Brady, Nature **317** (1985) 804

[19] W. Fontana, P. Schuster, Biophys. Chem. **26** (1987) 123

[20] J. Grefenstette, in: J. Grefenstette, (Ed.), *Proc. Int. Conf. on Genetic Algorithms and their Applications*, Carnegie Mellon University, Pittsburgh, 1985, 42

[21] H. Mühlenbein, M. Gorges-Schleuter, O. Krämer, Par. Comp. **7** (1988) 65

[22] Q. Wang, Biol. Cyb. **57** (1987) 95

[23] B. Lewin, *Genes*, Wiley, New York, 1983

[24] R. Feistel, W. Ebeling, BioSys. **15** (1982) 291

[25] P. Rujan, Z. f. Phys. **B73** (1988) 391

[26] D. Sherrington, S. Kirkpatrick, Phys. Rev. Lett. **35** (1975) 1792

[27] S.A. Kauffman in: R. Livi et. al. (Eds.), *Proc. Workshop on Chaos and Complexity*, Torino, 1987, World Scientific, Singapore, 1988

[28] W. Banzhaf, BioSystems **22** (1989), 163

[29] W. Banzhaf, in: H. Haken, (Ed.), *Neural and synergetic computers*, Proceedings of the Elmau International Symposion on Synergetics 1988, Springer Verlag, Berlin, 1988, 155

[30] N. Carriero, D. Gelernter, J. Leichter, in: *Proc. ACM Symp. on Principles of Programming Languages*, Jan 1986, 236

TEN THESES REGARDING THE DESIGN OF CONTROLLED EVOLUTIONARY STRATEGIES

Paul Ablay

I. TARGETS

Practical application of evolutionary strategies for solving complex optimization problems

- Where other methods fail.

- Irrespective of the resulting complexity it should be possible to treat the problem complex in a model reflecting reality as closely as possible.

- The design of efficient evolutionary strategies to solve individual problems should be realizable with justifiable effort.

- Computer-aided problem solving should be possible in feasible running times.

II. WORKING THESES

The theses below are the result of findings obtained when applying evolutionary strategies to economic optimization problems in practice. It is exclusively their feasibility which makes their relevance.

1st Thesis
NATURE AS MODEL

Nature's evolutionary strategies are composed of a variety of successful basic processes, e.g. reproduction, mutation, selection, isolation, population, recombination, annidation, etc., and , if adequately applied, they can be applied successfully to solve complex technical, economic and social problems.

The following discussion of evolutionary strategies will be limited to applications which can be tested under the **controlled** conditions of given experimental conditions, specifically computer models, and represent general optimization problems.

2nd Thesis
BETWEEN DETERMINISM AND PURE RANDOM

Well-designed evolutionary strategies are characterized by a well coordinated interaction of chance and necessity

Each kind of problem requires its own approach how to apply evolutionary principles in order to find the optimal way between determinism and pure random with as many random trials as necessary and as much necessity as possible.

Example:

Let us put Shakespeare's well-known quote:

"TO BE, OR NOT TO BE: THAT IS THE QUESTION"

on paper by typing the letters at random.

The following methods show the differences in efficiency with which this problem can be solved:

Proceeding according to the **Monte-Carlo method** would mean that the touches per line are determined at random and the line is checked as to how many of its touches correspond to the target text.

A method in which the touches are arranged most chaotically from line to line. How many lines would have to be typed in order to type the quotation correctly?

With 41 touches and 34 possibilities per touch (26 characters and 8 special characters), the probability of typing it correctly is

$$\frac{1}{34^{41}} = 1.62 \cdot 10^{-63}$$

If one could type one line with 41 touches per second and if one had done so since the beginning of universe about 15 billion years ago, only $5 \cdot 10^{17}$ lines would have been typed!

An appropriately designed computer program shows the development of the Monte Carlo method.

MONTE CARLO METHOD

Steps	Target text	Number of conformity
	TO BE, OR NOT TO BE: THAT IS THE QUESTION	
	*	
1	.RQEYHISDTPP.ZUZT:NRLJKDDOMJAVY,UHUNWZPSR	1
	* *	
2	LCQ;UASONI?P RBC DO,WOOO!ATY.JS'CZA,D;RKV	2
	* * *	
4	G?F'NBXER,;OR C AJJTAUOXYOY?ZOGKO'T,AUQDH	3
	* * * *	
27	DL.AEFLYGNWI, SR,IDTDVHY;MUS;!S.ICCJDAE?Z	4
	* * * * *	
269	CYFSJVS?RJNGQFSF!?YXVGGA,SY JSTCOBOLMTVOM	5
	* * ** * * *	
2157	HZ UEV OY ERFAO!RQMPTT!BS:KRX ?SWMRTIZQ,	7
	* * ** * * * *	
56799	TNBBYE!'M,H, RTOXPDDAGOACS?LBSTETHMECCIXP	8
	* * * ** * * * *	
277929	SB KT;,OPJNL.TF,TM?GIHOXN,ISEQKC !UGSWI;Z	9

An **evolutionary strategy** tailored to this problem would aim at more balance between chance and necessity.

Instead of changing all touches anew in each line, only few touches of the starting line are changed at random (**mutation**). One could also say that the original line is reproduced with mistake in some places.

If the number of letters equal to the target text increases, this line becomes the starting line for the next one, otherwise variation is again done on the basis of the original line (**selection**).

By merely limiting the random component and linking the evaluation with a decision, a **mutation - selection strategy** designed in this way exhibits the merit of higher efficiency than is shown by the Monte Carlo method and illustrated in the print-out of the results of an evolutionary strategy programmed in this way (see next page).

EVOLUTIONARY STRATEGY

Steps	Target text	Number of conformity
	TO BE, OR NOT TO BE: THAT IS THE QUESTION	
1	LAWMUIU'TRWV,.NN UOSC.Y:ZRKP :X;!HAJGMDPT	2
2	LAWMUIU'TRWV,.NN UOSCTY:ZRKP :X;!HAJGMDPT	3
4	LAYMUIU'TRWV,.NN UOSCTY:ZTKP :X;!HAJGMDPT	3
6	LAYMUIU'TRVV,.NN UOSCTY:ZTKP :X;!HACGMGPT	3
7	LAWMUIU'TRVV,.NN UOSCTYPZTKP :X;!HACGMGPT	3
8	LAWMUIU'TRVVY.NN UOSCTYPZUKP :X;!HACGMGPT	3
9	LAWGUIU'TRVVY.NN TOSCTYPZUKP :X;!HACGMGPT	3
11	LAWGUIU'SRVVY.NN TOSCTYPZUKP :X;!HACGMGPT	3
12	LAWGUIU'SRYVY.NN TFSCTYPZUKP :X;!HGCGMGPT	3
20	LAWGUIU'STYVY.NN TFSCTYPZUKP :X;!HGCGMGPT	3
26	LCWGUIU'STYVY.NN TFSCTYGZUKP :X;!HFCGMGPT	3
28	LCWGUIU'STZVY.NN TFSOTYGZUKP :X;!HFCGMGPT	3
32	LEWGUIU'STZVY.NN TFSOTYGZUKP :X;!HFCGMGPT	3
34	LEWGUIU:STZVY.NN TFSOTYGZUKP :X?!HFCGMGPT	3
41	LEWGUIU:STZVY.NN NFSOTYGZUKP :X?!MKNGMGPT	3
46	LEWGUIU:STZVY.NN NFSOTYGZZKP :X?'MKNGMGPT	3
47	LEWGUIV:STYQY.TN NFSOTYGZZKP :X?'MKNGMGPT	4
48	LEWGUIV:STYQY.TN NFSOTYGZZKP :X?'MNNHMGPT	4
49	LEWGUIV STYQY.TN NFSOTYDZZKP :X?'MNNHMGPT	4
55	LEWGULT STYQY.TN NFSSTYDZZKP :X?'MNQHMGPT	4
65	LEWGULT STYQY.TN NFSSTYDXZKP :X?'MNQHMGPT	4
66	LEWGULT RTYQY.TN NFSSTYDXTKP :X?'MNQHMIPT	6
78	LEWGULT RTYQY.TN NFSSTYDXTKP :X?'MNQHMIPM	6
81	LEWGULT RTYQY.TN NFSSTYDXTKP :X?'MNQHMIOM	7
83	LEWGUKT RTYQY.UN NFSSTYDXTKP :XZ"MNQHMIOM	7
85	LEWGUKT RTYOY.UN NFSSTYDXTKP :XZ"MNPHMIOM	7
86	LEWGULT RTYOY UN NFSSTYDXTKP :XZ"MNPHMIOM	8
89	LEXGULT RTYOY UP NFSSTYDXTKP :XZ"MNPHMIOM	8
99	LEXGULT RTYOY UP NFSXTYDXTKP XZ"MLPHMIOM	8
100	LEXGULT RTYOY UP NFSXTYDXTKP XZ"MLPIMIOM	8
101	PEXGULT RTYOY UP NFS.TYDXTKM XZ"MLPIMIOM	8
104	PEXGUOX RTYOY UP NFS.TYDXTKM XZ"MLPIMIOM	8
111	PEXGUOX RTYOY .O NFS.TYDXTHM XZ"MLPJMIOI	9
112	PEXGUOX RTYOY .O NFS.TXDXTHM XZ"MLPJMIOI	9
115	PEXGUOX RTYOY .O NFS.TXDXTHM WZ"MLPJMIOI	9
122	PEXGUOX RTYOY .O NFS TXDXTHM W "MLTJMIOI	10
	.	
	.	
	.	
556	TN BE. PRUKOT RO BE? TUCQ IW VHE QTNSUISN	25
560	TN BE. PRUKOT RO BE? TUCQ IU VHE QTNSUISN	25
561	TT BE. PRUKOT RO BE? TUCQ IU VHE QTNSUISN	26
565	TT BE. PRUKOT TO BE? TUCQ IU VHE QTGSUISN	26
591	TT BE. PRUKOT TO BE, TUCQ IU SHE QTGSUISN	26
595	TT BE. PRUKOT TO BE, TMCQ IU SHE QTGSUISN	26
608	TT BE. PRUKOT TO BE, TMCQ IU SHE QTGSUILN	26
612	TS BE. PRUKOT TO BE, TMCQ IU SHE QTGSUILN	26
626	TS BE. PRUKOT TO BE, TMBQ IU SHE QTGSUILN	26
633	TS BE. PRUKOT TO BE, THBQ IU SHE QTGSUILN	27
639	TS BE. PRUKOT TO BE, THBV IU SHE QTGSUILN	27
653	TS BE. PRVKOT TO BE, THBV IU SHE QTGSUILN	27
683	TS BE. PRWLOT TO BE, THBV IU SHE QTGSUILN	27
688	TR BE. PRWLOT TO BE, THBV IU SHE QTGSUILN	27
699	TR BE. PRWLOT TO BE, THBU IU SHE QTGSUILN	27
755	TR BE. PRWLOT TO BE THBU IU SHE QTGSUILN	27
799	TR BE. PRWLOT TO BE THBU IU SHE QTGSUION	28
814	TR BE, PRWLOT TO BE THBU IU SHE QTGSUION	29
829	TR BE, RR LOT TO BE THBU IU SHE QTGSUION	30
837	TR BE, RR LOT TO BE THBT I, SHE QTGSUION	31
861	TR BE, RR LOT TO BE THBT I, SHE QUGSUION	32
924	TR BE, RR LOT TO BE THAT I, SHE QUGSUION	33
941	TR BE, KR LOT TO BE THAT IW SHE QUGSUION	33
947	TP BE, KR LOT TO BE THAT IW SHE QUGSUION	33
1070	TP BE, KR KOT TO BE: THAT IW SHE QULSUION	34
1164	TP BE, KR KOT TO BE: THAT IW THE QULSUION	35
1173	TP BE, KR KOT TO BE: THAT IW THE QUHSUION	35
1222	TP BE, KR KOT TO BE: THAT IV THE QUHSUION	35
1255	TP BE, KR KOT TO BE: THAT IT THE QUHSUION	35
1269	TP BE, KR POT TO BE: THAT IT THE QUHSUION	35
1284	TP BE, KR POT TO BE: THAT IT THE QUFSUION	35
1285	TP BE, OR POT TO BE: THAT IT THE QUFSUION	36
1394	TP BE, OR NOT TO BE: THAT IT THE QUFSVION	37
1543	TP BE, OR NOT TO BE: THAT IS THE QUFSTION	38
1547	TP BE, OR NOT TO BE: THAT IS THE QUFSTION	39
1663	TO BE, OR NOT TO BE: THAT IS THE QUFSTION	40
1785	TO BE, OR NOT TO BE: THAT IS THE QUESTION	41

BASIC STRATEGY

The **basic principle of evolution** established by Charles Darwin - the further development of species via mutation and selection - can be schematized in two phases for our purpose

> **1st phase:**
> Effecting a random variation respectively a weighted random step
> **-mutation-**
>
> **2nd phase:**
> Evaluation of this random step and decision on whether this step becomes the starting point of the next step or is to be cancelled.
> **- selection -**

For comparison: pattern of a **deterministic way** of procedure:

> **1st phase:**
> Analysis of the possible steps and selection of the most favorable step in line with a given rule.
>
> **2nd phase:**
> Implementation of the selected step.
>
> *e.g. simplex method*

and pattern of a **erratic way** to proceed:

> **1st phase:**
> A random step
>
> **2nd phase:**
> Does not exist (evaluation of the random step does not entail any consequences).
>
> *e.g. monte carlo method*

The two-phase pattern of mutation and selection is most apt as a **basic strategy** for controlled evolutionary strategies. The schematized procedure of a mutation - selection strategy is particularly suited for finding **solutions by steps** of a given optimization problem.

We assume that an **optimization problem** is defined by n independent variables, each point of the solution space being definitely evaluated using an objective function, and the solution space further being limited to certain constraints.

The basis is a feasible starting vector, which is called parent vector ($\overline{X}$) in analogy to biology. An evolutionary step is initiated, a **mutation operator** varies the variables of the parent vector at random. Thereby, a new vector is created - the daughter vector (X^I). The evolutionary step is completed by the **selection operator**, which at first examines whether the daughter vector is qualitatively superior to the parent vector and whether it represents a feasible solution.

If both checks are positive, the daughter vector becomes the parent vector of the next generation; otherwise, the daughter vector dies out, and the original parent vector, in turn, becomes the basis for a new mutation step.

A step sequence designed this way will approach the optimum successively (convergence towards the optimum). How fast - and thus how **efficiently** - the strategy will lead to the optimum, is determined by:

- externally **given** necessities defined by the order structures of the objective function and the constraints.

- necessities **susceptible** to procedure-inherent factors, i.e. depending on the selection of mechanisms directing the random process to success.

  In this context, it is important to transfer the **mutation-selection process** from a pure random behavior **to random weighted step sizes and step directions tailored to the local order structures.**

SCHEME OF EVOLUTIONARY STRATEGY

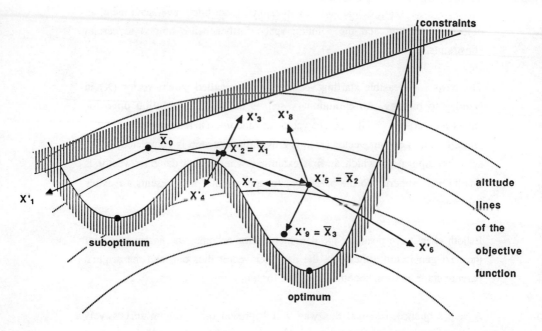

$$\overline{X} \quad : \quad \text{parent vector}$$
$$X' \quad : \quad \text{daughter vector}$$
$$Z \quad : \quad \text{random vector}$$
$$\text{start} \quad g, r = 0$$

Mutation:

$$X'_{g+1} \quad = \quad \overline{X}_r + Z_{g+1}$$

Selection: (here : minimization)

$$\overline{X}_{r+1} \quad = \quad X'_{g+1} \quad \text{if} \quad f(X'_{g+1}) < f(\overline{X}_r)$$

$$\text{and} \quad X'_{g+1} \text{ feasible}$$

$$\overline{X}_{r+1} \quad = \quad X_r \quad \text{if} \quad f(X'_{g+1}) >= f(\overline{X}_r)$$

$$\text{or} \quad X'_{g+1} \text{ unfeasible}$$

3rd Thesis
LIMITING RANDOM PROCESSES

In line with its order structures almost any type of problem suggests criteria for enhancing the probabilities of success for a mutation according to which the random process can then be weighted.

It is of no importance in which way these criteria are obtained - by way of analytical deduction or empirical induction, by way of learning processes or exchange of information (crossing over). The only thing that counts is to **reduce the quantity of degrees of freedom** so that successful mutations become feasible.

1st example: **TRAVELLING SALESMAN PROBLEM**

Supposing a travelling salesman has to visit 10 towns, he has $(10 - 1)! =$ 326,880 possibilities per tour.

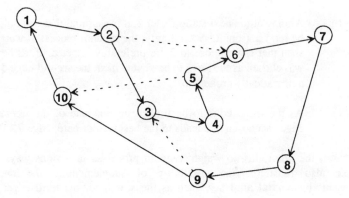

The tour
$$\overline{X} = (1, 2, 3, 4, 5, 6, 7, 8, 9, 10, 1)$$

shall be employed as a start vector within an evolutionary strategy.

To this end, mutations (permutations) can be performed, the **step size** of which can be determined by different metrics. It can, for instance, be determined how many positions or how many edges are to be changed.

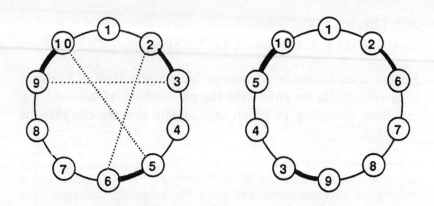

If **edge metrics** are chosen - as is appropriate for this type of problem - it is easy to see that the smallest step size for a mutation is not obtained before changing three edges. For mutations of this step size, most different weighting possibilities exist. The simplest may be to have the three edges determined through an equipartitioned random process. A **weighting** tailored to this type of problem may as well be as follows:

a) The selection of the **first** edge to be broken up is effected through a random process which can hit each of the 10 edges with the same probability (here for instance the edge 2-3).

b) Aiming at the next station after 2, it is essential that it should not be too far from 2. Accordingly, the random process is weighted such that towns close to 2 are preferably selected. If, as here, 6 was chosen as the town to be visited next, the **second** edge 5-6 is automatically broken up.

c) As the **third** edge to be broken up that one of the remaining edges is chosen that leads to the best result (here: edge 9 - 10).

Apart from the possibility to weight random processes in various ways, this example also illustrates the possibility of supplementing the random components by **partial analyses,** such as the search for the third edge. The convergence behavior of such a weighted mutation-selection strategy offers considerable advantages over unweighted strategies (see example of thesis 7).

Nature seems to **limit random processes** in a similar way by mutating specific combinations of genes by means of regulator and operator genes. Often, one could speak of a real "wiring network" of the genes.

2nd example: **NONLINEAR OPTIMIZATION**

Almost all algorithms for solving optimization problems with a non-linear and discrete objective function as well as with non-linear restrictions are bound to fail. Especially fringe areas, points of discontinuity or "narrow gorges" put the strategy strictly to the test. Here, successes can frequently only be attained via variations of few variables.

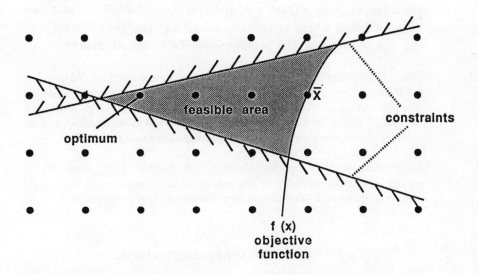

Thus, in this two-dimensional example, a mutation can only be achieved, if only one variable is varied.

When dispensing with local partial analyses, the chances of the mutation can be increased by introducing a weighting which tends to vary few variables. The **mutation weighting** could, for example, again be effected in three stages:

a)   At first, it is important to determine the number of the variables to be varied. Here, the random process is to be weighted in a way that tends to generate **few variables**. This can be effected by means of a binomial or a Gaussian distribution.

b)   The decision on **which of the variables** is to be mutated can be aided by an equi-distributed random process

c)   The **variation of each individual variable selected** is again effected through a weighted random process, whose probability distribution favors points in the local neighborhood. The local neighbor-hood can be defined here via the selection of the standard deviation of a corresponding distribution function.

It goes without saying that the probability profile of mutation can as well be applied to nonlinear problems with real variables.

4th Thesis
LEARNING BY TRIAL AND ERROR

The evolutionary strategy represents a counter-balance to the deficiency of being forced to do without an extensive problem analysis by using the capability of learning. With the mere trial as to how many and which mutations were successful or ineffi- cient respectively within a step sequence, adaptation rules can be established which are able to adapt any control parameter of the step sequence to the problem-specific order structures.

The simple **mutation-selection strategy** incorporates quite an old learning mechanism. Here, the progress of the step sequence is based on the successes of past mutations. Once achieved, no successful mutation is lost. It is "forgotten" within the step sequence only when it can be replaced by a better mutation. **Here, learning means: maintaining of advantages obtained.**

Quite contrary to this is the **Monte-Carlo method.** Here, each step is considered without reference to any preceding step. Basically, all steps could occur simultaneously. With this method, learning by trial is impossible.

ADAPTIVE STEP SIZE CONTROL

With many problems, it is useful to keep the step sizes of a mutation variable within the step sequence. At the beginning, in most cases large step sizes are recommended in order to reach near-optimum in a speedy way, whilst close to an optimum, smaller step sizes are adequate in order not to exceed the optimum too often.

But how does the strategy know which step sequence is the right one and when to change it? For this purpose, we have **adaptation rules**, which - in the case of a non-linear and discrete problem - could read as follows:

- If the hit rate in 50 mutations is lower than 2%, reduce the standard deviation by 50%.

- If the hit rate in 50 mutations is over 6%, increase the standard deviation by 100%.

These or similar adaptation rules make it possible that the dimensions of all control parameters adjust to each local order structure.

It is, however, not possible to determine in advance the optimal dimension selection for the adaptation rules in case of every problem. These can best be found out through empirical test series or even through a built-in adaptation of the control parameters for adaptation rules.

5th Thesis:
SYSTEM OF MODULES

The number of evolutionary strategy methods is an open set. They can be conceived as a collection of process modules, from which those modules are combined which support the solving of the specific problem via an interaction of chance and necessity.

Accordingly, the quantity of currently applied evolutionary process modules is rather incomplete and requires continuous updating.

Examples of modules of controlled evolutionary strategies:

a)   BASIC MODULES

are the principles of mutation and selection with manifold variation possibilities:

*   **Mutation weighting:**

    Varying distribution of the stochastic process to enhance the chances for successful mutation on the basis of feasible local order structures (limiting random processes).

*   **Selection rules:**

    -   Under the **strict** selection rule, mutation is not considered successful, unless it results in a feasible improvement in quality.

    -   The **neutral** selection rule also allows a step sequence with feasible mutations that neither improve nor deteriorate the quality.

    -   The **soft** selection rule allows to continue the sequence of generations also with mutations that, though resulting in a deterioration of quality, do not exceed a given band width relative to the former best value.

    The two latter rules in particular enhance the chances for further developing the step sequence near problematic fringe areas or points of discontinuity.

b)   **ADAPTATION RULES**

to adapt the control parameter of the process modules to the local order structures of the respective step sequence, the hit rate of past mutations serving as an orientation.

- **Step size-based adaptation:**

  Adaptation to step sizes offering better chances of success; in most cases the step size is reduced, the higher the number of ineffective mutations becomes.

- **Step direction-based adaptation:**

  Preference is given to those directions of mutation or to the selection of those variables that offer better chances.

- **Selection width-based adaptation:**

  Fixing of the selection width to support further development of the step sequence by the appropriate selection width.

c)   **PARTIAL ANALYSES,**

which suggest step sizes and/or step directions the probability of success of a mutation can be increased in order to weight the random process on the basis of these criteria.

d)   Modules for the fast localization of suboptima can be integrated in **OVERRIDING STRUCTURES** which aim at finding the global optimum.

- **Population concept,**

  where the sequence of generations consisting of several "parents" and "daughters" basically develops "simultaneously" in different local points.

- **Recombination concept,**

  which provides for the exchange of information between the elements of a multi-membered evolutionary strategy (e. g. crossing over).

- **Fluctuation concept,**

  which opens up new chances in the controlled sequence of stabilization and destabilization phases (see 6th thesis).

Although each evolutionary module has a relatively simple structure, their combination may introduce a hardly predictable behavior as far as the step sequence of the evolutionary strategy is concerned. This behavior can range from

- the step sequence **"getting stuck"** in a local area far away from the global optimum and

- via some kind of **fluctuating step development** which changes its neighborhood from time to time

- to an absolutely **erratic development** of the step sequence (random walk).

6th Thesis
CHANCES ARISING FROM FLUCTUATION

Fluctuations within the step sequences contribute considerably to the fact that new possibilities of development chances open up for an evolutionary strategy.

Therefore, it is important to determine the dynamics inherent in the process of the step sequence so that phases of long stability are followed by a short destabilized step sequence - involving a loss in quality - in order to be able to then restabilize itself in other local areas.

Example: STABILIZATION UND DESTABILIZATION PHASES

Apart from the basic fluctuation, i.e. the mutation as "basic noise" of the step sequence, **superim-posed patterns** can be designed **from fluctuations.**

For instance in the **change of the target orientation** of the step sequence. Long phases of an orientation towards an optimum - **stabilization phases** - can be interrupted by short-term instabilities - **destabilization phases**.

Modules of the destabilization phase:

- adaptation rules for starting and ending the instable phase

- larger step size

- adaptively controlled selection widths

- prohibition rules for mutations in order to prevent a return to the preceding stage

- tunneling of infeasible areas

- minimization according to the number of constraints in order to reach the feasible area.

FLUCTUATING EVOLUTIONARY STRATEGY

A typical characteristic of a step sequence containing **stabilization and destabilization phases** is illustrated by the convergence development of a fluctuating evolutionary strategy for the traveling salesman problem:

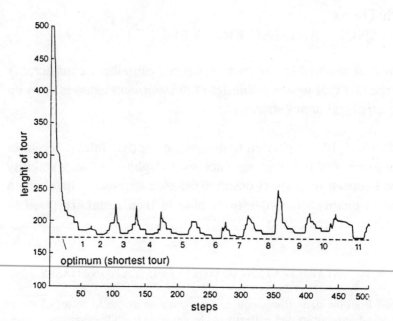

Any tour may be the start. Within few steps only the stabilization phase succeeds in approaching the optimum. After a given number of ineffective mutations the destabilization phase begins which evaluates poor-quality mutations as successful, as long as they are within a certain (adaptive) band width. In the example above the global optimum is reached in the sixth stabilization phase.

Proceeding like this, the exact solution to a 120 town-problem was found within a calculation time of one minute in the presence of ca. 6×10^{196} possibilities.

7th THESIS
DESIGN ACCORDING TO EVOLUTIONARY PATTERNS

The most promising method of designing efficient evolutionary strategies for the purpose of solving complex problems is the application of evolutionary procedures themselves. Instead of employing extensive mathematical analyses to evaluate the various strategy modules regarding their contribution to convergence, the advantages and disadvantages of single modules are relatively easy to establish by way of empirical induction based on the statistical evaluation of test series on their convergence development.

If evolutionary strategies are composed of a variety of random-based components that are moreover controlled by different adaptation processes, mathematical tools of analysis are likely to be inadequate. But comparatively little effort is required to prepare an average convergence characteristic with high informative value from a variety of empirical test series. For the illustrated case - a symmetrical travelling salesman problem touching 25 towns - the convergence characteristics were averaged from 50 test runs for different evolutionary strategies.

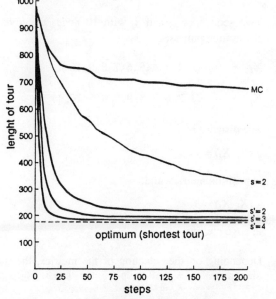

Weighted mutation-selection strategies exhibiting step sizes of s'= 2, 3 and 4 edge mutations show the best convergence (due to the symmetrical metrics also 2 edge mutations are permitted, since direction reversals of the edges are not evaluated). Clearly inferior is the convergence behaviour of the unweighted strategies. Only the best of all with step sizes of s=2 is illustrated above. The convergence behavior achieved with the Monte-Carlo method (MC) shows the poorest results.

Since in most cases a problem can be described with several models, which in addition can have different metrics, application of an evolutionary strategy requires identification of the model, the metrics of which suggests optimal convergence behavior.

8th THESIS
PROBLEM-ORIENTED METRICS

In order to solve a specific problem with evolutionary strategies, the best-suited metrics are those which during empirical test runs shows the highest unweighted mutation-selection strategy-based convergence at given step sizes.

Here, for any metrics, step sizes out of the spectrum of possible distances are to be selected.

Example: **SEQUENCE PROBLEMS**

These problems can be described by a decision tree model, as an 0 - 1 optimization problem, or by models the metrics of which are based on the number of edge or position differences.

Two sequences x_i and x_j with 10 nodes are given below for illustration purposes:

X_i =   (1, 2 ,3 ,4 ,5 ,6 ,7 ,8 ,9 ,10 ,1)

X_j =   (1, 2, 6 ,7, 8, 9, 3, 4, 5, 10, 1)

The distance is

$p(X_i, X_j) = 7$

in position metrics, and

$k(X_i, X_j) = 3$

in edge metrics.

Depending on the selection of the metrics, the mutation step size from X_i to X_j will be $p = 7$ or $k = 3$.

There is, however, a great difference as to the probability of exactly hitting X_j in one single mutation step.

With $p = 7$, the chances are 1 : 66744,
with $k = 3$ the chances are 1 : 120.

In the illustrated case - an asymmetrical touring problem touching 25 towns - the convergence characteristics were averaged from 50 test runs for different mutation-selection strategies.

PROBLEM-ORIENTED METRICS

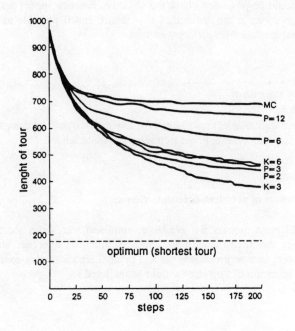

When comparing the convergence characteristics of the edge metrics with those of the position metrics, it can clearly be seen that the edge-metrics (k) are superior with comparable distance values.

The strategy with k=3 mutation step sizes shows the best convergence behavior, while the Monte-Carlo method (MC) is at the bottom of the league.

III. GENERAL DESIGN PRINCIPLES

1. Generation of models

In general, there are several options how to describe a real problem by formulating models.

It should be possible to have the objective function model defined without reservations at the desired level of detail, and if possible to extend their variable space over different metrics.

2. Base algorithm

The mutation-selection strategy for any metrics shall be designed such that the step sequence at given step sizes can materialize.

3. Selection of problem-oriented metrics

If different metrics are available, empirical test series shall be used to select the metrics with the highest mutation-selection strategy-based convergence at given step sizes. (Typical representative examples out of the spectrum of step sizes should be included.)

4. Strategy modules

The base algorithm with the selected metrics should be complemented by those strategy modules which increase both convergence speed and reliability.

 The selection is again based on empirical test series, which evaluate the convergence development of the mutation-selection strategies complemented by the respective modules, and which can also be used to "adjust the control parameter" of active modules.

All modules that further limit random trials and thus facilitate the search for the optimum, are suited to **increase convergence speed**. They include

- **special types of mutation weighting** which create specific preferences with regard to the probability distribution of the random process.

- **adaptation rules** that quickly adapt the step sequence to local order structures.

- **partial analyses** providing information about step sizes and directions which will increase the probability of a successful mutation.

Regarding **improvement of the convergence reliability**, there are modules which are to assure that the step sequence is not delayed or stopped by suboptima. They include:

- the **population concept**, in which the step sequence practically develops over several points in different local areas simultaneously (a lot of parent vectors with multiple daughter vectors).

- the **recombination concept** permitting forms of information exchange (e. g. crossing over) among the elements of a population which is subsequently evaluated in order to facilitate the search for the optimum.

- the **fluctuation concept**, where stabilization and destabilization phases confer process-like dynamics to the step sequence which facilitate the spotting of all promising areas.

5.   Evaluation

The type of evolutionary strategy, the harmonized modules of which are closest to the desired criteria of convergence speed and reliability, shall be applied to solve the initial problem.

IV. THESES REGARDING THE SELF- ORGANIZA-
TION OF EVOLUTIONARY APPROACHES

Since, as shown above, efficient evolutionary strategies can be designed on the basis of a module system supported by empirical test series without having to employ problem-specific mathematical analyses, it is a logical consequence to incorporate this procedure in a computer program.

9th THESIS
CONTEST OF STRATEGIES

It should be possible to have a program organized so as to provide for calculus-type access to a variety of metrics and evolutionary modules. In such a program evolutionary strategies are compiled, the convergence quality of which can be evaluated for any optimization problem input by means of program-inherent test series. Without a mathematical analysis of the input optimization problem, strategy contests between the individual varied strategy designs following the pattern of evolution, are to take place which create progressively improving convergence properties as a result of mutation and selection pressure. The winner of this meta-evolution from evolutionary strategies would finally be used for optimization of the initial problem.

If one day such a program materialized, its efficiency would above all be determined by the volume of predefined metrics and strategy modules. If moreover the self-organization of problem-solving were to be excluded from these restrictions, the thesis below would have to be feasible in practice:

10th THESIS
EVOLUTION OF EVOLUTIONARY RULES

A program which organizes strategy contests on the basis of available metrics and evolutionary principles should also provide the possibility of subjecting, in turn, the elements resulting from the collection of procedures to evolutionary rules, i.e. generate variations of the metrics and evolutionary rules and make a selection based on their contribution to problem-solving.

Such self-organizing programs could well be designated as "creative", since they would be able to find the solution to a so far unknown optimization problem on their own.

V. CASE EXAMPLE

When planning to apply evolutionary strategies for solving concrete practical problems, one will realize that the input for designing a model of the objective function model including all its restrictions will be much higher than the input required for tests with evolutionary modules.

The following case example particularly illustrates the complexity of the model if all relevant influencing factors occuring in practice are taken into consideration. The example refers to a project which is currently carried out in cooperation with *The Boston Consulting Group* for a client in the automobile industry.

Background:

An engine plant has a variety of manufacturing lines for engine parts. The target is to provide the required quantities of engine variants at the **required deadlines and the lowest possible costs.**

The most important **restrictions** to be considered with regard to availability of the **production capacity alone**, are

- The cycle time of each assembly station of the line depends on the version of the respective engine.

- The availability of the line is to be determined months in advance by taking into account when and exactly for which shift the individual assembly station will be ready (each assembly station may have its own shift schedule).

- Buffer possibilities between the assembly stations.

- Retooling times as a function of the sequence of engine versions to be assembled.

- Rejects as a function of the sequence of engine versions to be assembled.

- Consideration of idle times during retooling and assembly.

- Channeling pre-fabricated parts to any assembly station in the line.

Disposition tasks are rendered even more difficult by the fact that exact requirements are not available but until up to two weeks in advance and that within the weekly production planning which is subject to steady changes all further requirement data may change.

Given a restrictive stock policy, disposition personnel is confronted with the dilemma of creating more retooling and idle times when poduction lots decrease, and consequently is not in a position to meet delivery deadlines.

SOLUTION APPROACH

It is planned to provide the disposition personnel with a tool to facilitate their daily work, which presents optimized proposals for the planning of production lots.

The core of this model is an objective function model which records all relevant influencing factors of reality with regard to exact times, quantities and costs and which supports optimization by applying evolutionary strategies. The optimization targets may be selected freely.

In principle, all cost variants which can be influenced by disposition can be included in the objective function, e. g. cost of inventory, variable labor cost from retooling, cost of rejects and tool wear, transport and storage, as well as cost of machinery-related idle times from retooling, malfunction, or nonproductive times, or even extra cost from night shifts. First test runs considering all relevant cost variables have suggested a savings potential of several DM 100,000 p.a. and line.

VI. RESUME

Two issues highlight the differences between this kind of strategy design and other evolutionary approaches:

1) The evolutionary processes of nature have model character only. It is **not aimed at exactly simulating nature's mechanisms** with the computer (cf. genetic algorithms). Any attempt at exactly simulating natural processes with the computer would require too much effort in terms of calculating and thus not be in line with our target.

   It is rather aimed at transferring nature's evolutionary mechanisms to a computer program so that they produce clear advantages in terms of efficiency. I. e., selection is done under very pragmatic aspects, and if nature does not provide any patterns for the specific problem, the designer is called upon to create new evolutionary rules.

2) The design of controlled evolutionary strategies is **not limited by the level of complexity** which still allows mathematical analysis of the strategy's convergence quality.

   The module approach rather allows the design of strategies - based on empirical tests - wich employ 5 and more adaption processes at one time and which are surely too complex to be analysed with mathematical tools.

There is reason to believe that particularly the complex optimization problems occuring in practice can be approached with controlled evolutionary strategies.

Author:

Dr.rer.pol. Dipl.-Phys. Paul Ablay
ABLAY OPTIMIERUNG
Römerschanzweg 5

Tel. 089/8500666

D - 8035 München - Gauting   Fax 089/8500680

Some exhibits are produced from "Spektrum der Wissenschaft" by kind permission of the publishers of Scientific American.

Literature:

ABLAY, PAUL: *Optimieren mit Evolutionsstrategien.* Spektrum der Wissenschaft, Juli 1987, S. 104-115.

ABLAY, PAUL: *Konstruktion kontrollierter Evolutionsstrategien zur Lösung schwieriger Optimierungsprobleme der Wirtschaft.* In: Evolution und Evolutionsstrategien in Biologie, Technik und Gesellschaft. Hrsg. Jörg Albertz, Freie Akademie, Wiesbaden, 1989.

DAWKINS, RICHARD: *Der blinde Uhrmacher (The blind watchmaker).* Kindler Verlag, München 1987.

GOLDBERG, DAVID E.:   *Genetic Algorithms.* Addison-Wesley, 1989.

HOFSTADTER, DOUGLAS: *Gödel, Escher, Bach: an Eternal Golden Braid.* Basic Books, New York, 1979.

RECHENBERG, INGO: *Evolutionstrategie - Optimierung technischer Systeme nach Prinzipien der biologischen Evolution.* Frommann-Holzboog, Stuttgart, 1973.

RIEDL, RUPERT: *Die Strategie der Genesis.* Piper Verlag, München, 1986.

SCHWEFEL, HANS-PAUL: *Numerische Optimierung von Computer-Modellen mittels der Evolutionsstrategie.* Birkhäuser Verlag, Stuttgart, 1977.

INCOMMENSURABILITY OF LIOUVILLEAN DYNAMICS AND INFORMATION DYNAMICS

H. ATMANSPACHER

Max-Planck-Institut für Physik und Astrophysik

D-8046 Garching, FRG

1. Introduction

The notion of evolution, although originating in and mostly referred to biological issues, is highly significant for any processual world view. Thus the concept of a temporal evolution of variables of a system should be considered of very general relevance. As a consequence, a formal approach would be most desirable which leaves a high degree of freedom regarding the interpretation of the variables as well as the interactions between them. The present contribution aims toward the development of such a general evolutionary theory based on physical concepts.

In the framework of physics, there are two basically different types of evolution which might be classified into discontinuous and continuous ones. By a discontinuous evolution we think of phenomena such as thermodynamical phase transitions, bifurcations, and instabilities far from thermal equilibrium. The "synergetic" approach of Haken mainly concerns these fields. A huge number of examples has become popular. Just to give an arbitrary selection (ordered by decreasing common experience) we mention freezing water as a typical phase transition phenomenon, the formation of collective action as an instability in laser systems, and even such fundamental problems as that of quantum mechanical measurement.

Continuous ("smooth") processes constitute the main body of evolving physical systems in the sense of classical mechanics. During the last decade, a very general class of processes has gained a lot of interest in this field. They provide an irregular temporal evolution of their variables although being described by deterministic equations (hence the catchword "deterministic chaos"). Historically, the first studies of such processes date back to Poincare.

The present article is devoted to various properties related to this class of smooth evolutionary processes. The reason for this interest originates in the fact that such processes can be shown to introduce a temporally directed flow of information, a crucial criterion for evolution in general. This point is obviously related to one of the main unresolved mysteries of contemporary sciences: the problem of irreversibility. The formalization of a time – directed information flow is thus extremely important in this context.

An additional paradigm of evolution is that of increasing complexity. From a physical and mathematical viewpoint, the number of approaches toward complexity increases self – similarly. In order to pay an artistic tribute to biological complexity as a specific phenomenon of a general evolutionary theory we refer to Escher's inspiring image "Verbum" reproduced in Fig.1.

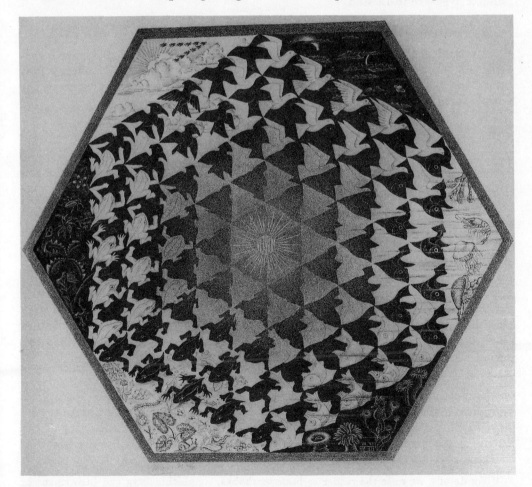

Figure 1: M.C.Escher's lithography "Verbum" as a famous artistic creation on the paradigm of evolution (© 1989 M.C.Escher Heirs / Cordon Art – Baarn – Holland).

The central objective of this article is the derivation and discussion of the differences between two different dynamical concepts used to describe evolution in the sense of evolving variables of a system. Both types of dynamics are defined based on the stability properties of dynamical systems (Sec.2), and on resulting considerations on information flow (Sec.3). In Sec.4 we show that there is a fundamental discrepancy between both dynamical concepts. Contrary to Liouvillean dynamics, the concept of information dynamics will turn out to be suitable for the description of a directed temporal evolution. The formal structure of this discrepancy is that of an incommensurability.

The remaining sections essentially represent an attempt to interpret this incommensurability by a comparison with quantum mechanical incommensurable quantities. A propositional calculus with respect to both types of incommensurability is suggested (Sec.5) and investigated using tools of lattice theory (Sec.6). Eventually the main results are summarized and possible perspectives are indicated.

2. Stability of dynamical systems[1]

Dynamical systems as they occur in nature are generally dissipative systems, i.e. they are not closed and there is a non — vanishing net energy flow out of the system. This means that the temporal evolution of a system of the simple form

$$\dot{\mathbf{x}} = \mathbf{F}(\mathbf{x}, \eta) \tag{1}$$

provides a shrinking phase space volume,

$$\text{div } \mathbf{F} < 0 \tag{2}$$

whereas the volume would be preserved for Hamiltonian systems. The vector $\mathbf{x}$ in (1) characterizes the variables $(x_1, ..., x_n)$ of the system, and $\mathbf{F}$ is a matrix describing the generally nonlinear coupling among the variables. The number of variables defines the dimension n of the phase space. Note that this definition is not restricted to the Hamiltonian formalism of canonically conjugate variables for each particle within a system. The dimension n may thus be small if a many particle system is described by collective variables (e.g., temperature, pressure, etc.). The quantity η refers to a so–called external control parameter (in principle η is also conceivable as a function of $\mathbf{x}$).

Since the phase space volume of a solution of (1) decreases during its temporal evolution, it is quite suggestive that the motion of this solution (the flow F_t) in phase space is asymptotically restricted to a finite subspace of the entire phase space. This subspace will be called an attractor for the flow F_t.

---

[1]For details, we refer the reader to basic textbooks, e.g. Lichtenberg and Lieberman[1] or Guckenheimer and Holmes[2]. We also recommend the compact review article of Eckmann and Ruelle[3].

This illustrative concept can be obtained as a rigoros reoult from a lineai stability analysis of the flow in phase space. It is possible to define a set of n characteristic exponents λ_i for F_t determining the stability properties along the trajectory within the attractor on a temporal average. These exponents (also called the Ljapunov exponents) are derived from the eigenvalues $\Lambda_i(t)$ of a time dependent stability matrix $\mathbf{L}(t)$ whose eigenvectors $\delta\varsigma$ form a comoving coordinate system for F_t (see Fig.2):

$$\mathbf{L}(t)\delta\varsigma = \Lambda(t)\delta\varsigma \tag{3}$$

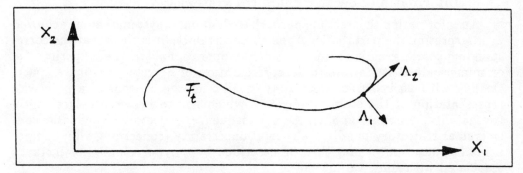

Figure 2: Schematic illustration of the eigenvalues Λ_i for a two dimensional flow F_t in phase space. Since Λ_i is defined locally, it is a time dependent quantity. In contrast, the Ljapunov exponents λ_i are given by the temporal average of Λ_i.

The Ljapunov exponents are obtained as the temporal average of Λ_i:

$$\lambda_i = \lim_{T\to\infty} \frac{1}{T} \int_0^T \Lambda_i(t)\,dt \tag{4}$$

The size of small perturbations $\delta\varsigma$ in the comoving frame thus shows an average evolution according to:

$$\delta\varsigma_i(t) = \delta\varsigma_i(0)\, e^{\lambda_i t} \tag{5}$$

Using the Ljapunov exponents, the dissipation condition (4) can be reformulated as

$$\sum_i \lambda_i < 0 \tag{6}$$

whereas this sum has to vanish for Hamiltonian systems. Moreover, the λ_i are a convenient tool for a classification of different types of attractors. Fixed points, corresponding to stationary solutions of (1) require that $\lambda_i < 0 \; \forall i$. For a limit cycle, corresponding to a periodic system, there is one vanishing exponent into the direction of the flow. This direction is locally accessible by a vanishing vector product $\delta\varsigma \times \dot{x}$. In order to guarantee the stability of the cycle, all the other exponents have to be negative. In case of a k − periodic system the attractor is the surface of a k − torus with k vanishing Ljapunov exponents. For systems with at least three degrees of freedom $n \geq 3$, condition (6) can be satisfied by a combination of positive and negative exponents. This situation defines a chaotic attractor producing chaotic motion due to the sensitive dependence of the evolution on small perturbations. The different types of attractors and some of their properties are summarized in Table 1 below. A random process is not confined onto an attractor, since (6) is not satisfied.

3. Information flow in dynamical systems[2]

An information theoretical approach to dynamical systems can be achieved by interpreting the pertubation $\delta\varsigma$ as an uncertainty. It is clear that the corresponding coarse graining is inherent in any operationalistic (i.e. experimental or numerical) attitude toward a representation of a system in phase space. Dealing with an intrinsic uncertainty invalidates the concept of a pointwise representation of the state of a system which has to become replaced by a (probability) distribution $\rho(x)$. As a consequence, trajectories have to be considered as trajectory bundles. The total uncertainty (concerning all n components of phase space) may then be described by a hypervolume (for instance associated with some variance of ρ)

$$\delta V = \prod_i \delta\varsigma_i \tag{7}$$

which evolves (on the average) according to:

$$\delta V(t) = \delta V(0) \, e^{\sum \lambda_i t} \tag{8}$$

The size of δV with respect to a coarse graining characterized by a resolution ϵ can now be used to define an information measure

$$I = \log_2 \frac{\delta V}{\epsilon^n} \tag{9}$$

where $\delta V / \epsilon^n$ is simply the number of (uniformly distributed) hypercubes of ϵ^n required to cover δV. Equation (9) provides the Shannon information[7] which is obtained by specifying the state of a system within one of the hypercubes.

If one is interested in the temporal evolution of I in a coarse grained phase space, one has to consider each individual component $\delta\varsigma_i(t)$. Starting with $\delta\varsigma_i(0) = \epsilon$ (experimental or numerical resolution for initial conditions) it is easy to see that for all components with non−positive λ_i the decreasing

---

[2] The idea to ascribe an information flow to the evolution of dynamical systems has gained a lot of interest subsequent to the extensive articles of Shaw[4] and Farmer[5] in the early 80s. For more recent studies see the Ph.D. thesis of Fraser[6].

uncertainty will become irrelevant, since for these components $\delta\varsigma(t) \lesssim \epsilon$. For this reason only those components for which $\lambda_i > 0$ will contribute to a change in information.

Having this in mind, a suitable quantity to determine the rate of information change in the system is given by the sum of positive Ljapunov exponents:

$$K = \sum_i \begin{cases} \lambda_i, & \text{if } \lambda_i > 0 \\ 0, & \text{otherwise} \end{cases} \tag{10}$$

This quantity is the famous *Kolmogorov* entropy[8] or *dynamical* entropy. The equality (10) has been found[9] to hold for a huge class of systems giving rise to a natural measure in phase space. Using only those components with positive λ_i, the informational content of the system is given by the subvolume

$$\delta V_+ = \prod_i \begin{cases} \lambda_i, & \text{if } \lambda_i > 0 \\ 1, & \text{otherwise} \end{cases} \tag{11}$$

with a temporal evolution

$$\delta V_+(t) = \delta V_+(0) \, e^{Kt} \tag{12}$$

Analogous to (9) an information measure refering to δV_+ can be defined by

$$I = \log_2 \frac{\delta V_+}{\epsilon^{n_+}} \tag{13}$$

where n_+ is the number of positive Ljapunov exponents. The information I will then increase according to:

$$I(t) = \log_2 \left(\frac{\delta V_+(0)}{\epsilon^{n_+}} \, e^{Kt} \right) \tag{14}$$

$$= I(0) + \frac{1}{\ln 2} Kt \tag{15}$$

For $\delta V_+(0) = \epsilon^{n_+}$, one has $I(0) = 0$. In addition, it should be noted that the factor $\frac{1}{\ln 2}$ would disappear if a thermodynamical terminology with natural logarithms were used. We shall drop this factor for our subsequent argumentation. With respect to the meaning of of Eq.(15) two further major remarks are in order concerning the linearity and the sign of $I(t)$.

The positive sign of $I(t)$ originates in the fact that the intrinsic "instability" of the system due to positive Ljapunov exponents enhances initial uncertainties. From an *internal* point of view, this is equivalent with an increasing amount of information within the system. From an *external* (observer) point of view, the information about the actual state of the system decreases to the same degree. It is of general importance to distinguish between these two viewpoints. Information in the first sense (on the level of the system) will therefore be called *potential information*. It can be gained by an observer who then obtains *actual information* about the state of the system by measurement. As a fundamental idea behind this two–level description one might see an – informal – principle of information conservation. A more detailed elaboration on these matters in the context of elementary information transfer between two levels has been given elsewhere.[10]

The linearity of $I(t)$ in Eq.(15) is a consequence of the linear stability analysis providing the Ljapunov exponents. It is thus clear that it is a first order approximation, only valid for small $t > 0$. Although K is a *global* invariant of the system in the sense of being a temporal average, this does not at all imply an overall linear information increase. On the contrary, it is well known that the information flow is generally nonlinear. The dynamical entropy K is to be considered as an averaged *local* rate of information production in the system. Its inverse, $1/K$, measures the time interval for which the temporal evolution of the system can reasonably well be predicted. Since $K > 0$ is a sufficient and necessary condition for deterministic chaos (see Table 1), the temporal predicition of chaotic systems is generally limited. As mentioned above, this limitation occurs in a completely deterministic manner due to an intrinsic instability acting on the coarse graining of phase space.

There are a lot of examples showing that the structure of chaotic attractors may be highly complicated. In many cases this complexity is of a type known from fractal objects (see Mandelbrot[11]). Fractal objects can be characterized by a dimension which is not restricted to integer numbers. The first definition of such a dimension dates back to Hausdorff[12] in 1919. Within our information theoretical approach, the information dimension D describes how the information I_ϵ about the state of the system (its location in phase space) scales with varying spatial resolution ϵ. The dimension D is defined by

$$D = \lim_{\epsilon \to 0} \frac{\log I_\epsilon}{\log 1/\epsilon} \tag{16}$$

thus assuming a power law scaling of Shannon's information I_ϵ as an inherent property of fractal objects. In the last column of Table 1, the dimension D is indicated for the typical attractors discussed above.

It is of utmost importance that algorithms exist which allow for a determination of D and K for experimental situations. Based on work of Takens[13] and Packard *et al.*[14] different procedures have been proposed[15] to estimate these quantities. For a summary of details and recent developments we refer the reader to the proceedings of the 1985 workshop on *Dimensions and Entropies in Chaotic Systems*[16]. Early examples of applications cover areas from hydrodynamic[17] and acoustic[18] turbulence over neurophysiology[19] and climatology[20] to laser dynamics[21] and astrophysics[22].

	Ljapunov spectrum	K entropy	dimension
fixed point	$\lambda_i < 0\ \forall i$	$K = 0$	$D = 0$
limit cycle	$\lambda_1 = 0;\ \lambda_{i>1} < 0$	$K = 0$	$D = 1$
k - torus	$\lambda_1 = ... = \lambda_k = 0;\ \lambda_{i>k} < 0$	$K = 0$	$D = k \in N$
chaotic attractor	$\exists j : \lambda_j > 0;\ \sum \lambda_i < 0$	$0 < K < \infty$	$D \in R, D < n$

Table 1: Specific types of attractors, properties of their Ljapunov spectrum, the dynamical entropy K, and the dimension D of the attractor as a characterization of its geometrical properties.

4. Liouvillean dynamics versus information dynamics[3]

At the beginning of this section it is helpful to reconsider two central pairs of terms used to characterize dynamical systems. Adopting a view with a strong emphasis on realistic systems, dissipation is regarded as a mechanism of very general relevance. In contrast, conservative (Hamiltonian) systems represent a limiting situation characterized by $\mathrm{div}F = 0$ (cf.Eq.(2)) which is equivalent to Liouville's theorem of the invariance of ρ. (Note that this invariance does not refer to any coarse graining.)

In addition to the "dichotomy" of conservative and dissipative systems one has to distinguish regular $(K = 0)$ and chaotic $(K > 0)$ systems. Both types of behavior are possible in Hamiltonian as well as dissipative systems, since a positive Ljapunov exponent can occur for $\sum \lambda_i = 0$ as well as for $\sum \lambda_i < 0$.

Hamiltonian and dissipative systems essentially differ by the fact that the canonical formalism to describe the temporal evolution of Hamiltonian systems has no rigorous counterpart for dissipative systems.For Hamiltonian systems the temporal evolution of ρ can be described by the Liouville equation

$$i\partial_t \rho = L\,\rho \tag{17}$$

where L is the Liouville operator acting on ρ and $i = \sqrt{-1}$. The concept of an evolution operator acting on distribution functions in phase space has been introduced and extensively applied by I.Prigogine[24]. The solution of Eq.(17) is given by:

$$\rho(t) = e^{-iLt}\rho(0) \tag{18}$$

It represents a unitary evolution based on the unitary operator $U_t = e^{-iLt}$. The corresponding unitary dynamical group is said to be generated by L. The

[3]The present section strongly refers to an essential part of an earlier article[23] on information flow in dynamical systems.

existence of an inverse element in this group ensures its invariance under time reversal. A temporal evolution of ρ as given by L is temporally reversible. The dynamics determined by Eq.(18) will be called *Liouvillean dynamics* in the present article.

Apart from the dynamical formulation based on L we now consider the information flow as defined in Eq.(15). This expression is valid for Hamiltonian and dissipative systems, since it merely contains the dynamical entropy K as an independent quantity. (It does not include the Hamiltonian of the system as L does.) Defining an information operator M with eigenvalues $I(t)$ according to Eq.(15) provides the eigenvalue equation:

$$M\rho = I(t)\rho = (I(0) + Kt)\rho \qquad (19)$$

The operator M is time independent. It acts on a time dependent distribution ρ, yielding a temporally increasing eigenvalue $I(t)$ if $K > 0$. (A more detailed discussion of the eigenvalues of information has been given elsewhere[23].) In contrast to L, the information operator M introduces the notion of irreversible evolution in the sense that M provides a semigroup representation (lacking inverse element) whenever $K > 0$. The dynamics according to Eq.(19) will subsequently be denoted as *information dynamics*.

The operators L and M constitute two dynamical formalisms acting on distribution functions in phase space. It has been shown[23] that the commutator of these operators is just given by the Kolmogorov entropy K:

$$i[L, M] = K\mathbf{I} \qquad (20)$$

($\mathbf{I}$ is the identity operator.) This incommensurability of Liouvillean dynamics and information dynamics reflects the fact that the information theoretical description of a system (M) and its description by means of its Liouvillean (L) are not equivalent except for $K = 0$. Predictions based on the evolution operators L and M are not commensurable. This result may be interpreted as a consequence of the increasing uncertainty in predicting the state of a system as time proceeds. Whenever $K > 0$ the state $\rho(t_o)$ of a system cannot be predicted as accurate as initial conditions have been measured (or otherwise fixed) at $t < t_o$.

At this point it would be a serious lapse not to mention the most important work of I.Prigogine and his Brussels school on the same subject[25]. Starting from statistical mechanics and thermodynamical arguments, their purely theoretical derivation provided a commutation relation between the Liouville operator and an entropy operator (instead of our information operator). The approach described here differs from Prigogine's in its degree of operationalism. It provides the experimentally available quantity K as the commutator of L and M which can be directly identified for concrete systems (see the concluding remarks of Sec.3 and Refs.21 for experimental examples).

From a rigorous viewpoint, the relevance of the incommensurability of L and M is restricted to conservative systems (Refs.21 concern dissipative laser systems). This means that the evolution operators L and M are "equivalent only for conservative systems with $K = 0$. A system with positive dynamical entropy requires an evolutionary semigroup constituted by M in order to be appropriately described. The transfer of these concepts to dissipative systems is nontrivial, since L is rigorously defined only for vanishing dissipation.[4] An intuitive motivation for such a transfer might be possible by the "smoothly" varying phase space representation of systems with increasing (weak) dissipation[27].

Whatever rigorous results concerning this issue might turn out in the future, the information dynamical formalism applies to conservative and dissipative systems with $K \geq 0$. In this sense it is superior to the Liouvillean dynamical formalism. With a view toward evolution in a broader context than that of physical systems, information dynamics is able to account for temporally directed information flow, an essential feature of dynamical systems as they occur in nature.

5. Incommensurability of facts and of theories

The preceding interpretational remarks on the incommensurability of Liouvillean and information dynamics did already offer a picture to imagine a reason for the non–vanishing commutator. However, the fundamental meaning of the incommensurability as such has not yet been touched. The present and the following sections represent an attempt to inquire into this issue.

The formal structure of the commutation relation

$$i[L, M] = K\mathbf{I} \tag{20}$$

is isomorphic to the quantum theoretical commutation relation between momentum and position:

$$i[P, Q] = h\mathbf{I} \tag{21}$$

This correspondence will be used as the interpretational "core" of the matters to be discussed in the following. Before we start, let us still introduce some linguistic details which will be required. Concerning the commutation relations (20) and (21), we shall subsequently speak of incommensurability of propositions, not of quantities. This terminology clearly aims toward an investigation of the logical structure of the propositional calculus related to (20) and (21). A formalization of this logical structure, i.e. its lattice theoretical formulation, will be given in Sec.6.

---

[4] Note that the Brussels school[24,26] uses a "dissipativity" condition based on a phase space behavior more complicated than phase mixing (see Sec.14.1 in Ref.24 and the remark on p.421 in Ref.26). This is different from the condition $\text{div} F < 0$, distinguishing conservative and dissipative systems in our sense. Probably the Brussels condition of dissipativity comes close to our condition $K > 0$.

Talking about propositions implies that (20) and (21) can be unified in a certain sense: both relations express incommensurabilities of propositions. While (21) characterizes propositions concerning *facts* (as ontological bases for measurement), relation (20) characterizes propositions concerning *theories* (as ontological bases for prediction).[5] Both types of propositional objects are formalized by the eigenvalues of the corresponding operators.

It is now important to realize that facts are always defined within the framework of a theory. Thus there is a hierarchy involved in both types of incommensurability. Nevertheless, the relations between both pairs of incommensurable propositions are quite sophisticated. The incommensurability concerning facts can be considered as the syntax of a theory. On the other hand the incommensurability of theories bears some relevance concerning the semantics of facts. Moreover, the aspects of measurement and prediction express a pragmatic component with respect to both, facts and theories.[6] Due to their propositional formulation both types of incommensurability with their mutual relationships can, however, be formalized in a unified manner. The logical structure of this unified description (the corresponding lattice) may be viewed as part of a syntax of a metatheory of facts and theories.

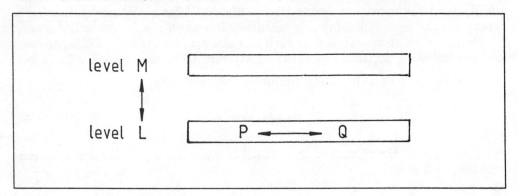

Figure 3: Illustration of the pairwise dichotomy of facts ($P \leftrightarrow Q$) and theories ($M \leftrightarrow L$). See text for possible incommensurable facts on the level of M.

The pairwise dichotomy of facts and theories according to (20) and (21) is visualized in Fig.3. The incommensurability concerning P and Q is of course defined on the semantic level of L, since L is a differential operator based onto partial derivatives of the Hamiltonian with respect to momentum and

[5]In order to indicate a tentative relation to modern developments in the philosophy of science[28] we point out that facts in this sense can be viewed as being related to Elkana's *corpus of knowledge*, while theories would correspond to his notion of *conceptions of knowledge*.

[6] For those readers with some semiotic background it will be easy to imagine that further detailed studies of these relationships can be expected to be highly interesting and relevant. More information on language and reality in modern physics is available in a recently published text of P.Mittelstaedt[29].

position. The semantic level of M is superior to that of L: M accounts for systems with non–vanishing information production $K > 0$ in addition to $K = 0$.

It might be interesting to speculate about the non–commuting operators on the level of M. In this respect one could think of an information dynamical extension of Prigogine's approach of a dynamics of correlations[24]. The corresponding operators accounting for creation and destruction of correlations would then have to be modified to related information source and information sink operators.

In the same context, the concept of subdynamics developed by the Brussels school (see Secs.14-17 in Ref.26) is probably relevant. Since relation (21) means that no common eigendistributions exist for L and M if $K > 0$, one could suppose that L and M refer to eigendistributions which become different for $t > 0$ (Fig.16.2.1 in Ref.26 might indicate a way of formalization).

6. Lattice theoretical investigation

We are now going to study the logical structure of the propositional calculus which is related to relations (20) and (21), i.e. to the incommensurability of propositions concerning both facts and theories. It has become common practice to investigate such a logical structure in the framework of lattice theory. Readers interested in lattice theory as a mathematical field should consult the extensive monograph of G.Birkhoff[30]. A very compact overview of important definitions and theorems is listed in the appendix of M.Jammer's book[31]. In the present article we content ourselves with those details relevant for our purposes.

In our context a lattice $V(G, \geq, \leq)$ consists of a set $G = \{a, b, c, ...\}$ of propositions for which partial ordering relations are defined by the symbols $\geq$ and $\leq$. With respect to propositions these partial ordering relations represent implications $\rightarrow$ and $\leftarrow$, and = corresponds to $\leftrightarrow$. In addition the operations meet ($\wedge$) and join ($\vee$) are defined in G.

A partially ordered set constitutes a lattice if the following algebraic properties are satisfied:

Idempotency :	$a \wedge a = a$	(22)
Commutativity:	$a \wedge b = b \wedge a$	(23)
Associativity:	$a \wedge (b \wedge c) = (a \wedge b) \wedge c$	(24)
Absorption:	$a \wedge (a \vee b) = a \vee (a \wedge b) = a$	(25)

together with the corresponding dual properties which are obtained by interchanging $\wedge$ and $\vee$.

A lattice V is *complemented* if for each a an element a' exists such that:

$$a \wedge a' = 0 \tag{26}$$

and (dually)

$$a \vee a' = I \tag{27}$$

Here I (unity) and 0 (zero) represent universal upper and lower bounds in G, respectively.

A lattice V is *distributive* with respect to the complement if for each pair $(a, b) \in V$ the relations

$$a = (a \wedge b) \vee (a \wedge b') \tag{28a}$$
$$\vee \quad b = (b \wedge a) \vee (b \wedge a') \tag{28b}$$

and (dually)

$$a = (a \vee b) \wedge (a \vee b') \tag{29a}$$
$$\wedge \quad b = (b \vee a) \wedge (b \vee a') \tag{29b}$$

are satisfied. If V is non–distributive (28) and (29) have still to satisfy the distributive inequalities, obtained by replacing $=$ by $\geq$ in (28) and replacing $=$ by $\leq$ in (29). The logical significance of a non–distributive lattice consists of the non–uniqueness of the complement (invalidity of the *tertium non datur*). In this case V as well as the logical structure of the propositional calculus are non–Boolean.

There are additional important properties of lattices (e.g. modularity, metricity, etc.) which will not be treated in this article. These properties might nevertheless be of particular interest concerning practical problems of constructing evolutionary models. For the following theoretical discussions we shall restrict ourselves to complementation and distributivity.

Let us now first consider propositions about incommensurable facts which are measurable as eigenvalues of non–commuting quantum mechanical operators P and Q. The corresponding lattice theoretical formulation relates to the issue of a quantum logic, a field of great interest and lively debate since the pioneering work of Birkhoff and von Neumann in 1936[32]. Section 8 in M. Jammer's book[31] presents an excellent review on quantum logic.

As an essential point Birkhoff and von Neumann have shown that a quantum mechanical propositional calculus is characterized by a complemented (more exact: orthocomplemented) non–distributive lattice. Denoting incommensurable propositions about facts (P, Q) as (a, b) the non–distributivity relations quoted in this context correspond to (28):

$$a > (a \wedge b) \vee (a \wedge b') \tag{30a}$$
$$\wedge \quad b > (b \wedge a) \vee (b \wedge a') \tag{30b}$$

These inequalities can be supplied with an interpretational meaning, namely a limitation of the simultaneous availability concerning propositions about P

and Q. Although the dual property of (30) is formally satisfied if (30) is satisfied, the quantum mechanical incommensurability concerning facts does not provide any interpretational clue to the dual of (30). It is the objective of the remaining part of this section to check whether an incommensurability of propositions about theories might provide an interpretational possibility corresponding to (29). As a specific example we intend to study the incommensurability of Liouvillean dynamics and information dynamics.

We consider propositions (a, b) as propositions concerning theories $(\mathcal{L}, \mathcal{M})$ related to L and M. Let $\mathcal{L}$ be characterized by a vanishing dynamical entropy,

$$\mathcal{L} := \{K | K = 0\} \tag{31}$$

whereas $\mathcal{M}$ is characterized by

$$\mathcal{M} := \{K | K \geq 0\} \tag{32}$$

In order to check the complementation and distributivity of the corresponding lattice we apply the affirmative propositional calculus introduced by P.Lorenzen[33] which has also been adopted in Mittelstaedt's discussions of quantum logic[29]. Affirmative propositions are object to a dialogical (proponent – opponent) demonstration of their validity.

For our theories $\mathcal{L}$ and $\mathcal{M}$ this means that the complement $\mathcal{L}'$ of $\mathcal{L}$ is easily characterized by

$$\mathcal{L}' := \{K | K \neq 0\} \tag{33}$$

However, the only way to find $\mathcal{M}'$ is to falsify a proposed $K = K_o$ for a specific system. If an opponent proves $K \neq K_o$, he has found $\mathcal{M}'$:

$$\mathcal{M}' := \{K | K \neq K_o\} \tag{34}$$

If an opponent found $K = K_o$, $\mathcal{M}$ would not be "verified" but corroborated. There would be infinitely many more cases to be checked in order to "prove" $\mathcal{M}$. This argument gives an idea of the impossibility of any positive verification.

In order to study the distributivity of the lattice, we analyze the distributivity relation dual to (30). (Remember that we hope to provide a meaningful interpretation of this relation for $\mathcal{L}$ and $\mathcal{M}$.) The algebraic term $(\mathcal{M} \vee \mathcal{L}) \wedge (\mathcal{M} \vee \mathcal{L}')$ is then characterized by:

$$\{K | K \geq 0 \vee K = 0\} \wedge \{K | K \geq 0 \vee K \neq 0\} = \{K | K \geq 0\} \tag{35a}$$

such that

$$(\mathcal{M} \vee \mathcal{L}) \wedge (\mathcal{M} \vee \mathcal{L}') = \mathcal{M} \tag{36a}$$

The second relation to be checked concerns $(\mathcal{L} \vee \mathcal{M}) \wedge (\mathcal{L} \vee \mathcal{M}')$ which provides:

$$\{K|K = 0 \vee K \geq 0\} \wedge \{K|K = 0 \vee K \neq K_o\} = \{K|K \geq 0, K \neq K_o\} \quad (35b)$$

such that

$$(\mathcal{L} \vee \mathcal{M}) \wedge (\mathcal{L} \vee \mathcal{M}') > \mathcal{L} \quad (36b)$$

Relation (36b) satisfies the distributive inequality. Since according to (29) distributivity would require the equality of both (36a) and (36b), the logical structure related to the incommensurability of propositions concerning theories is non–distributive, hence non–Boolean.

We have thus shown that the logic of incommensurable propositions concerning facts and theories may be interpreted as dual properties in a corresponding lattice theoretical formalism. This result implies two basic programmatic issues: 1. The incorporation of the evolutionary theories based on L and M into quantum theory. A series of papers by Prigogine and Petrofsky[34] is strongly relevant to this point. 2. Additional lattice theoretical properties should be checked for $\mathcal{L}$ and $\mathcal{M}$ and compared with corresponding properties concerning incommensurable facts.

From the viewpoint of the philosophy of science, the presented approach offers a way to introduce the idea of "two level thinking" formally into the conceptual structure of scientific method. For more details on this idea which is tightly related to relativistic realism as well as to constructivism we refer the reader to the text of Elkana[28] and to a collection of articles on the constructivist approach[35].

Last but not at all least the work of Krueger[36] on the aspect of evolution in physics and on the significance of evolutionary theory in general has to be mentioned. In fact the incommensurability of L and M has been found to represent an example for the conceptual structure proposed in Ref.35. This structure provides the "splitting" of Boolean truth values into the non – Boolean logic corresponding to the dual unity of theories and facts. It is shown in Fig.4 and suggests a unification of evolutionary logic (also of scientific discovery[37]) and of quantum logic[32].

7. Summary and perspectives

The main content of this article refers to the development of evolutionary theory based on physical concepts. In this context, two types of evolution operators, L and M, have been studied. Although the Liouville operator L is suitable for the description of an evolution in p, q – space, it misses an essential point of evolution: the directed flow of information. This point is introduced using an information operator M. The commutation relation between L and M has been derived, providing the information production rate K as the commutator. The dynamics according to L and M are incommensurable, if $K > 0$.

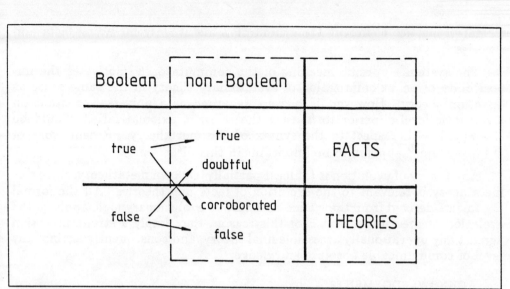

Figure 4: Schematic illustration of the splitted Boolean truth values into the mutually interlocked truth values "true" and "doubtful" (for facts) and "false" and "corroborated" (for theories) [36]. These modified truth values (*modalities*) correspond to the dual formulations of non – distributivity in Eqs.(30) (quantum logic) and (36) (evolutionary logic).

An investigation of common properties of the incommensurability of L and M and that of the quantum mechanical operators P and Q has been carried out. It provided a structural approach concerning the lattice theoretical equivalent of the propositional calculus concerning (P, Q) and (L, M). With regard to $\mathcal{L}$ and $\mathcal{M}$ as evolutionary theories related to L and M, a formal analogy between the logics of *facts* in quantum theory and of evolutionary *theories* has been found.

The following points are of particular interest:
Both lattices are complemented, but non – distributive, hence non – Boolean. Consequently, the complements are not unique. This non – uniqueness relates to the central paradigm of diversity in evolution which is beautifully illustrated in Fig.1. The Boolean truth values split into those shown in Fig.4, providing a scheme of mutual interlocking between truth values and facts/theories.

It is not hard to prove that $\mathcal{L}$ and $\mathcal{M}$ are modular for a finite number of propositions (compare the modularity of quantum logics in a finite – dimensional Hilbert space[32]). The modularity of evolutionary theories might be of relevance to the development of evolutionary models and strategies as addressed in the article of P.Ablay (especially his thesis #3)[38]. Additional formal properties as metricity, valuation functionals, etc. are also of interest.

As a further point, the logics concerning facts and theories have been incorporated into a common scheme. This scheme allows for a formal unification of both using the dual properties of the corresponding lattice. It has indeed turned out that incommensurable facts and theories may be interpreted according to the dual formulation of non – distributivity of this lattice. Its formal properties represent the syntax of a metatheory including facts and theories.

The system – specific meaning of the commutator K motivates the impossibility of an axiomatization of evolutionary theory in the same sense as quantum theory. However, if K were regarded as a universal constant on a semantic level superior to h, then the relevant axiomatization should be attempted with respect to the syntax of the metatheory. Recent work of D.Finkelstein[39] appears to be important in this direction.

Any theory of evolution is (at least partially) its own metatheory, since this metatheory is relevant to the evolution of facts *and* theories. All the formal arguments derived from the lattice structure of evolutionary logic apply to the evolution of theories as well. For this reason, they imply a strong objection against any operationally accessible final theory, and consequently against any level of communicable absolute knowledge.

ACKNOWLEDGEMENTS

It is a great pleasure to thank F.Mündemann and J.Becker for the hospitality during the workshop for which this article has been prepared. The work on the presented matters has profitted from discussions with F.Krueger and H.Scheingraber which are gratefully appreciated.

REFERENCES

1. A. J. Lichtenberg and M. A. Lieberman, *Regular and Stochastic Motion* (Springer, Berlin, 1983)

2. J. Guckenheimer and P. Holmes, *Nonlinear Oscillations, Dynamical Systems, and Bifurcations of Vector Fields* (Springer, Berlin, 1983)

3. J.-P. Eckmann and D. Ruelle, *Rev. Mod. Phys.* **57**, 617 (1985)

4. R. Shaw, *Z. Naturforsch.* **36** a, 80 (1981)

5. J. D. Farmer, *Z. Naturforsch.* **37** a, 1304 (1982)

6. A. M. Fraser, Ph.D. thesis, University of Texas at Austin, 1988

7. C. E. Shannon and C. Weaver, *The Mathematical Theory of Communication* (Univ. of Illinois Press, 1962)

8. A. N. Kolmogorov, *Dokl. Akad. Nauk.* **98**, 527 (1959)

9. J. B. Pesin, *Russ. Math. Survey* **32**, 455 (1977) (*Usp. Math. Nauk* **32**, 55 (1977))

10. H. Atmanspacher, *Found. Phys.* **19**, 553 (1989)

11. B. B. Mandelbrot, *The Fractal Geometry of Nature* (Freeman, San Francisco, 1982)

12. F. Hausdorff, *Math. Ann.* **79**, 157 (1919)

13. F. Takens, in *Dynamical Systems and Turbulence, Lecture Notes in Mathematics* **898**, eds. D. A. Rand and L. S. Young (Springer, Berlin, 1981), p.366

14. N. H. Packard, J. P. Crutchfield, J. D. Farmer, and R. S. Shaw, *Phys. Rev. Lett.* **45**, 712 (1980)

15. Among the pioneering publications are P. Grassberger and I. Procaccia, *Phys. Rev. Lett.* **50**, 346 (1983); P. Grassberger and I. Procaccia, *Phys. Rev.* **A 28**, 2591 (1983);

Y. Termonia and Z. Alexandrovitch, *Phys. Rev. Lett.* **45**, 1265 (1983); R. Badii and A. Politi, *J. Stat. Phys.* **40**, 725 (1985)

16. G. Mayer - Kress, ed., *Dimensions and entropies in dynamical systems* (Springer, Berlin, 1986)

17. A. Brandstater, J. Swift, H. L. Swinney, A. Wolf, J. D. Farmer, E. Jen, and P. J. Crutchfield, *Phys. Rev. Lett.* **51**, 1441 (1983); J.-C. Roux, R. H. Simoyi, and H. L. Swinney, *Physica* **8 D**, 257 (1983)

18. W. Lauterborn and J. Holzfuss, *Phys. Lett.* **115 A**, 369 (1986)

19. A. Babloyantz, J. M. Salazar, and C. Nicolis, *Phys. Lett.* **111 A**, 152 (1985)

20. G. Nicolis and C. Nicolis, *Proc. Ntl. Acad. Sci. USA* **83**, 536 (1986)

21. A. M. Albano, J. Abounadi, T. H. Chyba, C. E. Searle, S. Yong, R. S. Gioggia, and N. B. Abraham, *J. Opt. Soc. Am.* **B 2**, 47 (1985); H. Atmanspacher and H. Scheingraber, *Phys. Rev.* **A 34**, 253 (1986)

22. M. Auvergne and A. Baglin, *Astron. Astrophys.* **168**, 118 (1986); W. Voges, H. Atmanspacher, and H. Scheingraber, *Ap. J.* **320**, 794 (1987)

23. H. Atmanspacher and H. Scheingraber, *Found. Phys.* **17**, 939 (1987)

24. I. Prigogine, *Non - Equilibrium Statistical Mechanics* (Wiley, New York, 1962)

25. I. Prigogine, *From Being to Becoming* (Freeman, San Francisco, 1980)

26. R. Balescu, *Equilibrium and Nonequilibrium Statistical Mechanics* (Wiley, New York, 1975)

27. G.M. Zaslavskii, *Chaos in Dynamical Systems* (Harwood, London, 1985) App.3

28. Y. Elkana, in *Sciences and Cultures. Sociology of the Sciences*, Vol.5, eds. E. Mendelsohn and Y. Elkana (Reidel, Dordrecht, 1981) pp.1-76

29. P. Mittelstaedt, *Sprache und Realität in der modernen Physik* (BI Verlag, Mannheim, 1986.)

30. G. Birkhoff, *Lattice Theory*, 3rd ed. (AMS Coll. Publ. Vol.25, Providence, Rhode Island, 1979)

31. M. Jammer, *The Philosophy of Quantum Mechanics* (Wiley, New York, 1974

32. G. Birkhoff and J. von Neumann, *Ann. Math.* **37**, 823 (1936)

33. P. Lorenzen, *Metamathematik* (BI Verlag, Mannheim, 1962)

34. I. Prigogine and T. Petrofsky, *Physica* **147 A**, 33, 439, 461 (1987)

35. P. Watzlawick, ed., *Die erfundene Wirklichkeit* (Piper, München, 1981)

36. F. Krueger, *Physik und Evolution* (Parey, Berlin, 1984); *Theorie der Emergenz* (unpublished manuscript, 1988)

37. K. Popper, *The Logic of Scientific Discovery* (Hutchinson, London, 1959)

38. P. Ablay, *this volume*

39. D. Finkelstein, *Finite Physics* (preprint, 1988)

Selforganization by Evolution Strategy in Visual Systems

-Reinhard Lohmann-

Department "Bionik und Evolutionstechnik", TU Berlin,
Ackerstraße 71-76, D-1000 Berlin 65

Summary

Pattern recognition usually distinguishes two fields in its systems, feature detection and classification. There is a certain similarity to vision in man where feature detection partly takes place in the retina while the process of assigning the features to the proper concepts is supposed to be located in the cortex. With respect to a certain set of pictures one may ask whether the features being detected are the most advantageous ones for the solution of the classification problem. Within the system of parallel distributed pattern recognition being developed at the Department for Bionics and Evolution Techniques a certain feature is given by a local filter which is described by its structure and by some parameters contained in the structure. The locally detected signals are added and the result is a global value of that particular feature. If the structure of the local filter is well defined there will be no problems with the determination of parameters by applying evolution strategy.

In this paper, however, we will show how the structure of a local filter can be developed by means of evolutionary self-organisation. The optimization of structure is superimposed on the optimization of parameters. The task of the local filter, treated as an example, is counting the coherent areas in binary pictures. The self-organization by means of evolution strategy is carried out with 24 pictures and the result will be tested on 12 different pictures.

Selforganisation under biological and psychological aspects of perception

Visual perception in biology is the result of the evolution. Psychology describes the capabilities of biological systems of perception. On one hand these capabilities are often astonishing, on the other in most cases self-evident. It is easy, for example, to distinguish a maple leaf from an oak leaf (see figure 1). But the common aspects in both figures become evident if we add a geometrical triangle. All the differences in the figures, however, seem to vanish if we perceive that there are three figures. We assume that some or even most of the features which enable us to perceive, to distinguish, to generalize and to count figures are not conscious and are not represented in language.

Fig. 1: Leaves and triangle.

Physiology gives some information about structures of the nervous system in which perception takes place. There is no doubt that a number of features are extracted locally from the pattern of physical excitation of the receptors in the retina, combined with considerable data reduction, and that the extracted features are processed into specialized information in the cortex.

Although the retina is one of the best investigated parts of the brain (Dowling 1987) there is only little information about what the features particularly are like and which specific neural circuits do extract them from the pattern of excitation.

One access to neuronal structures which are able to do what the retina is supposed to do is the method of self-organisation by evolution strategy, i.e., by means of computer simulations. Self-organization means that the structure of a system, a local filter in this case, is developed only by applying the rules of evolution, mutation and selection, in order to solve a given task. To count the figures in a binary picture seems to be a sufficiently complex task for a first experiment with self-organization in the field of neuronal structures. The stimuli for the selforganization of the local filters are 24 pictures given in figure 2. The success of the procedure will be discussed on 12 different pictures at the end of this paper.

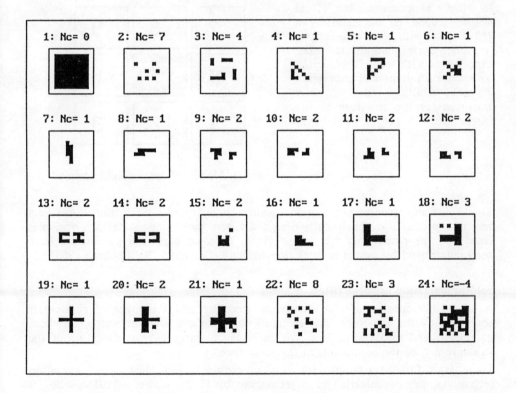

Fig. 2: 24 pictures used in the selforganization of the counting filter, correct number of figures Nc written above each.

Structure of technical simulation of perception in biology

Biological perception is simulated in this technical system as far as it concerns local feature detection in the retina. The receptors are replaced by the points of the monitor, the possible excitation of a point being restricted to values of one or zero.

The circuits of neuronal cells in the retina are represented by local filters which cover the whole monitor like a homogeneous layer and which work with a small amount of neighbouring points, which correspond to the receptive fields in biology. The output of the local filter depends on the activation pattern in the neighbourhood of the particular point where the operation is carried out. The signals of all local filters are added and the result is the global value of that feature measured in a certain picture.

We can choose freely between a more or less large neighbourhood of a point to bo integrated in the local operation of the filter, and, further, there are many possible ways to construct the filter. Computer technologies of today are poorly suited for the simulation of neurons in all details like time response or the influence of geometrical form on the response. The formula choosen for the neuronal filter should be the most appropriate one for the computer medium.

In this experiment the local neighbourhood is described by a 3*3-matrix with the elements a_i (i=1...9). Corresponding to other experiences with local filter design we choose a polynomial expression P of arbitrary degree and extension for the structure of the filters:

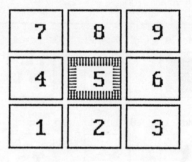

$$P = F (a_i^g , K_j)$$

Fig 3: Matrix of local neighbourhood

The K_j are any parameters, g is any positive integer and the function F can be a sum or a product. In heuristically designed filters the mathematical description is fixed and the parameters are well defined. In some cases the parameters have been determined by means of evolution strategy (Lohmann, Rechenberg 1988).

Self-organization of local filters

A local filter consists of a mathematical structure and a number of parameters contained by the structure. The structure is of no use without well-determined parameters and vice versa. Thus it is necessary to carry out the evolution of the structure and of the parameters at the same time.

There is a famous example for this procedure in the history of evolution techniques, the optimization of a jet engine by H.-P. Schwefel (1968). The jet engine consists of a number of segments whose radii are determined by mutation out of the set of real numbers. The structure is given as the number of segments and the stock of possible structures is described by positive integers. Adding or removing one segment is the smallest step in structure mutation.

503

In the case of local filters the stock of structures is the set of all polynomial expressions. The problem now is to change such expressions by mutations without violating the principle of strong causality which is a necessary demand for evolution strategy applicability (Rechenberg 1973). The replacement of an addition operation by a multiplication or the removal of a bracket are very small alterations if one looks at the written version of the expression, but they will result in a drastic change of its value. There is a possibility, however, to perform mutations in the expression under the restriction of strong causality in the following ways: The addition of any other polynomial, provided it is multiplied with a very small number e, does not change the value of the former expression very much. The second way uses the multiplication of the expression with another one whose value is close to one. Let us call the expression to be mutated F and the random polynomial P. Then the mutants F_1 and F_2 of F are performed in the following way.

$$F_1 = F + e * P$$
$$F_2 = F * (1 + e * P), \quad \text{with } e \ll 1$$

In order to simplify this procedure the filter is assembled by a number of elementary operators - according to the segments of the jet engine. Three basic types of operators have been defined for this purpose.

$$O_1 = c_1 * K_1 + c_2 * K_2$$
$$O_2 = c_1 * K_1 + K_2$$
$$O_3 = c_1 * c_2 * K_1$$

Here c_i is the adresse of an element in the local neighbourhood, calling for values of one or zero, or the adress of an other operator in the filter. K_j are real numbers , called the parameters of the system. In this formulation of the filter it is possible to carry out mutations of the filter structure while complying with the principle of strong causality.

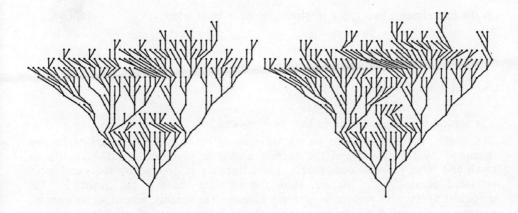

Fig. 4: Graphical illustration of filter structures; each end of a branch calls for one of the elements in the local filter matrix.

Demonstrations of filter structures are not as vivid as those of the structures of the jet engine. The most common way is simply to write down the complete expression which is not very instructive. Another possibility is a list of elementary operators assembling the filter structure. I prefer a graphical illustration of the structure as a binary tree (Fig. 4). The graphics of two different filters are plotted which are obviously of great similarity. One represents a mutation of the other.

Strategies

Concerning the evolution strategies being applied we point out once more that two optimizations have to be carried out at the same time, one of structure and the other adjusting the parameters. In the case of parameter evolution a (μ,λ)-strategy with mutative step width control was choosen. The parameters are optimized exclusively for one particular structure. A second and different structure is treated in a second isolated population and accordingly all of the following μ' structures. For the selection among the structures the convergence velocity in the parameter evolution of each structure is of most importance in addition to their qualities at a certain time. In order to measure convergence velocity and to involve it in the selection process of structures the parameter evolution of each structure has to be carried out for a certain number of generations while the populations with their individual structures are kept isolated during this time. This procedure refers to the phenomenon of isolation observed as well in biological evolution.

Schwefel (1977) and Rechenberg (1978) have suggested a certain form of characterizing different types of evolution strategies for one or more populations.

The phenomenon of isolation needs to be incorporated in the formal description of strategies. Isolation can be expressed by the number γ , written like an exponent to the round brackets. γ denotes the number of generations during which the populations are isolated until the next cycle of structure alteration and selection takes place. On the level of structures we have μ' parants- and λ' descendant populations. The formal description of strategies is:

$$[\mu' +, \lambda' \ (\mu , \lambda) ^{\gamma}] - ES$$

In the experiments two types of strategies have been tried:

$$[4 + 1 (5 , 20) ^{10}] - ES$$

$$[1 + 1 (5 , 20) ^{5}] - ES$$

The second one seemed to have had more success.

The quality function depends on the differences between the numbers of figures measured by the filters and the correct numbers. In order to make results on small and large pictures comparable this difference is divided by the number of activated points in the picture. Here the absolute value is the quality of the particular filter with respect to a single picture. The quality according to a set of pictures is evaluated by the squareroot of the mean square of all "single picture qualities". The goal of the evolution strategy in this experiment is defined as the minimization of the quality.

Results

In figure 5 quality and filter length are plotted against generations. 4000 seems to be a rather large number of generations but remember that a new structure occurs only every 5 generations. That means that we have 800 generations in structure evolution which is not too much. Alterations of structures are plotted in the same figure using the number of elementary operators, called filter length, as an indicator. The filter length increases over all despite stagnation in some periods. Increasing of filter length is probably caused by the fact that there are no appropriate mutation operators implemented for the execution of associative and distributive law.

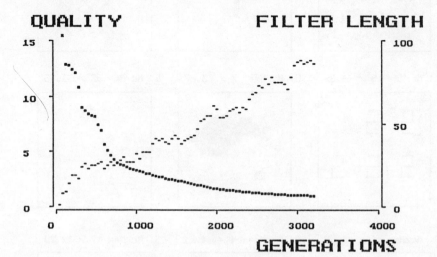

Fig. 5: Quality and filter length versus generations.

Nr.	Nc.	Nm.	SPQ/%	R.
1	0	-0.001	0.007	0
2	7	6.991	0.130	0
3	4	4.016	0.122	0
4	1	1.000	0.004	0
5	1	1.003	0.027	0
6	1	0.998	0.023	0
7	1	0.999	0.011	0
8	1	0.999	0.006	0
9	2	2.003	0.029	0
10	2	2.004	0.047	0
11	2	1.994	0.070	0
12	2	1.999	0.007	0
13	2	2.004	0.030	0
14	2	1.981	0.136	0
15	2	2.001	0.012	0
16	1	1.008	0.085	0
17	1	0.986	0.068	0
18	3	3.003	0.012	0
19	1	0.984	0.127	0
20	2	2.057	0.283	0
21	1	0.954	0.178	0
22	8	8.021	0.165	0
23	3	2.980	0.089	0
24	-4	-3.991	0.020	0

Now let us examine the results of the filter developed by evolutionary self-organisation acting on the 24 pictures mentioned above. Deviations of the numbers of figures measured by the filter compared with the correct numbers are extremely small and *rounding gives us the correct numbers for each picture. This is not a surprising result (see figure 6).*

Fig. 6: Table of results for 24 pictures being involved in the optimization process. The columns contain the numbers of the pictures - Nr -, the correct - Nc - and the measured - Nm - numbers of coherent areas, the single picture qualities - SPQ - and the round - R - of the differences between Nc and Nm.

But the success of self-organisation becomes evident, when the filter acts on the set of 12 test pictures (figure 7).

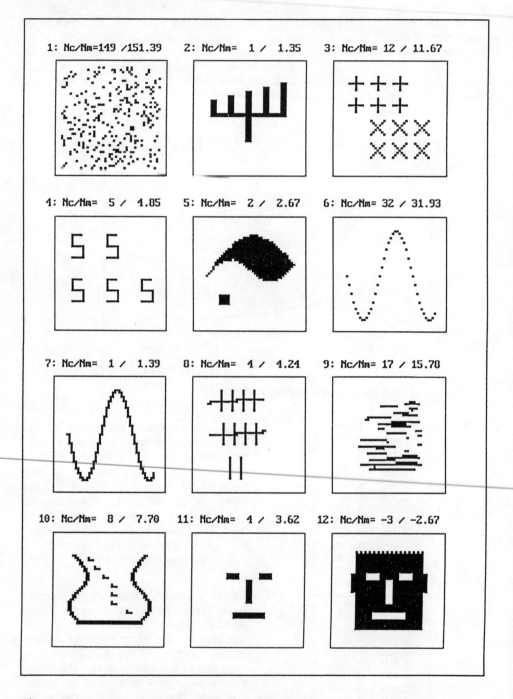

Fig. 7: Twelve test pictures; at the top of each picture you read correct - Nc - and the measured - Nm - number of coherent areas.

There is no reason to complain about the mistakes made by the filter for the first picture. Human perception would not hit the right number as precisely as the filter. Asked for an estimate I would say 80 or 100. Even in an attempt to count correctly I would expect to miss the right number by 5 or 10. Concerning the eleven pictures left in most of them the measured results can be adjusted by rounding. Despite this deviations seem to be greater than in the first set of 24 pictures. If we have a look at a comparison of the normalized single picture qualities in figure 8 all qualities turn out to be of similar magnitude.

The comparison allows the careful conclusion that something like the ability of generalisation has been developed by means of evolutionary self-organisation.

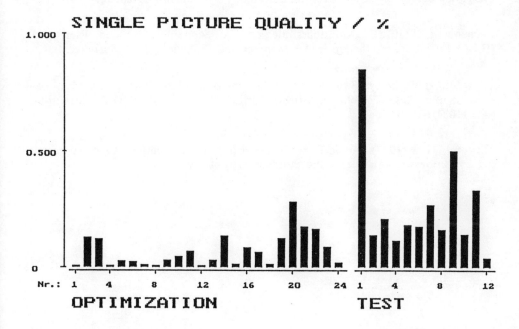

Fig. 8: Single picture qualities.

Outlook

The method of self-organisation by means of evolution strategy has been quite successful in this first experiment which was meant to consolidate the procedure. After this application in the field of feature detectors in visual systems we planned to develop neuronal networks by self-organisation, networks which will have to solve classification problems. Development of filters for features without expressible concepts will be possible, in my opinion, if selforganisation is applied to a complete visual system including local filters as well as classification system which can be flexibly structured.

Moreover, the tools of self-organization by means of evolution strategy are worth being applied in other problems which can be technically formulated.

Literature

Dowling, J. E.(1987): The Retina - An approachable Part of the Brain. Cambridge, Massachusetts: Belknap Press of Harvard University.

Lohmann, R. & Rechenberg, I.(1988): Parallelstrukturierte Zeichenerkennung in schnellbewegten visuellen Systemen. In: Levi, P., Rembold, U. (Hrsg.): Autonome Mobile Systeme, 4. Fachgespräch. Karlsruhe.

Rechenberg, I.(1973): Evolutionsstrategie - Optimierung technischer Systeme nach Prinzipien der biologischen Evolution. Stuttgart: Frommann-Holzboog.

Rechenberg, I.(1978): Evolutionsstrategien. In: Schneider, B. und Ranft, U. (Hrsg.): Simulationsmethoden in der Medizin und Biologie. Berlin: Springer.

Schwefel, H.-P.(1968): Experimentelle Optimierung einer Zweiphasendüse. Bericht 35 des AEG Forschungsinstituts Berlin zum Projekt MHD-Staustrahlrohr (Nr. 11034/68). Berlin.

Schwefel, H.-P.(1977): Numerische Optimierung von Computer-Modellen mittels der Evolutionsstrategie. Basel und Stuttgart: Birkhäuser.

On Steiner Trees and Genetic Algorithms

J. Hesser, R. Männer, O. Stucky

Physics Institute, University of Heidelberg, Heidelberg, West Germany
bd0@dhdurz2.bitnet

Abstract

In this paper the application of a Genetic Algorithm (GA) to the Steiner tree problem is described. The performance of the GA is compared to that of the Simulated Annealing Algorithm (SA) and one of the best conventional algorithms given by Rayward-Smith and Clare [1] (RCA).

Particular attention has been paid to find an optimal setting of the parameters and operators of the GA. A mutation probability P_M=0.01 and a crossover probability P_C=0.5 have been obtained according to the values found by Grefenstette [2]. An optimal population size has been chosen using a heuristic comparable to that of Goldberg [3]. If, according to Goldberg, arbitrary combinations of bits are used as genetic material, a population size of 10^{19} is obtained. Instead, we chose problem inherent structures yielding a population size of 50.

In addition the application of two problem specific heuristics is discussed, the removal of Steiner points of degree <3 and a local relaxation of the tree. A speedup of ≈20 was obtained if these heuristics were applied to the computation of the fitness only without changing the individuals themselves. For a fair comparison, these heuristics have been applied to the GA, the SA, and, in contrast to its original definition, to the RCA.

According to our results, all three algorithms find the optimum and converge equally fast, i.e., GA and SA need not more function evaluations than the RCA, that has even been improved by the relaxation heuristic. The GA reaches the optimum in a small number of generations which is considered the reason why it did not show an even better performance.

1. Introduction

The construction of minimal Steiner trees (MStT) is a problem that should be suited very well for applying Genetic Algorithms. Despite of the fact, that we selected this problem just to compare the efficiency of GA's to other algorithms, it is of high practical importance. One example is the design of large networks, e.g., for telecommunication, power distribution, or oil pipelines. Here, the bulk of the costs lies in the connection between nodes. A number of given nodes has therefore to be connected so that the minimal connection length is obtained, resulting in a minimum spanning tree (MSpT). A practical example has been given by Gilbert [4]. Computing the costs for connecting the main cities in the USA he obtained savings of 240 million dollars when using a MStT instead of a MSpT only. A similar problem arises, e.g., in the layout of PC boards and VLSI chips. Here again, a given number of device terminals has to be connected so that the interconnections are as short as possible. This maximizes the device speed and minimizes the device costs.

More precisely, the MStT is defined as follows. A number of n given nodes in the plain can be joined by edges such that all nodes are connected to form a tree. Numbers can be attached to all edges defining their Euclidian lengths. The length of the tree is the sum of the length of these edges. Among all possible trees one or more can be found with minimal length. They are called the MSpT's. Optimally, a MSpT with n nodes can be found in $O(n \log n)$ operations [5].

However, a still shorter tree can be found if additional nodes are introduced, so called Steiner points. The principal effect of shortening the MSpT by introducing Steiner points is shown in figs. 1 and 2. The additional nodes may be placed at all positions inside the convex hull of the given nodes. Typically, the MStT that can be formed using the additional Steiner points is shorter than the MSpT by about 10%. In the applications mentioned above, an improvement in this order of magnitude yields very important cost savings. Since an optimization has to be done only once during the design phase, a high computational effort is justified.

Figure 1: MSpT of 4 nodes.
Length = 3.00

Figure 2: MStT of 4 nodes.
Black nodes are Steiner points.
Length = 2.73

For finding the MStT it is necessary to know the number of Steiner points and their positions in the plain. This problem has been shown to be discrete and NP complete [6]. Its discreteness follows from the fact that a large number of possibilities for choosing Steiner points can be omitted by using information about the optimal tree without knowing the precise solution. This makes the search space finite. Thus the problem is reduced to a combinatorial optimization. Nevertheless the number of possible configurations which constitute the search space increases exponentially with the problem size. The computational effort to construct the optimal solution is so high that only problems with up to 17 nodes have been solved exactly for any arrangement of points [7]. Some 30 point problems have also been solved. In order to get near-optimal solutions on larger problems, heuristics are used. One of the best known heuristical algorithms for the construction of Steiner trees was given by Rayward-Smith and Clare [1]. It was applied to several problems with up to 80 nodes.

2. The Genetic Algorithm

For the construction of the MStT, a conventional GA was applied. We used the operators crossover, mutation, and selection which operate on a number of individuals forming a population. The individuals are represented by chromosomes which consist of a binary string of length l. A chromosome is an assembly of the x,y-positions of a fixed number of Steiner points. Therefore, each chromosome represents one Steiner tree. The fitness of the individual belonging to the chromosome corresponds to the length of the MStT that can be constructed by using the original fixed points and the encoded Steiner points. This length is a function of the positions of the Steiner points as specified by the chromosome.

According to Holland´s theoretical work [8] the GA should be well suited for the construction of Steiner trees. This is due to the fact that MStT´s consist mainly of structures as shown in fig. 3. The selection operator leads to an accumulation of good structures and the crossover operator combines them to form even better individuals. The mutation operator ensures that the algorithm does not converge to the same structures in all individuals in the population. For a better understanding the exact form of the operators used is explained below.

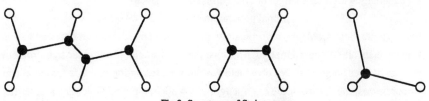

Fig.3: Structures of Steiner trees.

2.1 Parameter setting
Crossover

The crossover operator exchanges a part of a chromosome between two individuals. This part begins at its first bit and has a length between 1 and l that is chosen with uniform probability. Crossover is applied to all individuals with a probability $P_C/2$. The second individual required for each crossover is chosen from the population with uniform probability. This form of crossover was selected with respect to a later implementation on a multiprocessor system [9]. There, the crossover can be executed concurrently and asynchronously.

The crossover probability P_C was chosen at 0.5 according to the value given by Grefenstette, who optimized the parameter values of GA´s with another GA [2]. It was pointed out that this setting can successfully be used in most applications. This setting has been shown to be optimal in our case (see below).

Mutation

The mutation operator inverts a number of bits in every individual. The inversion is done with a fixed probability P_M for each bit in every chromosome. P_M was chosen at 0.01, again in accordance with the optimal value given by Grefenstette.

Selection

For our problem the selection algorithm introduced by Holland [8] has been used. Here, an individual I appears in the subsequent population with a frequency $h(T+1)$, which depends on the current frequency $h(T)$ and the ratio of its current function value $f(I)$ to the mean of the function values in the population $<f>$,

$$h(T+1) = \frac{f(I)}{<f>} \, h(T).$$

Since this selection procedure has been proposed for maximization problems, the MStT problem has been converted to a maximization problem by subtracting the values of the considered function from an offset, the largest function value in the population. According to Grefenstette [2] this offset should be optimal.

Population size

Goldberg [3] presented a heuristic that in principle allows to determine an optimal population size. This heuristic tries to find a good compromise between a large amount of available genetic material and a small population size. Therefore, the number of different combinations of bits per individual is maximized. This ratio grows exponentially with the bit length l of the chromosomes, since

$$\text{optimal population size} \approx 1.65 \cdot 2^{0.21 \cdot l}.$$

For most practical problems, unfortunately, the optimal population size obtained in this way exceeds all computational possibilities. As an example, if a Steiner tree problem with 25 Steiner points is to be solved and we choose a resolution of 6 bits for the x- and y-coordinates, the chromosome has a length of $l \approx 300$. Applying Goldberg's heuristic, the optimal population size is $\approx 1.5 \cdot 10^{19}$, a number that is clearly not usable. However, a smaller optimal population size is obtained by using the following argument. Since a binary representation of the x,y-coordinates of each Steiner point is used, the individual bits are of different significance for the length of the encoded Steiner tree. This is in contrast to the presupposition of Goldberg that assumes an identical weight of all bits. A lower limit for the optimal population size can therefore be estimated, if only the most significant bit per x- and y-coordinate is assumed to be relevant for each Steiner point. This yields an effective length l_{eff} of 50 bits. According to the formula given above, this chromosome length leads to a population size of 2240 individuals. However this number is still too large to work with, as can be seen later.

For this reason we determined a reasonable population size in the following way. In analogy to Goldberg's argument, we optimized the amount of available genetic material that is required to assemble good Steiner trees. Since the trees are built of structures like the ones shown in fig. 3, the number of different structures in populations of various sizes has been investigated. Fig. 4 shows the number of different structures vs. the population size. The largest increase in the number of different structures is in the range from 30 to 110 individuals. This means that within this range an increased population size leads to more different structures per individual than outside. Based on this results it was expected that a population size of 50 individuals is a good compromise.

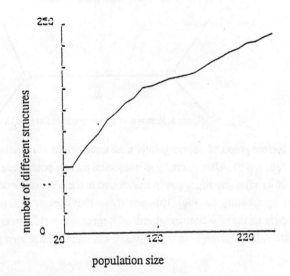

Figure 4: Number of different structures vs. population size.

2.2. Heuristics

Due to the immense number of suboptima and the small difference in length between the respective Steiner trees, the GA as described so far converges so slowly that practical Steiner tree problems cannot be solved in reasonable time. In order to speed up the algorithm, heuristics have to be applied. By using the heuristics described below, the computing time is reduced by a factor of ≈20.

First we used the fact that, according to Gilbert et al. [6], the number of possible Steiner points is limited to n-2 for n given nodes. Second, we exploited the fact that in a MStT the number of edges connected to a Steiner point, called its degree, is three and the angle between the edges 120 degree.

Consequently, Steiner points of degree 1 and 2 were removed and Steiner points of degree >3 were resolved into multiple Steiner points of degree 3 as follows. Steiner points of degree 1 can clearly be dropped without further action. In the case of Steiner points of degree 2, the edges to both adjacent neighbors are removed. To construct the new minimal tree, the two separate subtrees would have to be linked by the shortest possible edge. Instead, we connected both subtrees by linking the former neighbor nodes directly to reduce the computational effort. In most cases, the MSpT of the smaller problem is obtained in this way. The few exceptions do not influence the convergence rate of the algorithm significantly.

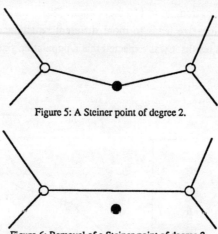

Figure 5: A Steiner point of degree 2.

Figure 6: Removal of a Steiner point of degree 2.

Steiner points of degree greater than three can be eliminated in the following way (figs. 7, 8, and 9). After inserting an additional Steiner point close to the considered one, some of its adjoining edges can be transferred to the additional one. The two subtrees are joined by creating an edge between these two Steiner points. Unexpectedly, this procedure to resolve Steiner points of degree >3 yields worse results because the algorithm often converges to suboptima. It has therefore not been used in our GA.

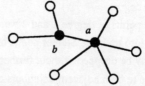

Figure 7: Changing the degree of a Steiner point of degree >3 (a): Starting tree.

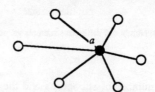

Figure 8: Two edges from Steiner point a are disposed to the new Steiner point (b).

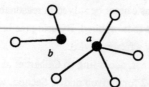

Figure 9: Connection between both subtrees.

The Steiner trees found so far can further be shortened by a local optimization. Every Steiner point is moved locally so that its edges are arranged symmetrically. This was realized by relaxing the tree.

For each Steiner point a displacement vector was computed. The direction of the displacement was given by Gilbert et al. [6]. Considering all edges as unit vectors, the vector sum points into the direction where a shift leads to a shorter tree. The amount of the displacement can be found in analogy to the damped Newton procedure [10] for finding zeroes of a function. Starting from a fixed distance, it is tested whether shifting the Steiner point by this distance shortens the tree. If so, the shifting distance is accepted. Otherwise, the distance is divided by two and the test is repeated until the position of the Steiner point results in a shorter tree (figs. 10 and 11).

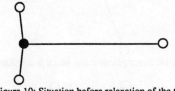

Figure 10: Situation before relaxation of the tree.

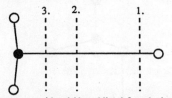

Figure 11: The different distances considered (dotted lines) for relaxing the tree. An originally large distance (1) was divided repeatedly by 2 until the new position of the Steiner point results in a shorter tree. This is the situation with the distance 3.

This procedure to rearrange Steiner points is applied until the tree is relaxed. Since some of the modifications caused by mutation and crossover are cancelled by the relaxation procedure, with relaxation less new structures are created than without (figs. 12, 13, and 14). This leads to a slower convergence.

On the other hand, the GA with relaxation finds the locally optimal positions of Steiner points much faster than the GA without. To exploit this speedup provided by relaxation and to avoid the reduction in the number of new structures, relaxation was used only to compute the fitness of each individual. The chromosome itself remained unchanged in order not to loose genetic material.

For large Steiner tree problems, the heuristics described above are only useful if their complexity is $\leq O(n \log n)$, which is the complexity of the MSpT algorithm. Otherwise for certain problem sizes the heuristics would require more computing time than spanning the MSpT. Our theoretical studies have shown that these heuristics are of time complexity $O(n)$. Therefore the function evaluations per step for spanning the Steiner tree still require only $O(n \log n)$ operations.

By applying these heuristics, the convergence speed is increased by more than a factor 20. They are only inherent to the Steiner tree problem and are independent of the optimization algorithm used. Consequently, for a fair comparison, they have also been implemented in the other two algorithms the GA was compared to. These two algorithms are a Simulated Annealing Algorithm (SA) and the best known conventional algorithm, proposed by Rayward-Smith and Clare[1].

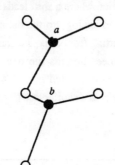

Figure 12: Starting Steiner tree which is not relaxed.

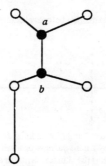

Figure 13: Steiner point *b* of fig. 12 is shifted slightly upward, resulting in another structure.

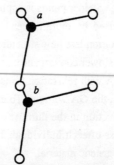

Figure 14: The tree of fig. 12 is relaxed. A small change of the positions of Steiner points *a* or *b* does not result in another structure.

3. Other algorithms
The Simulated Annealing Algorithm

Simulated Annealing Algorithms are function optimizers which are very effective for problems like the construction of MStT´s [11]. For this problem, the function to be optimized was chosen the same as the one introduced for GA´s. Starting from arbitrary function arguments it is tested, whether modifying these arguments improves the function value. In this case the modification is accepted. Otherwise it is accepted with a probability which is proportional to a parameter called temperature. This probability is given by

$$P_{accept} = e^{(new\_value - old\_value)/temperature}.$$

An exploration of new arguments, the positions of the Steiner points, has been implemented using mutation as described above. A mutation probability of ≈0.08 has been shown to be the most efficient.

A repeated application of these steps leads to the optimum if the temperature is adjusted correctly. It has to be be chosen such that the algorithm can escape from a suboptimum with non-negligible probability. If, e.g., the number (*new_value - old_value*)/temperature is 1, an escape probability of $P_{accept} = 1/e$ is obtained. In our problem, suboptima differ only by ≈2% from the function values in their surroundings. So a suitable temperature should have a value equal to this difference in order to fulfill the condition described above.

In conventional SA´s this temperature is lowered during the optimization process. One cooling scheme is that of Kirkpatrick [12]], where the temperature is lowered by a factor of 0.9 each 100 function evaluations. According to this scheme, the temperature would decrease by a factor of three during 1000 function evaluations. However, this does not change the order of magnitude of the probability P_{accept}. If this scheme is applied to our problem, no significant improvement of the convergence speed or quality can been obtained compared to the simpler algorithm where the temperature is held constant. For more complex problems requiring more function evaluations, nevertheless this annealing strategy has been found to give good results [11].

The Rayward-Smith/Clare Algorithm

In contrast to the methods mentioned above, the problem of MStT´s is usually tackled in the following way. Starting with the given nodes, a heuristic function determines which one of a given set of Steiner points is selected. It is accepted only if the new MSpT that can be constructed by using this node is shorter than the former MSpT. These steps are repeated until all possible Steiner points are tested. The positions of the Steiner points of this set are chosen with uniform distribution. Such an algorithm was described by Rayward-Smith and Clare[1].

To approximate the minimal Steiner tree, the RCA needs $O(k\,n^2)$ operations, with k as the number of Steiner points in the given set and n as the number of original nodes. For our problem with n=25, we chose k= 100 which represents the upper limit regarding the available CPU time.

4. Results

To compare the three algorithms with each other, the Steiner tree problem of a 5×5 grid of nodes was chosen, as described by Gardner [13]. It satisfies two conditions. First, the problem is complex enough to be comparable to the technical applications described in the introduction. Second, the optimal solutions are known so that the overall quality for each algorithm can be measured. The grid scaling was chosen such that the distance between two adjacent grid points in x and y direction was 1. A typical optimal solution is shown in fig. 15. It consists of structures of the type shown in fig. 3.

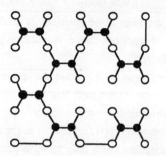

Figure 15 An optimal Steiner tree on the 5×5 grid.
Length = 22.15

In order to compare the three algorithms, the number of function evaluations was defined as a performance measure. It represents the computational effort, since the major part of the CPU time is required for spanning the Steiner tree and not for the optimization algorithm itself. Additionally, it is independent of the effectiveness of the implementation.

Comparison of the algorithms

Figs. 16, 17, and 18 show the convergence of the SA, GA, and RCA respectively, in particular the minimal length of a Steiner tree which can be expected within a certain number of function evaluations. The three curves present the mean value and the standard deviations for 5 trials. Fig. 17 shows the length of the best individual in the population for the GA. Since the SA operates only on one Steiner tree at a time, the length of this tree is given in fig. 16. For the RCA, that needs 100 function evaluations per trial, fig. 18 shows the best found solution so far.

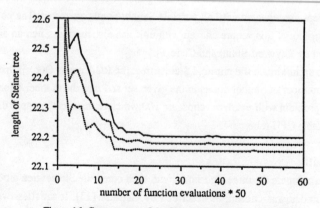

Figure 16: Convergence of the Simulated Annealing Algorithm.

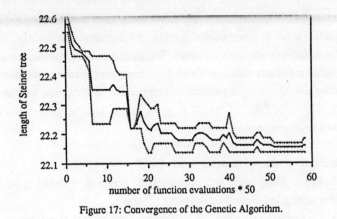

Figure 17: Convergence of the Genetic Algorithm.

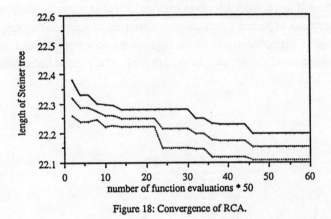

Figure 18: Convergence of RCA.

An algorithm is called converged, whenever it has reached the optimum within the variance of its results. The results of two different algorithms at any number of function evaluations are considered not to be significantly different, if their variances overlap. According to this the three algorithms find the optimum in about 750 to 1200 function evaluations. The SA operates on a single set of Steiner points, so the length of the first Steiner tree is still close to the MSpT. However, the length of the Steiner tree is reduced fast enough that after 750 function evaluations the algorithm has converged. On the other hand, the GA starts with a better value, because the best length out of an initial population is taken. In subsequent generations its variance increases dramatically between 250 and 750 function evaluations. During this period the algorithm selects good structures and combines these to the optimal solution after 1000 function evaluations. The RCA finds an even better suboptimum after one trial. In contrast to the other algorithms it constructs a good suboptimum after one trial already. Subsequent trials of the RCA lead to the optimal solution after 1200 function evaluations.

At first sight the SA finds the optimum faster than the other algorithms. Nevertheless, there is no significant difference from a statistical point of view since their variances overlap. This result is in contrast to our assumption that the GA should be very well suited for the Steiner tree problem. We think the reason is that the GA requires a reasonable number of generations to combine good structures. It can show its full power only for sufficiently complex problems. We assume that the investigated problem of a 5×5 grid is too small although a very high computing power is required due to the expensive tree spanning algorithm. In comparable applications, e.g., the Traveling Salesman problem [14, 15], the GA requires several hundred generations to converge compared to 20 generations in our case.

Verification of the parameter setting

Before tackling larger and more complex problems with GA's, we verified the parameter setting and the genetic operators chosen.

The crossover operator used breaks up chromosomes at arbitrary bit positions. Since there is no correlation between exchanged Steiner points and their position in the grid, structures of Steiner points are often destroyed. It was therefore supposed that an exchange of nearby structures would improve the convergence speed. However, the convergence speed was not affected at all (fig. 19). This effect is not understood yet.

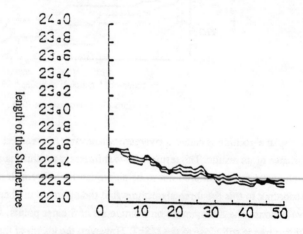

number of function evaluations *50

Figure 19: Convergence of the GA with crossover of nearby structures only.

For a verification of the optimal parameter setting, several different crossover probabilities (P_C) and mutation probabilities (P_M) have been investigated. The larger these probabilities, the larger the search space exploited and, consequently, the probability to find a better solution. On the other hand, crossover and mutation can destroy good structures in the population if their probabilities are chosen too large. For understanding these effects, different mutation probabilities have been tested, as shown in figs. 20, 21, 22, and 23. P_M was varied from 0.001 to 0.03. In fig. 23, a mutation probability of 0.03 destroys too many good structures, since the best function value in

521

the population is lost from time to time. On the other hand, a small value of P_M results in a smaller number of new structures compared to higher mutation probabilities. Therefore the mutation probability has to be maximized under the condition that not too many good structures are lost. The same holds true for the crossover probability for which we performed similar experiments.

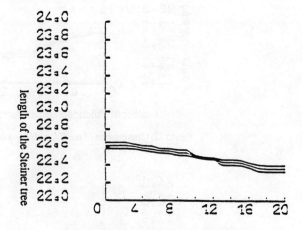

number of function evaluations *50

Figure 20: Convergence of the GA with $P_M = 0.001$.

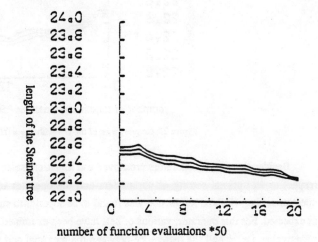

number of function evaluations *50

Figure 21: Convergence of the GA with $P_M = 0.003$.

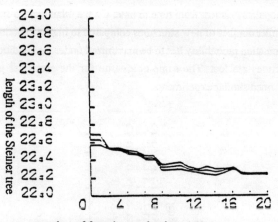

number of function evaluations *50

Figure 22: Convergence of the GA with $P_M = 0.01$.

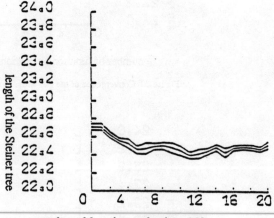

number of function evaluations *50

Figure 23: Convergence of the GA with $P_M = 0.03$.

Besides the genetic operators crossover and mutation, selection was examined in respect to an optimal setting of parameters and the operator itself. Concerning the operator setting, Holland´s selection was found to be superior to simple ranking schemes as expected. For this operator, various offsets have been examined. Large changes in the offset value, i.e., ± half the difference between the maximal and the minimal length in the population, can affect the speed of the convergence. This can be understood, since for a large offset, i.e., a weak selection pressure, good individuals are preferred less in the reproduction process and the algorithm converges slower. On the other hand, a strong selection pressure leads to a faster convergence, but the probability to converge into suboptima increases. This results in a larger variance as shown in figs. 24 and 25. For our setting, an acceptable compromise between convergence speed and minimal variance is obtained (fig. 17).

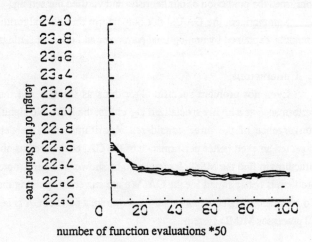

Figure 24: Convergence of the GA with smaller selection pressure

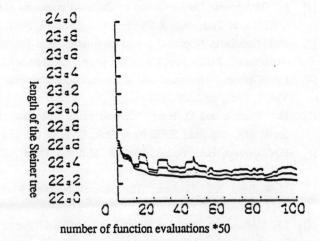

Figure 25: Convergence of the GA with higher selection pressure

The sensitivity of the convergence speed to the population size has been investigated. We chose the population size to meet the following conditions. Increasing the population size increases the probability to reach the optimum. On the other hand, a large population size requires a large number of function evaluations per generation. For a given number of function evaluations, this would only allow a few generations to be computed. But as pointed out above, a reasonable number of generations is required for a good performance. We found an optimal population size between 30 and 100 individuals. These limits represent the lower bound where the GA reaches the optimum and the upper bound where its performance drops significantly. These results confirmed the prediction of our heuristic and verified our setting.

Summarized, the GA did not outperform the other algorithms as expected. This cannot be explained by non-optimal parameter and operator settings.

5. Conclusions

Even not problem specific algorithms as the GA and the SA show the same performance as a highly specialized algorithm, the RCA. No significant difference in the convergence of the three considered algorithms could be detected. However, we expected an even better performance of the GA, because of its ability to combine good structures to find the MStT. It remains to be shown whether a more complex Steiner tree problem is better suited for the GA. We are currently applying the GA to a 17x17 grid. To provide the high computing power required, a parallel GA is being implemented on a 30 processor MIMD system.

References

[1] V.J. Rayward-Smith, A. Clare, "On finding Steiner Vertices", *Networks*, Vol. 16, 1986, pp.293-294

[2] J.J. Grefenstette, "Optimization of Control Parameters of Genetic Algorithms", *IEEE Trans. Syst., Man & Cybern.*, Vol. SMC-16, No.1, 1986, pp.122-128

[3] D.E. Goldberg, "Optimal initial population size for binary coded Genetic Algorithms", TGCA Rep. No.85001, Univ. of Alabama, AL, 1985.

[4] E.N. Gilbert, "Minimum cost communication networks", *Bell Sys. Tech. J.*, Vol. 9, 1967, pp.2209-2227

[5] M.I. Shamos and D. Hoey, "Closest point problems", 16th Ann. Symp. on found. of Comp. Sci., IEEE, New York, 1975, pp.151-162

[6] E.N. Gilbert, H.O. Pollak, "Steiner Minimal Trees", *SIAM J. Appl. Math.*, Vol. 16, No.1, 1986, pp.1-29

[7] M.W. Bern, R.L. Graham, "The Shortest-Network Problem", *Sci. Am.*, Jan. 1989, pp.66-71

[8] J.H. Holland, "Adaption in Natural and Artificial Systems", Univ. of Michigan Press, Ann Arbor, MI, 1975

[9] A. Genthner., R. Hauser, H. Horner, R. Lange, R. Männer, "NERV - A Simulation System for Neural Networks", will be publ. in: Proc. Int'l Symp. Connectionism in Perspective, Zürich, 1988

[10] W.H. Press, B.R. Flannery, S.A. Teukolski, W.T. Vetterling, "Numerical Recipes", Cambridge University Press, Cambridge, MA, 1986, pp.258-259

[11] L. Davis, "Genetic Algorithms and Simulated Annealing", Pitman, London, 1988

[12] S. Kirkpatrick, C.D. Gelatt, M.P. Vecci, "Optimiuation by simulated annealing",

525

Science, Vol. 220, No.4598, 1983, pp.671-680

[13] M. Gardner, "Casting a net on a checkerboard and other puzzles of the forest", *Sci. Am.*, Aug. 1986, pp.14-17

[14] H.Mühlenbein, M.Gorges-Schleuter, O.Krämer, "Evolution algorithms in combinatorial optimization", *Parallel Computing*, Vol. 7, 1985, pp.65-85

[15] M. Gorges-Schleuter, "ASPARAGOS - An Asynchronous Parallel Genetic Optimization Strategy", subm. to 3rd Int'l Conf. on Genetic Algorithms, 1989

Lecture Notes in Artificial Intelligence (LNAI)

Lecture Notes in Computer Science